TIME IN CONTEMPORARY INTELLECTUAL THOUGHT

AZimuth

Volume 1 Color for Science, Art and Technology
edited by **Kurt Nassau**

Volume 2 Time in Contemporary Intellectual Thought
edited by **Patrick Baert**

Cover illustration (top left-hand corner): Astrolabe, with permission from Museum Boerhaave, Leiden, The Netherlands.

TIME IN CONTEMPORARY INTELLECTUAL THOUGHT

EDITED BY

PATRICK BAERT

Faculty of Social and Political Sciences
University of Cambridge
Cambridge, U.K.

N·H

2000

ELSEVIER

Amsterdam - Lausanne - New York - Oxford - Shannon - Singapore - Tokyo

ELSEVIER SCIENCE B.V.
Sara Burgerhartstraat 25
P.O. Box 211, 1000 AE Amsterdam, The Netherlands

First edition 2000

Library of Congress Cataloging-in-Publication Data

Time in contemporary intellectual thought/edited by Patrick Baert. -- 1st ed.
p. cm. -- (Azimuth ; v. 2)
Includes bibiliographical references and index.
ISBN 0-444-82903-2
1. Time. I. Baert, Patrick, 1961– . II. Series: Azimuth
(Amsterdam, Netherlands) ; v. 2.
BD638.T563 2000
115--dc21 99-23944
CIP

ISBN: 0-444-82903-2

The paper used in this publication meets the requirements of ANSI/NISO Z39.48-1992 (Permanence of Paper).

Transferred to digital printing 2005

To the memory of my parents, Suzanne Laureys and Georges Baert, and my brother, Emmanuel

Acknowledgements

I would like to thank Arjen Sevenster from Elsevier for his remarkable patience and useful editorial guidance. I would also like to thank numerous anonymous referees who have helped in the refinement of the contributions. This book was completed during a sabbatical term, for which I thank the cooperation of my colleagues at New Hall and King's College, Cambridge. All contributions in this volume are new with the exception of chapter 4 which appeared in 'Time in Science, Language and History: An Interdisciplinary Research Seminar', edited by Peter Ohrstrom, and published by the Department of Communication of the University of Aalborg. I thank Peter Ohrstrom for allowing it to be included in this volume.

Alfred Gell died while the book was in the later stages of editing. We would like to pay tribute to his cooperation on this project and his exceptional contribution to anthropology.

I thank Julia for her almost unwavering support in the final stages of this project.

CONTENTS

INTRODUCTION

INTRODUCING TIME

by Patrick Baert and Alan Shipman

In this book, fifteen authors from a wide spectrum of disciplines (ranging from the natural sciences to the arts) offer assessments of the way time enters their work, the definition and uses of time that have proved most productive or problematic, and the lessons their subject can offer for our understanding of time beyond the classroom and laboratory walls. The authors have tried, without sacrificing analytical rigour, to make their contribution accessible to a cross-disciplinary readership.

Each chapter reviews time's past and present application in its respective field, considers the practical and logical problems that remain, and assesses the methods researchers are using to escape or resolve them. Particular attention is paid to ways in which the technical treatment of time, for problem-solving and model-building around specific phenomena, call on – or clash with – our intuitive perceptions of what time is and does. The spans of time considered range from the fractions of seconds it takes unstable particles to disintegrate to the millions of years required for one species to give way to another. Like all central conceptual words, time is understood on several levels. By inviting input from a broad range of disciplines, the book aims to provide a fuller understanding of those levels, and of the common ground that lurks at their base. Much agreement emerges – not only on the nature of the problems time presents to modern intellectual thought, but also on the clues that recent discoveries may offer towards possible solutions.

Section I of this introduction distinguishes several components to the recent debates on time. It attempts to draw out complementarities and common themes from the collection. Section II summarises the individual papers.

I

Four major components emerge from the recent debates on time. One controversy concerns the precise nature of time, one deals with the notion of directionality of time, one with the precise status of the future, and one with the appropriation of the past.

Regarding the nature of time, several ontological questions are raised within that debate. Probably the most tantalising question, and one prior to all other queries, is which language or vocabulary to use to answer the fundamental problems of time. Is it necessary, desirable or even possible to resort to the language of philosophy, or does one need to draw upon science, linguistics or the social sciences, or maybe something different again like, say, intuition (Chapters 3 and 4)? In the philosophical tradition, perennial questions are whether time really exists and whether the present has duration. The problem of existence famously pre-occupied the British philosopher McTaggart about a century ago, but this does not make it *vieux jeu.* The puzzle still haunts us, and even McTaggart's famous distinction between A- and B-series has, in and outside the realms of philosophy, remained useful until today (Chapters 1, 9 and 13). Another conceptual issue is whether time is independent or relational. Newton for one conceived of time as independent, as did most of his contemporaries, but several current thinkers claim that it is relational; time is at least partly, if not *in toto*, constituted by something else. In particular, they often argue that the concept of time emerges out of social practices or discourses (Chapters 2, 3, 11 and 13). Relatedly, some authors wish to ascertain whether time is a singular unity or whether there are many times. In the philosophy of science, some hermeneutic approaches stress the differences in temporality between the social and the natural realm, whilst recent developments with regard to non-linear dissipative structures might suggest otherwise (Chapters 2 and 5). Furthermore, some would contend that it is only a small step from the claim that time is at least partly constituted by society to the assertion that there are as many times or temporalities as there are cultures or forms of life (Chapters 2, 11, 13, 14, 15).

A second debate surrounds the issue of directionality. This issue has especially puzzled several generations of natural scientists and philosophers of natural science. Whilst various laws in the natural sciences are time-symmetric, our intuitions based on daily experience suggest otherwise. Furthermore, recent developments in physics and chemistry put forth macroscopic laws that are sensitive to temporal direction (Chapters 5 and 6). A large number of contributions to nineteenth century social and political theory postulated directionality: the reconstruction of patterns in the past would enable one to prophecy, justify and steer the future (Chapters 11 and 12). But directionality was soon to be lost. In the course of the twentieth century, economists and sociologists tended to emulate the earlier natural sciences and searched for lawlike generalisations that are equally time-symmetric. The models can be and have been defended by citing their simplicity or predictive power (Chapters 9 and 10). A less common, though equally powerful, justification is that, although the models are incurably 'unreal', they can nonetheless inform one about what would occur if the initial conditions and assumptions were true. The most influential of these 'atemporal' models is probably rational choice theory – originated in economics but now underpinning several strands of political theory (Chapters 9, 10 and 12). But quite a few social scientists and economists now shed doubts over that enterprise, and appeal for evolutionary models that are situated in Bergsonian 'real time'. The processes described are not time-symmetrical in that they cannot be run backwards (Chapters 9 and 15).

A third, though related, debate refers to the precise status of the future. Late eighteenth century scientific endeavours tended to portray the future as relatively closed, and even the then-emerging social sciences anticipated discovering 'laws of dynamics' that took the future as somehow given – contained in the past or, reversing the order, entailing it (Chapters 5, 11 and 12). Darwinian biological evolution later became one of the first scientific stepping stones towards the recognition that it is possible to explain without being able to predict (Chapters 7 and 8). A similar notion of a relatively open future emerges as a central implication of quantum physics and more recently chaos theory. The former retreats from predictive certainty to probabilities, and the latter underscores that infinitesimal changes to initial conditions have overwhelming repercussions so that any prediction will be well off the mark (Chapter 5). Likewise, the social sciences are now attempting to recognise the intrinsic openness of the future. Drawing upon Bergsonian principles of emergent novelty and the evolutionary paradigm, it is possible to regard social systems as regularly bringing about and continuously facing creativity (Chapters 9 and 11).

A fourth debate concerns the appropriation of the past. Common sense suggests that there is an asymmetry between the past and future in that the past has been and the future is not yet. While it may be true that the past has occurred by definition, this does not put one's beliefs or reconstructions of the past beyond dispute. Physics and cosmology provide a good example. Contrary to what some scientists purport to have 'discovered', there is little empirical evidence to infer what happened exactly at the very beginning of the universe (Chapter 4). Biologists who attempt to unravel 'lines of descent' face similar problems; for instance, they often rely upon records which are already the outcome of the very same process (natural selection) which they intend to investigate (Chapters 7 and 8). In the social sciences, selective marshalling of past phenomena plays at least an equally important role. This is firstly because the social world is pre-interpreted; hence the people, who are the subject of investigation, adopt a particular notion of the past, and act upon that notion (Chapters 11 and 15). Secondly, the appropriation of the past ties in with power, as in the case of political theorists who, for a long time, have sought to justify their practical guidelines through alleged continuity with the 'past' (Chapter 12). More recently, the post-modern condition can be characterised as a power struggle between diverse narratives or distinct appropriations of the past, and no yardstick can be provided to compare and judge between them (Chapter 14).

These four themes by no means exhaust the issues raised by the contributors to this book, but they can be detected as recurrent concerns in what follows. A summary of each chapter is given below.

II

Roger Teichmann (Chapter 1) assesses time from the angle of analytical philosophy. His starting point is J.M.E. McTaggart's attempt to demonstrate that time is unreal. In this context, McTaggart drew a distinction between A- and B-series: the A-series refer to past, present and future, whereas the B-series denotes 'before' or 'after'. There are two steps to McTaggart's 'proof' that time is unreal. First, he argued that the A-series cannot be real, for its existence would imply incoherence. Secondly, McTaggart asserted that it follows from the unreality of the A-series that time is unreal: if time exists, it would involve change, and the B-series on its own fails to capture change. Teichmann analyses scrupulously both stages of McTaggart's argument, and reviews the ongoing debate. His main verdict is that the existing or possible assaults on the A-series are unconvincing, and that past, present and future are objective features of reality.

Mike Sandbothe (Chapter 2) considers time in continental philosophy. His initial proposition is that there are three ways in which time is discussed in contemporary philosophy. First, some, like Ilya Prigogine, conceive of time as a unifying principle. A new conception of time is thought to transcend the previous opposition between the study of the natural world and the study of the social realm. Second, some conceive of time as pluralizing. For instance, Paul Ricoeur argues that there is a discontinuity between the historical and the natural sphere since the former relies upon phenomenological time. Third, some relativize and historicize time. For instance, Richard Rorty contends that the concept of time emerges out of people's interaction with the world. Time is thus culturally specific and variable. Sandbothe's contribution traces back the intellectual antecedents of this tendency to relativize time. He argues that this genealogy leads us to Martin Heidegger and even Immanuel Kant. Sandbothe shows convincingly that several interpretations of Heidegger and Kant have failed to appreciate the distinctive and original nature of their notions of time.

Lars Löfgren (Chapter 3) traces the implications of time being 'self-referential'. By this is meant that any description of time makes use of words in which time is already implicit. It is perfectly possible for individuals to convey their 'temporal vocabulary' to others without having to describe time explicitly. People can do so because they share language. For Löfgren, shared language allows for a 'complementaristic understanding of time': that is, individuals 'know' each other's temporal vocabulary without having to describe it. Löfgren analyses the main features of such linguistic complementarity, leading to a proof of the incompleteness theorem for descriptions of time. From this perspective, he redefines the Gödel-Turing concept of a formal system. These insights throw new light on various philosophical discussions on time. For example, McTaggart's, Bergson's and Gödel's conceptions of time are reassessed from the perspective of linguistic complementarity.

WH Newton-Smith (Chapter 4) grapples with the issue of time in contemporary cosmology. His main objective is to expose the speculative nature of current discussions about the origin of the universe. Modern philosophy tends to eradicate metaphysics, whereas Newton-Smith argues that contemporary physics is unwittingly embracing it. Stephen Hawking and others make several empirically unsubstantiated claims under the guise of science. In particular, Newton-Smith argues that several aspects of Roger Penrose and Hawking's theory of Big Bang Singularity are problematic. For example, not only is the meaning of some of the parameters in the theory unclear, but alternative hypotheses fit the data equally well. Newton-Smith's central point is that the evidence available is insufficient to make claims about what happened or did not happen before the Big Bang. Furthermore, this domain of cosmology is in constant flux; new findings regularly arise. Small changes in science can affect the general world-view which it is supposed to support, and *a fortiori* Hawking's theoretical inferences lack solid empirical foundation.

Peter Coveney's contribution (Chapter 5) focuses on the direction of time in physics. His starting point is the irreversibility paradox: whereas the laws of physics at the microscopic level are symmetric with regard to time, the behaviour of macroscopic systems is irreversible. The aim of this chapter is to demonstrate how microscopic reversibility can be reconciled with macroscopic irreversibility. There seem to be two ways in which scientists have tried to solve this paradox. First, some scientists have treated the notion of irreversibility as illusory. Second, others recognise irreversibility, and they have tried to use statistical techniques at the microscopic level to accommodate the phenomenon. Coveney shows that the second route is more promising. In the course of his discussion, he touches on a variety of fundamental notions in contemporary science, including non-linear dissipative systems, self-organisation and chaos.

Like Coveney, de Hemptinne (Chapter 6) explores the notion of irreversibility in physics and chemistry. In sympathy with Newton-Smith, de Hemptinne argues that scientific discussions in this area have suffered from vague definitions and hasty conclusions. They both warn against the many leaps or invalid extrapolations in the scientific literature. This chapter engages with central notions in the debate *vis-à-vis* the arrow of time in macroscopic systems: that is, isolation, relaxation, equilibrium, and finally irreversibility itself. de Hemptinne suggests a very different perspective than traditionally conceived. For example, he demonstrates how the results of Joule's experiment have been misrepresented in the scientific literature. Joule's experiment is only one of the wide range of topics covered by de Hemptinne; he also touches upon issues concerning Hamiltonian dynamics of isolated systems, Boltzmann's and Gibb's equations for entropy, and Heisenberg's uncertainty principle.

Problems of time enter into evolutionary biology in a variety of ways. One is in the reconstruction of evolutionary history, the subject of Adrian Friday's contribution (Chapter 7). Evolutionary history depicts 'trees of life', which indicate origin, genetic separation and structural divergence of lineages through time. There are various ways in which attempts have been made to establish this history. Initially, many scientists attempted to infer knowledge about evolutionary history from the study of embryological development. However, there is now enough evidence to cast some doubts on the assumption that the short time-span of embryological development 'matches' the long time span of geological history. Others instead tried to derive insights into trees of life from detailed investigation of fossil records. One of the problems here is that the tools of research (fossil records) have already been subjected to the very selection process which might be the object of investigation. The reconstruction of evolutionary history enables Friday to introduce aligned debates, for example between Darwinian 'gradualism' on the one hand, and Gould and Eldredge's 'punctuated equilibrium' on the other.

Ian Tattersall (Chapter 8), like Friday, investigates time and evolution. But Tattersall's contribution takes a broader canvas. Whereas Friday focuses on issues of relationship, Tattersall writes about the history of evolutionary theory in general. That history starts, so we learn, with Jean-Baptiste de Lamarck. Before Lamarck, naturalists like Linnaeus already classified the rich diversity of life, but they failed to put the listed items into a temporal order. Lamarck was ingenious in doing so, but made the fatal error of claiming that learned characteristics can be inherited. By drawing upon the notions of mutation, variation and selection, Darwin and Wallace pioneered the basics of modern evolutionary theory. Subsequent progress in genetics allowed for further development of the Darwinian paradigm, first via an 'evolutionary synthesis', and later via the theory of 'punctuated equilibrium'. Tattersall's paper ends with his own version of evolutionary theory, which builds on the theory of punctuated equilibrium. New species emerge quite suddenly and then remain relatively invariable for long periods of time.

Mario Rizzo (Chapter 9) is concerned with the role of time in economics. He notes that the notion of relative indeterminacy underlies the way in which philosophers and natural scientists think in the twentieth century. In this context, Rizzo discusses, for example, Henri Bergson's reflections on time and Heisenberg's 'uncertainty principle' at microphysical level. Both are indicative of the growing awareness of indeterminacy. The irony is, however, that whilst philosophers and natural scientists are increasingly drawn to this Bergsonian picture, social scientists are not. In particular, contemporary economic theory tends to present a static picture. Like their nineteenth century predecessors, economists today draw upon analogies with the science of mechanics, and thus inherit the latter's bias towards equilibrium and statics. Rizzo is of the opinion that economics has much to learn from a Bergsonian concept of time. He shows how some Bergsonian notions, like 'real time', can be employed to handle more effectively the complexity of the economic sphere. Some non-mainstream economists, like Hayek and Mises, have already made successful steps in this direction.

Like Rizzo, Frank Hahn (Chapter 10) addresses time in economics. But whereas Rizzo tackles the philosophical dimensions of the debate, Hahn chooses not to. Instead, Hahn demonstrates how time enters into recent attempts by economists to explain and predict people's choices regarding goods and services. The explanation, provided by economists, assumes a rational individual: that is, somebody who has coherent preferences (with regard to goods and services), has rational beliefs about the means to obtain the preferences, and acts accordingly. Time comes into play in that the goods and services, which are the subject of one's preferences, might not be available at the moment of making the choice. The same goods and services might be more or less valuable at a later stage. Furthermore, as the future is not yet and so is not known with certainty, the individual (producer or consumer) is reduced to assigning probabilities to future states. Hahn builds on these simple examples to show how mathematical formulae can be used to take into account these temporal dimensions.

Patrick Baert (Chapter 11) presents the history of social theory, particularly in the twentieth century. He distinguishes five aspects of temporality. The first component refers to the synchrony versus diachrony axis. Synchronic analyses take a snapshot, whereas diachronic studies deal with processes across time. The second dimension relates to the status of the future. For instance, to what extent is the future already given, and if it is, can it be foreseen? The third component involves the relationship between what is changing and what is constant. To what extent is the invariant more significant than the temporal flux? The fourth aspect concerns the relationship between the various so-called temporal levels. Should one favour the longer temporal spans of institutional change or the shorter temporal *durée* of daily interaction? The fifth element raises a methodological issue: how can one explain or understand in the present social practices that happened some time ago? Baert's essay traces back the history of social theory by taking into account these five components. He concludes with an appeal for a temporalised social theory, drawing upon analogies with biological evolution.

Melissa Lane (Chapter 12) traces back the notion of time in political theory. Both past and future are central to practical reasoning. Indeed, concerns about political action are often informed by a philosophy of history. Likewise, political theory addresses values and goals, and the means to obtain them, and so it is coloured by one's time-horizon, and one's expectations about the future. Lane's concise history shows, surprisingly, that political theorists have until recently put less emphasis on the future than the past. Current debates in this area centre around the right balance between the various temporal modes. In particular, John Rawls' *Theory of Justice* is indicative of recent attempts to construct normative political theory independently of the course or direction of history. This contemporary alliance between political theory and 'present-tense' rational choice, as exemplified in Rawls, has been subject to substantial criticisms. From contrasting angles, libertarians and communitarians unite in criticising Rawls and likeminded political theorists for failing to acknowledge the role of history. Others criticise Rawls for disregarding the dialogical nature of decision making across time.

The notion of temporality in anthropology is the subject of Alfred Gell's contribution (Chapter 13). This angle allows Gell to express, more generally, his discomfort with a tendency amongst anthropologists to treat every feature of the observed world as 'socially constructed'. From Emile Durkheim to Edmund Leach, anthropology looks for difference across cultures. Not only do anthropologists stress the variability of collective representations or narratives, they also treat categories, like time, space and number, as variable across cultures. Gell's position is that although different cultures might entertain radically different representations, these rely upon unchanging categories, like time, space and number. So Gell is willing to promote what he calls 'cultural relativity', but not 'temporal relativity': it is simply an error to talk about different temporalities, because time is a logico-cognitive universal. With this Kantian framework in mind, Gell discusses a number of empirical researches in anthropology. In particular, he pays close attention to some of Pierre Bourdieu's field work.

Marissa Quie (Chapter 14) examines the role of time in contemporary culture, particularly in so-called post-modernism. The project of modernity endeavours, amongst other things, to establish a univeral epistemology which would allow human beings to develop rational self-determination. Post-modernism challenges central presuppositions of the Enlightenment project; for example, the assumption that science is epistemologically superior and that it is value-neutral. The so-called post-modern condition also differs from its predecessors with regard to time. For example, whereas the avant-garde and futurism attempt to destroy or negate the past, post-modern culture chooses to revisit the past in an ironical and eclectic fashion, often relying upon assemblage and pastiche. Whereas Enlightenment philosophers assert conceptual continuity between past and present, post-modernists consider history as an amalgamation of mere narratives, stories or fictions. Quie does not simply delineate the temporal dimensions of post-modern culture and philosophy, she also uses this characterisation to voice her scepticism regarding the political and moral value of post-modernism.

William Friedman (Chapter 15) introduces the psychology of time, which he defines as the empirical study of human temporal experience. Friedman distinguishes four components to the psychology of time. First, there is the research on 'time perceptions'. Psychologists, working in this field, study how people judge time intervals. Second, some psychologists investigate memory. They seek to identify which factors influence the accuracy of people's recollection of the past. Third, some investigate the development of children's and adolescents' experience of time. For example, a sense of 'temporal sequencing' or the ability to distinguish between the different temporal modes (past, present, future) are not given, but gradually developed throughout childhood. Fourth, research is carried out on the representation of time. This involves establishing how people learn, relate and adjust to the recurrent temporal patterns of their environment. The week is an example of such a recurrent pattern.

By way of a conclusion, Alan Shipman and Patrick Baert consider what can be learnt from the cross-disciplinary study of time, by drawing together some common components from the foregoing chapters. The atemporalities introduced by axiomatic natural-scientific enquiry are identified as a possible corrective to the tendency for more experiential, social-scientific enquiry to squeeze past and future into an extended present. But definite conclusions must remain tentative, in view of the linguistic and conceptual limitations on the discussion of time noted in earlier contributions. Despite perhaps ending on a lighter note, it is hoped that the areas of common concern – and agreement – identified will show that perusal of this collection has been time well spent.

LIST OF CONTRIBUTORS

Patrick Baert is Fellow at New Hall, Cambridge, and Director of Studies in Social and Political Sciences at King's College, Cambridge

Peter Coveney is Professor of Physical Chemistry and Director of the Centre for Computational Science, Queen Mary and Westfield College, University of London

Xavier de Hemptinne is Emiritus Professor of Chemistry at the Catholic University of Leuven

Adrian Friday is University Lecturer in Zoology at the University of Cambridge

Bill Friedman is Professor of Psychology at Oberlin College

The late Alfred Gell was Reader in Anthropology at the London School of Economics, he was awarded a Professorship posthumously by the same institution

Frank Hahn is Emiritus Professor of Economics at the University of Cambridge, and Professor of Economics at the University of Siena

Melissa Lane is University Lecturer in Intellectual History at the University of Cambridge, and a Fellow at King's College, Cambridge

Lars Löfgren is Emiritus Professor of System Theory at the University of Lund

W.H. Newton-Smith is Fairfax Fellow in Philosophy at Balliol College, Oxford

Marissa Quie was Fellow at Gonville and Caius College, Cambridge

Mario Rizzo is Professor of Economics at New York University

Mike Sandbothe is Assistant Professor of Philosophy at the Friedrich-Schiller University of Jena

Alan Shipman is a researcher at the Judge Institute, University of Cambridge

Ian Tattersall is Curator of the Natural History Museum, New York

Roger Teichmann is Lecturer in Philosophy at St Hilda's College, Oxford

chapter 1

THE COMPLETE DESCRIPTION OF TEMPORAL REALITY

ROGER TEICHMANN
University of Oxford

Time in Contemporary Intellectual Thought
Patrick Baert (Editor)

CONTENTS

1.1. McTaggart

When a philosopher produces a proof of something extraordinary, one's reaction may well be the same as one's reaction to a fairground magician. The rabbit *can't* have appeared inside the hat by magic; so where was the sleight of hand? The philosopher's conclusion *can't* be right – or can't have been *proved*; so where, in the philosopher's reasoning, was the sleight of hand?

More than one sleight of hand may have occurred during the trick, of course. Different critics may discern different faults in a philosophical 'proof', and they may all be right. The ontological proof for God's existence is perhaps a case in point: is it invalid because it treats existence as a predicate? or because it presupposes its own conclusion by the use of a name ('God')? or because of a confusion about 'conceive of' ? . . . or for all these reasons?

J.M.E. McTaggart is probably best known now for his 'proof' that time is unreal (McTaggart, 1927). The 'proof' has been much discussed since he produced it, and indeed it has been defended in one way or another. These defences, however, cannot be said to have involved defending the argument's conclusion. On the whole, they have been defences of certain parts of McTaggart's reasoning – *other* parts of his reasoning being taken to be responsible for landing us with the outrageous conclusion. Michael Dummett's well-known 'defence of McTaggart's proof' is not quite of this form, but it too is more of a diagnosis than a true defence, as we shall see.

Before we can evaluate others' reactions to McTaggart's argument, however, we had better give the outlines of that argument itself.

1.2. The argument for the unreality of time

McTaggart prefaces his discussion with a distinction: the distinction between the A-series and the B-series. The A-series is the series of positions in time, as they are past, present or future, or have tenses derived from these three; the B-series is the series of positions in time, as they are related to one another by the relations 'earlier than' and 'later than', and relations derived from these. The argument then proceeds as follows. The predicates 'is past', 'is present' and 'is future' are clearly incompatible predicates: no two of them can be true of one thing (event).

However, if the A-series is real, we must admit that every event starts as future, becomes present, and finally becomes past. In that case, every event will have each of three incompatible properties (pastness, presentness, futurity). This is impossible; so the A-series cannot be real. But the reality of the B-series would not on its own be enough for the reality of time; for time (if it were real) would involve change, and change cannot be captured in the terms of the B-series. The B-series gives temporal relations between events in, as it were, a 'static' manner, thus failing to express true 'becoming'.

Since the reality of time would require the reality of the A-series, and the A-series involves incoherence, it follows that time is unreal.

Many have attacked the argument by attacking what it says about the A-series – and here the multiplicity of possible criticisms is perhaps most apparent. One thing that quite a few commentators seem to agree on, however, is this: that McTaggart's argument poses the threat (to a defender of the A-series) of one's falling into an infinite regress. The question then appears to be whether that regress is harmless or vicious. The moves that lead to the alleged regress are essentially as follows:

A: You must admit that every event is past, present and future.

B: No. A past event, for example the Battle of the Bulge, *was* future, then *was* present, and *is* past. A present event *was* future, *is* present, and *will be* past. And a future event *was* future, *will be* present, and *will be* past. No event *is* past, present and future.

A: You've just introduced such complex tenses as 'will be past', 'was past', etc. But you must admit that every predicate for a complex tense applies to every event; including incompatible predicates, such as 'past in the past' and 'present in the future'.

B: No. It may, for some event E, be true that: (it *is* the case that it was the case that E was the case), and that: (it *was* the case that it is the case that E will be the case). It can't be that E *is* 'past in the past' and *is* 'present in the future'.

Person A then points out that yet more complex tenses have been invoked, and that every predicate for such a complex tense will apply to every event. And so on.

Is the squabble one about whether 'is' can, or must, be tensed, or tenseless? Certainly, the two parties to the above dialogue seem to be using 'is' in two different ways: a tensed way and a tenseless way.

Let us try to remain neutral on this last question, by first of all seeing what follows if the 'is' is tenseless, and then seeing what follows if it is tensed. Assume that our 'is' is tenseless. Then why can't it be admitted that 'Every event is past, present and future' is *true*? For this sentence just amounts to: 'Every event is at some time past, is at some time present and is at some time future'; and this is compatible with the axiom that 'past', 'present' and 'future' are incompatible predicates. (The incompatibility of these predicates resides in this: that more than one of them can't be true of a given event *at a given time*.)

If, on the other hand, we assume that our 'is' is present-tensed, then clearly no one need admit that every event is (now) past, present and future – a proposition that *would* conflict with the axiom that 'past', 'present' and 'future' are incompatible predicates.

Thus on either reading of the copula, it would appear that the infinite regress shouldn't be embarked upon in the first place. For the first step – A's first proposition – is either true but harmless, or it is false.

1.3 The theory of relativity

McTaggart's attack on the A-series has been the part of his 'proof' to have attracted most attention among philosophers. Some, however, have gone along with McTaggart's view of the A-series, or at any rate have not found fault with it, instead taking issue with his claim that the reality of the B-series would not be enough for the reality of time. The reality of time, claim these philosophers, just *is* the reality of the B-series. Here we have the view that 'time is tenseless'.

As a matter of intellectual psychology, it seems probable that people are not generally attracted to the 'tenseless' view of time by the notion that the A-series involves incoherence, at any rate not in the first place; other attractions are more real. For example, the 'scientific worldview' encourages the idea that past, present and future are not objective features of reality. This is partly because of the (seeming) fact that science does not deal with tensed propositions at all, and partly because of what modern relativistic physics has to say, especially about *simultaneity*: to wit, that there is no such thing as absolute (reference-frame-independent) slmultaneity, which appears to mean that there can be no such thing as the Present (nor, it would therefore seem, such a thing as the Past, or the Future). Bertrand Russell's advocacy of the tenseless view was surely the result, in large part, of his having read Einstein.[1]

Now as to the denial of absolute simultaneity and of an 'absolute Present': a defender of the reality of tense may acquiesce in this denial, if he is willing to say that any given present-tensed statement with a sufficiently clear sense, for instance, "My cat is drinking her milk", will be objectively true or false, albeit only within some frame of reference. And it is hard to deny that many ordinary present-tensed statements of this kind *are* definitely true or false (and not only in a 'loose and popular' sense). The 'reality of the present', for such a philosopher, will simply reside in the objective truth of (many) present-tensed statements. Perhaps on that account the philosophical use of the phrase 'the Present' may indeed appear to be somewhat misleading, insofar as it can suggest something 'absolute'; but it will not be as misleading as the assertion, 'Time is tenseless'.

Moreover, it is not in any case obvious that the revolution in modern physics has, as is usually claimed, produced a corresponding revolution in our ordinary temporal concepts. Take 'simultaneous'. Has the Theory of Relativity shown that simultaneity is non-transitive? Has it, that is, shown that if A is simultaneous with B, and B is simultaneous with C, one may *not* infer that A is simultaneous with C? One can ask: from the point of view of what inertial frame would it ever be true that A was simultaneous with B, B with C, but A not simultaneous with C? The answer is: 'None'. The most that relativity shows is that 'A is simultaneous with B' may be true in A's reference-frame, 'B is simultaneous with C' true in *B's* reference-frame, and 'A is simultaneous with C' false in A's (or C's) reference-frame. The lesson to draw is surely that statements of simultaneity are only possible *within* some given inertial frame, not that simultaneity is non-transitive.

Did ordinary people ever think that one could infer A's simultaneity with C from A's with B and B's with C, in *all possible situations*? How are we to tell? Ordinary people only ever used (and only ever use) the concept in ordinary situations: i.e. as applied to events of sufficient duration as to count, for all intents and purposes, as being within a single inertial frame. And even if they did or do think these things, would that show that the ordinary *concept* of simultaneity was somehow at fault? Wouldn't it just show that there were commonly-held *beliefs* that were false? (If people once commonly held the view that leeching cured fevers, does this show that the concepts 'leech' or 'fever' were somehow at fault? No. People certainly knew what leeches and fevers were.)

These remarks are intended to encourage a certain caution: one should not hastily swallow the idea that ordinary notions (simultaneity, the present, etc.) have been 'overturned' by modern science. Modern science has mainly overturned older science, naturally enough. So if one is assuming that modern science must be largely true (a case, for those ignorant of its details, of being impressed by its practical successes), still one need not on that account infer that the onus of proof is on those who reject a 'tenseless' view of time. For that rejection is not a rejection of the Theory of Relativity, properly interpreted.

1.4. Scientific language

What of the point that science does not deal with tensed propositions? Does this cast doubt on past, present and future?

In fact, as it stands, the claim is too sweeping. Palaeontology is surely science, as is evolutionary biology, much of whose evidence, if not also its assertions, consist of past-tensed propositions. So perhaps by 'science' we should understand 'physics'.

Russell wrote that

> "A physicist will not say "I saw a table", but like Neurath[2] or Julius Caesar, "Otto saw a table"; he will not say "A meteor is visible now", but "A meteor was visible at 8 h 43 m GMT", and in this statement 'was' is intended to be without tense." (Russell, 1940.)

As a remark about the actual utterances of physicists, this is of course somewhat naive. Two of the main activities of physicists are (1) explaining phenomena, and (2) predicting phenomena. Each of these activities necessarily involves tense; the very explosion that could only be predicted *before* it occurred can only be explained *after* it occurs: 'An explosion will take place', and later on, "The explosion occurred because . . .". In explaining, and also in recording, phenomena, the physicist will typically use the past tense. Russell's astronomer would seem to be recording a phenomenon; and there seems no good reason to say, as Russell does, that the past-tensed 'was' is 'intended to be without tense'. (If 'A meteor is visible now' does not get written down, that is because of a preference for the past tense over the present. The same preference, and one that has the same rationale, is shown by the secretary who writes down the proceedings of a committee as they take place.)

It seems that physicists do go in for tensed statements. Some of these might even be said to be essential ingredients of certain physical *theories*, e.g. statements about the Big Bang, including ones about how long ago it took place. But what about the laws of physics? Aren't *these* at any rate tenseless? For don't they apply throughout time and space, rather in the way that mathematics does? 'F=ma' appears to be without a tense to just the same extent as '3+4=7'. Perhaps, then, we could make the point we are trying to make by refining it further: physics, insofar as it consists of general laws, eschews tensed propositions.

In fact, there is a case for saying that neither in physics *nor* in mathematics do we really find genuinely 'tenseless' propositions; the candidates for such propositions being actually present-tensed. One should recall that a present-tensed proposition need not be 'about' a short length of time (the time it takes to utter the proposition, for example). "Moscow is the capital of Russia" is present-tensed – it was once false, and came to be true – but it relates to a considerable period of time (if it is all right to talk of its 'relating' to any period of time). One could support the assertion by appeal to a Soviet guidebook published fifteen years ago.

A good test for whether a sentence is present-tensed, as opposed to 'timeless', is this: conjoin it to the past- or future-tensed version of its negation and see if you get a contradiction. If you don't, that suggests that the sentence *is* present-tensed. "Moscow is the capital of Russia – but it wasn't always"; "Pigs can't fly – but one day they'll be able to"; etc. And "Water boils at 100 degrees C. – but in a million years this will no longer be true", although it may be as *absurd* as the flying pigs proposition, is surely not a *contradiction*.

But it is not particularly my concern here to attack the view that physics deals with tenseless propositions. For even if physics *does* deal only or mainly with tenseless propositions, one can still ask: 'How does *that* show that time is tenseless?'.

A straightforward answer to this last question might be, "The true and complete description of the world is provided by physics". A number of possible presuppositions are buried in this sort of answer. Is it being said that a description of things that is couched in terms *not* derived from physics must be false?

This seems to have been the view of Èddington, when he wrote of his 'two tables', one of them being the solid, brown object spoken of in everyday discourse, the other being a conglomeration of minute colourless particles with gaps between them; physics told him, he thought, that only the second of the two tables really existed. The presupposition here would seem to be that since physics tells him that the second table exists, the first table cannot exist. (Eddington could not have supposed that physics *explicitly* told him that the first table did not exist – it clearly doesn't). And the reasoning here is: a table can't *both* be solid and brown *and* consist of colourless particles with gaps between. To which one should simply reply, 'Why not?'. As Wittgenstein was to put it:

> "Our perplexity was based on a misunderstanding; the picture of the thinly filled space had been wrongly *applied*. For this picture of the structure of matter was meant to explain the very phenomenon of solidity."(Wittgenstein, 1969)

Modern physics, in other words, tells us what solidity *consists* in: solid objects are actually made up of minute particles, etc. The most this can do is surprise us – it cannot show that there *are* no solid objects. Once again, if modern science overturns anything, it overturns older science.

1.5. Complete description

Turning back to time, one might indeed say, "Time cannot be *both* tenseless *and* tensed". This statement is not like "A table can't both be solid and consist of particles with gaps in between". On the contrary, the statement about time is tautologically true; but it involves a misstatement of the problem. A 'tenseless' description of things is one that does not have any tense, that somehow avoids being tensed. It is not one that says that things (events, time itself) *lack* tense. Time or history may or may not be tenselessly describable; but if it is tenselessly describable, still, a tenseless description of time is not the same thing as a description of time as tenseless. For the latter to be a correct description of time, the use of tenses would have to be somehow impugned. Compare: a description of a table that eschews colour-words is not a description of a table as colourless. For the latter to be a correct description, the use of colour-words would have to be somehow impugned.

One way to impugn tenses (or colour-words) might seem to be by insisting that a description of the world that did without tenses (or colour-words) could nevertheless be *complete*. If the complete description of the table makes no mention of its colour, then the everyday ascription to it of the predicate 'brown' is somehow redundant; it doesn't really *describe* the table.

This form of argument is very often used in support of the 'tenseless' view of time. An analogy with space may well be employed. One would not really add anything to a description of a region of space by tacking on a proposition like, "The bookcase is over there", or "The bookcase is further off than the electric kettle". For these propositions are only true from a certain perspective – a certain person's perspective, in fact – and so other, non-perspectival propositions stating the spatial relations of objects to that person would suffice to describe what the perspectival propositions aim at describing. In the same way, it is argued, propositions stating the temporal relations of events to persons suffice to describe what tensed propositions aim at describing. "The Battle of Hastings took place many years ago" could simply be replaced by "The Battle of Hastings is many years earlier than this utterance". There is no 'objective pastness' because a complete description of the world need not employ the past tense. Russell, immediately after the earlier-quoted remark, writes that "there can be no question that the non-mental world can be fully described without the use of egocentric [i.e. perspectival] words" (Russell, 1940, p. 102).

Our problem now is going to be: what counts as a full, or complete, description? To be sure, if one form of words is simply an abbreviation of, or synonymous with, another, one will not add anything to some propositions employing the first form of words by tacking on equivalent propositions employing the second. 'There is a wireless on the shelf' adds nothing to 'There is a radio on the shelf'; and a description of things that includes the first sentence may be quite complete though lacking the second. So can tensed propositions be seen as simply abbreviations of, or synonymous with, the tenseless propositions that we spoke of as 'replacing' them?

Well, if there were such a relationship of synonymy, this would hardly help the proponent of the 'tenseless' view. And this is just because synonymy is a symmetrical relationship. A proposition about a wireless adds nothing to an equivalent one about a radio; but similarly, the proposition about the radio adds nothing to the one about the wireless. Should we talk about radios or wirelesses when describing the world? A silly question. Should we use tensed forms or their tenseless equivalents? An equally silly question, if all that is at issue is synonymy. Either is possible, and there will be no grounds for saying that the tenseless form is more basic.

"But a description of the world using only tensed forms will leave something out, while one that uses only tenseless forms will not. To every tensed form there corresponds a tenseless one; but it is not the case that every tenseless form corresponds to a tensed one." Actually, this claim is false. A *metric tense logic*, of the kind delineated by Arthur Prior (Prior, 1967), can 'do' whatever a tenseless language with 'earlier' and 'later' can do. A simple example: "X is earlier than this utterance, and Y is earlier than X" can be re-written as: "For some n and m, it was the case n units ago that X, and it was the case (n+m) units ago that Y". (The units could be days, years, or whatever.) So the status of tensed discourse *vis à vis* tenseless discourse would seem to be quite analogous to the status of 'wireless'-talk *vis à vis* 'radio'-talk.

1.6. Facts

Of course, the idea that 'The Battle of Hastings took place' is synonymous with 'The Battle of Hastings is earlier than this utterance' is hardly believable in any case. The two sentences surely do not *mean* the same. And the proponents of the 'tenseless' view who have written since Russell wrote have tended to avoid talking of synonymy. Instead, they have very often alleged that tensed propositions are 'made true' by facts that are tenseless, not by ones that are tensed. 'Time is tenseless' has then been cashed out as meaning, "The only facts are tenseless ones; there are no tensed facts". Hugh Mellor is perhaps the best-known advocate of this sort of view (Mellor, 1981). A complete description of the world can now be characterised as one that mentions all the facts there are – and such a description will, it is argued, mention only tenseless facts. Even if it uses tensed *forms*, these propositions will in truth be 'about' tenseless facts.

But the problem that was encountered with synonymy has not really gone away. For that problem essentially had to do with *symmetry*. The 'tenseless' view will not make any headway if whatever is alleged of tenseless forms can with equal justice be alleged of tensed ones. Some sort of asymmetry between tensed and tenseless forms must be made out. In the present context, it is hard to see what could establish the existence of tenseless facts that would not also open the way to tensed facts.

Part of our difficulty, of course, is with the question: What is a fact? If it rained yesterday (and no one ought to be disputing the possible truth of such statements), then presumably I can say that it is a fact that it rained yesterday. This would be a very swift argument to the existence of a past-tensed fact. In using the term 'fact', it seems, the proponent of the 'tenseless' view must have something else in mind.

One can know facts, be annoyed by facts, marvel at facts. Arthur Prior, in a celebrated article (Prior, 1976), pointed out that we can thank goodness for certain facts. What, then, am I thanking goodness for when, in relief at the abatement of a toothache, I cry, "Thank goodness that's over!"? As Prior wrote of this exclamation:

> "It certainly doesn't mean the same as, e.g. "Thank goodness the date of the conclusion of that thing is Friday, June 15, 1954", even if it be said then. (Nor, for that matter, does it mean "Thank goodness the conclusion of that thing is contemporaneous with this utterance". Why should anyone thank goodness for that?)"

Not only does "Thank goodness that's over!" not *mean* anything tenseless along the lines given by Prior, what one who utters it thanks goodness *for* is surely not any tenseless fact. Rather, one thanks goodness that the toothache, say, is past. And this, apparently, is a tensed fact. So a complete description of the world ought to make mention of tensed facts – such as the fact that my toothache is over.

One can thank goodness for other 'perspectival' facts, of course. A businessman faced with impending bankruptcy, having just won the lottery, might think to himself: "Thank goodness that I won the lottery". And this isn't the same thought as: "Thank goodness that Henry Paunch won the lottery", even if the businessman is Henry Paunch (if only because Paunch might be suffering from amnesia, and have forgotten his name). Does this show that a complete description of the world should include mention of the fact for which Paunch thanks goodness, *as well as* of the fact that Henry Paunch won the lottery? This would seem absurd. After all, *such* a description of the world would be one that only Paunch could give: only he could include the statement, "I won the March lottery". And a *proper* description of the world, it might be said, should be one that anyone can give. So it should be one that simply includes the statement, "Henry Paunch won the March lottery".

Now, the tensed fact for which (apparently) I thank goodness when I exclaim, "Thank goodness that's over!", may be one that people other than me can know and can talk about. If I am thanking goodness that it has stopped raining, others too can say, "It has stopped raining". So a description of the world that included "It has stopped raining" would be one that anyone could give. However, it would be a description that could only be given at a certain time. A bit earlier, when it was still raining, that world-description would be false. It may now be claimed that a proper world-description, as well as being one that could be given by any person, should be one that could be given at any time.

1.7. Complete *vs* maximal

Michael Dummett thought that it was an implicit adherence to this last principle that lay behind McTaggart's attack on the A-series. He wrote:

> "The description of what is really there, as it really is, must be independent of any particular point of view. Now if time were real, since what is temporal cannot be completely described without the use of token-reflexive expressions [i.e. tensed expressions], there would be no such thing as the complete description of reality. There would be one, as it were, maximal description of reality in which the statement "The event M is happening" figured, others which contained the statement "The event M happened", and yet others which contained "The event M is going to happen".' (Dummett, 1978, p. 356.)

In Dummett's terms, the question can be put as: Is a maximal description of the world a complete description of the world? For a maximal description ot the world to be complete, one might think that it was enough for it to mention all the events and facts of history. And if "The event M is happening" alludes to the same fact or event as the subsequent statement, "The event M happened", it will surely be true that a maximal world-description *must* be complete. Recall that the typical proponent of the 'tenseless' view of time does think that both of the above tensed statements allude to (are 'made true by') the same fact; so that he must admit that a tensed maximal description of reality is complete, if completeness is as I have characterised it.

If, on the other hand, "The event M is happening" mentions a *different* fact from the one mentioned by the subsequent "The event M happened", then no maximal description of the world will be complete; but nor will any description of the world. Only a description that includes "The event M is happening" will mention *that* fact; but that description cannot include "The event M happened" – only the subsequent description can. And naturally, any tenseless description will leave out both of these facts.

When Dummett writes, ". . . there would be no such thing as the complete description of reality", it looks as if the point he is making is that any maximal world-description (and so any world-description) must be incomplete for the reasons outlined in the last paragraph. In the context of McTaggart's argument, the way to block the move to the unreality of time would then seem to be by asking: "Why *must* there be such a thing as a complete description of reality? Don't our considerations rather go to show that there cannot be such a thing?" McTaggart's 'proof' would in this case be no better than the following proof of the unreality of the physical universe:

(1) any description of the physical universe must be one that any normal adult can comprehend;
(2) an adequate description of the physical universe will be incomprehensible to some normal adults; so
(3) there is no physical universe.

The weak point in this attempted proof, of course, lies in the first premise. Premise (2), in fact, shows that premise (1) must be false. Analogously, if the viewpoint of the last paragraph is correct, that surely shows that the 'requirement' that reality be completely describable is a bogus one – not that time is unreal.

In fact, Dummett's remark ("there would be no such thing as the complete description of reality") follows on from his remark that "the description of what is really there, as it really is, must be independent of any particular point of view". This is rather odd. To start with, as we have just seen, the most likely reason for denying that a maximal world-description could be complete is the belief that a description of what is really there, as it really is, must *take account of* the particular point of view (i.e. point in time) from which it is made: it must *include* tensed statements such as "The event M is happening" and therefore exclude certain others, such as "The event M happened". To back up the claim: "there is no such thing as the complete description of reality", one should say: "a description of reality must be *dependent* on a particular point of view".

Secondly, it is unclear where the requirement of independence from any particular point of view could come from, in any case. It certainly doesn't derive from the meaning of 'complete', in 'complete description'.

Let us return to Henry Paunch. It was suggested that we need not include "I won the March lottery" in our description of the world; this statement, and others made from Henry Paunch's point of view, were to be excluded. But the main reason for this exclusion was that there was no need for "I won the March lottery" *in addition to* "Henry Paunch won the March lottery". For each of these statements (it was being presumed) alluded to the same fact. Actually, this observation would not support the claim that a proper description should be independent of any point of view: for Henry Paunch would, *ex hypothesi*, manage to mention the fact in question by saying, "I won the March lottery". (As he could talk of wirelesses instead of radios, if he wanted to.) We might require that the impersonal statement be used instead of the personal one on the grounds that it's good to have a description of things that we can all use. ("That would be simplest".) But this is surely nonsense; nobody is actually trying to formulate a complete description of reality. It follows that no pragmatic considerations, such as those that have to do with standardisation, can be relevant. Otherwise we might as well argue that the proper description of the world should be in Esperanto.

Similar remarks go for the alleged requirement that a description of the world should be one that could be given at any time. If tensed statements allude to the same facts as (putative) tenseless statements, why not use the tensed statements if one wants to? If, on the other hand, tensed statements correspond to their own, tensed, facts, then a proper description of the world will surely be one that (because tensed) cannot be given at any time whatever.

1.8. Conclusion

I have not here argued for a particular view of tensed statements, nor of putatively tenseless statements. Rather, I have tried to show that, whatever (reasonable) view of such statements one holds, it is difficult to mount a convincing attack on the A-series which relies on the notion of a true and complete description of reality.

The argument that science, or rather physics, provides reasons to reject the A-series can be faulted on a number of grounds: the argument from special relativity only works, if it works, against an 'absolute' conception of the present; the point that physics eschews tenses, if true, would show only that a tenseless description of reality was possible, not that a description of reality *as tenseless* was possible.

Could tensed forms be made out to be redundant? The view that tensed forms are synonymous with tenseless forms, as well as being implausible, would not establish this; nor would any view that failed to set up an asymmetry of the right sort between tensed and tenseless forms.

What of the idea that there are only tenseless *facts*? To begin with, there are *prima facie* problems here: e.g. if it will rain, then it is a fact that it will; and one can seemingly thank goodness for tensed facts. But in any case, the 'tenseless' view faces a dilemma: either "p" and the subsequent statement, "It was the case that p", record different facts, which ought then both to be included in a description of reality – or they record (are made true by) the same fact, in which case a world-description including such statements as "p", or "It was the case that p", will be *true and complete*. And admitting that "p" and the later "It was the case that p" both record the same fact need not be to admit that the fact in question is tenseless. If you ask me, "*What* fact, then, is it that both the present-tensed statement and the past-tensed one record?", I need not answer by giving a tenseless proposition; for I can say, "The fact that p", if it is still true that p when I give my answer, or alternatively, "The fact that it was the case that p", if it is no longer true then.

And it may actually be *impossible* for me to answer by giving a tenseless proposition: if, that is, it is impossible for there to *be* any genuinely tenseless propositions. That it is impossible I have argued elsewhere (Teichmann, 1995). But even if reality could be tenselessly described, this, as I hope to have shown, would not in any way undermine the view that past, present and future are objective features of reality.

Notes

1. This is not to say that Einstein alone was the source of such ideas, in Russell or in anyone else. The gist of 'On the Notion of Cause' (in Russell, 1918) is that we should replace the notion of a cause with the notion of a state of the universe's being a function of another state of the universe (earlier *or* later), an idea perhaps more reminiscent of Laplace than of Einstein.

2. Otto Neurath, one of the leading logical positivists, developed a theory of *Protokollsätze*, or 'basic propositions', from which all perspectival terminology was eliminated. The full protocol-version of, "Otto saw a table", would, according to Russell, be:

> "Otto's protocol at 3.17: {Otto's word-thought at 3.16 (In the room at 3.15 was a table perceived by Otto)"}.

Russell was not happy with the philosophical views which went with the theory of *Protokollsätze*; but he still seems to have thought that in physics, it is propositions of this sort, tenseless and impersonal, that are asserted.

References

Dummett, M.A.E., 1978, 'A Defence of McTaggart's Proof of the Unreality of Time'. In: Truth and Other Enigmas, Duckworth.

McTaggart, J.M.E., 1927, The Nature of Existence, **1**, Cambridge.

Mellor, D.H., 1981, Real Time, Cambridge.

Prior, A.N., 1967, Past, Present and Future, Oxford.

Prior, A.N., 1976, 'Thank Goodness That's Over'. In: Papers in Logic and Ethics, (ed). Geach and Kenny, Duckworth.

Russell, B., 1918, Mysticism and Logic, Longmans, Green and Co.

Russell, B., 1940, An Inquiry into Meaning and Truth, Allen and Unwin.

Teichmann, R.P.L., 1995, The Concept of Time, Macmillan.

Wittgenstein, L., 1969, The Blue and Brown Books, trans. G.E.M. Anscombe, 2nd ed., Blackwell.

chapter 2

THE TEMPORALIZATION OF TIME IN MODERN PHILOSOPHY

MIKE SANDBOTHE
Friedrich-Schiller University
Jena

Time in Contemporary Intellectual Thought
Patrick Baert (Editor)
2000 Elsevier Science B.V.

CONTENTS

2.1. Introduction

The subject of 'time' has occupied both scholars and the layman repeatedly throughout the course of the twentieth century, but it has acquired particular importance and a certain brisance over the last two decades. The current vogue for time is a multidisciplinary one, in fields ranging from the humanities, social natural sciences, history, literature, media theory and linguistics through to medicine, law, sciences and economics (Macey, 1991). This situation, characterized by a plurality of heterogenous concepts, lends particular significance to philosophical debate about the problem of time (cf. Wood, 1989; Baumgartner, 1993; Le Poidevin/McBeath, 1993, Zimmerli/Sandbothe, 1993, Gimmler/Sandbothe/Zimmerli, 1997). The central problem for contemporary philosophy is to relate to one another the varying conceptions of time developing in individual disciplines. The considerations in the following three sections seek to make a basic level theoretical contribution to this undertaking. To this end, in the first section (2.2), I extract three basic tendencies, representing differing strategies for solving the outlined task of relating the different concepts of time in a transdisciplinary manner. In the second and third sections I attempt to anchor one of these three tendencies – that which to me seems particularly tenable for the solution of the outstanding problems – in philosophical history with recourse to Kant (2.3) and Heidegger (2.4) and to argue for its plausibility (cf. Sandbothe, 1998).

2.2. The three basic tendencies in contemporary philosophy of time

The first basic tendency in contemporary philosophy of time may be described as the tendency to unify our understanding of time. The protagonists of this unification tendency are convinced that time's validity is that of being a new Archimedean point which unifies our everyday experience of the self and the world with our academic theories about nature and man. This point of unification, they contend, has been emphasized time and time again in philosophy (for instance by von Baader, Schelling, Bergson or Whitehead), but has been ignored for far too long by science and technology. It wasn't until the second half of this century that a global time concept was developed and mathematically implemented at the interface between physics, chemistry and biology within the framework of the so-called theories of 'self-organization' (cf. Griffin, 1986 and Krohn/Küppers/Nowotny, 1990). According to the proponents of the unification tendency this new conception of time enables the old duality between natural time and historical time to be overcome and resolves the conflict between physical, biological and philosophical approaches to time which had characterized the first half of the twentieth century. Against this background the German theoretician of time, Hermann Lübbe, observed, "that even the temporal structure of historicality, which, according to Heidegger and the hermeneutic theory which followed him, results exclusively from the subject's relationship to itself, which constitutes meaning, is in reality a structure belonging to all open and dynamic systems which is indifferent to the subject matter" (Lübbe, 1992, p. 30). Lübbe's convergence theorem can be supported by the deliberations of the Nobel prize-winning physicist, chemist and self-organization theorist Ilya Prigogine, who noted as early as 1973, in the light of his thermodynamic theory of irreversibility: "Whatever the future of these ideas, it seems to me that the dialogue between physics and natural philosophy can begin on a new basis. I don't think that I can exaggerate by stating that the problem of time marks specifically the divorce between physics on one side, psychology and epistemology on the other. (. . .). We see that physics is starting to overcome these barriers" (Prigogine, 1973, 590f. cf. Prigogine, 1980, 1997, Prigogine/Stengers, 1984, 1988, Sandbothe, 1998, pp. 7–73, 124–131).

The convergence theorem, which lies at the heart of the unification tendency in contemporary philosophy of time, remains, however, by no means undisputed. Paul Ricoeur, the French phenomenologist, opposes it for example with his diagnosis of an inpenetrable incommensurability between historical and natural time. Ricoeur's outset serves to illustrate the second basic tendency in contemporary philosophy of time, namely the tendency to split time into a multitude of mutually incompatible heterogenous concepts. As a representative of this pluralization tendency, Ricoeur regards "the break, on the level of epistemology, between phenomenological time on the one hand and astronomical, physical, and biological time on the other" (Ricoeur, 1988, p. 94) as being insurmountable. In the light of the underlying discontinuity which exists "between a time without a present [natural time – M.S.] and a time with a present [historical time – M.S.]" (Ricoeur, 1988, p. 91), Ricoeur describes the alleged coherence between the two heterogenous understandings of time as being "a phenomenon of mutual contamination" through which "the notion of history had been extrapolated from the human sphere to the natural sphere" (Ricoeur, 1988, p. 90).

From Ricoeur's perspective the "reciprocal overlapping of the notions of change (or evolution) and history" (Ricoeur, 1988, p. 90) is factually without foundation, and is, as such, to be refuted. Ricoeur reasons, "whatever the interferences between the time with a present and the time without a present, they presuppose the fundamental distinction between an anonymous instant and a present defined by the instance of discourse that designates this present reflexively" (Ricoeur, 1988, p. 91). Hence, to Ricoeur, "it (. . .) seems impossible (. . .) to include phenomenological time in the time of nature, whether it is a question of quantum time, thermodynamic time, the time of galactic transformations, or that of the evolution of species" (Ricoeur, 1988, p. 91). Phenomenological time consisting in a dimension of future, past and present is explained by Ricoeur as being appropriate only in the narrative medium; and time in the narrative 'refiguration' (Ricoeur, 1988, 180f. cf. Ricoeur, 1984) itself becomes comprehensible only up to a point. Finally, for Ricoeur, time marks the 'mystery' (Ricoeur, 1988, p. 274) in our thought which denies representation in that our existence irrevocably pervades our thinking. This negativist feature of Ricoeur's is also found in Emmanuel Levinas's (Levinas, 1979) and Michael Theunissen's (Theunissen, 1991) philosophical thoughts on time.

The third basic tendency in contemporary philosophy of time is best seen when one considers the common assumptions which relate the thesis of convergence and its antithesis of an irreconcilable divergence between natural and historical time. In both cases time is considered as being a basic universal structure which disavows itself of historical contingency and cultural change. Those who advocate the tendency towards unification regard the "ontological universality of the temporality aspect" (Lübbe, 1992, p. 31) as having been proved through the unity of historized natural time in theories of self-organization. Proponents of the incommensurability of time argue quite differently, but reach a similar result. According to them, the plurality of time refers to a negative unity in time, which even in principle defies representation, but which seems to be factually evident through the experience of its unrepresentability. Ricoeur for example sees the narrative 'totalization' (Ricoeur, 1988, pp. 193–206, 249–261) and the associated "universal ambition of the metahistorical categories of historical thought" (Ricoeur, 1988, p. 215) as being confirmed in the essentially irreconcilable phenomenological 'fundamental' (Ricoeur, 1988, p. 273) of our experience of time.

The third basic tendency which is of importance in the contemporary philosophy of time deviates from the two previously discussed with regard to the universality and ahistoricality of time presupposed by the first two tendencies. Supporters of the third tendency, the tendency to relativize and historize time, assume that the role time plays for human understanding of the self and of the world is an aspect of practical means of interaction with the world, one which is culturally divergent and, within individual cultures, subject to change over time. The American pragmatist Richard Rorty represents this basic idea with particular refinement. The fundamental premise of his thinking is "that a belief can still regulate action, can still be thought worth dying for, among people who are quite aware that this belief is caused by nothing deeper than contingent historical circumstances" (Rorty, 1989, p. 189). According to Rorty, a radical approach to time must do away with the conception, based on theology, that time and eternity come together in man (Rorty, 1995). Instead of this Rorty demands, "that we [should] try to get to the point where we no longer worship *anything*, where we treat *nothing* as a quasi divinity, where we treat *everything* – our language, our conscience, our community – as a product of time and chance" (Rorty, 1989, p. 22). According to Rorty we will only achieve this when we no longer mystify time, but understand it in a radically reflexive way as being a product of chance and human action.

The interrelations between the different conceptions of time currently being discussed in academia, as well as the question of the relationship between academic and everyday perceptions of time, are, on Rorty's account, to be dealt with pragmatically on the basis of the historization tendency. The convergence of different vocabularies of time is, from Rorty's perspective, by no means proof of an intrinsic coincidence between natural and historical time. The transfer of the vocabulary of historical time from the context of human self-description into the realms of the natural world, as well as the mathematical operationalization of time, illustrate only the historical ability to adapt, inner flexibility and contextual feedback even in a highly attuned vocabulary such as that found in physics and mathematics. In Rorty's view, the different vocabularies which we make use of for differing purposes and in varying contexts are to be understood as neither convergent in an intrinsic sense, nor as being essentially incommensurate in a phenomenological sense. Rather, they are themselves subject to change over time, through which they become related and disjoined in various ways according to the various historical situations which arise.

The radical temporalization of time expressed in these deliberations has already been outlined in literature by the Austrian novelist Robert Musil. In his novel *The Man without Qualities* he writes, "The train of events is a train unrolling its rails ahead of itself. The river of time is a river sweeping its banks along with it. The traveller moves about on a solid floor between solid walls; but the floor and the walls are being moved along too, imperceptibly, and yet in very lively fashion, by the movements that his fellow-travellers make" (Musil, 1954, p. 174).[1] The inner reflexivity in the modern apprehension of time, which Musil enounces here, was introduced within philosophy by the differing approaches of Kant and Heidegger respectively. The following parts of my considerations concern themselves with this twofold foundation, in which the debate between universality and ahistoricality of time on the one hand and relativity and historicality of time on the other hand is central.

[1] I am grateful to Wolfgang Welsch for his advising me of this reference, one of eminent appropriateness to the philosophy of time.

2.3. The reflexive temporalization of time in modern philosophy by Kant

The transcendental philosophy of time, from Kant's *Transcendental Aesthetic* in the *Critique of Pure Reason*, may be described as the Magna Carta of modern philosophy of time. In the *Critique*, Kant ordained time as being reflexive, i.e. with recourse to the basic constitution of human subjectivity as being "a pure form of sensible intuition" (Kant, 1985, p. 75 [B 47]). There is almost no single philosophical theory which has been misunderstood as often as Kant's designation of time as a pure form of sensible intuition. The standard misinterpretation is that Kant, with his theory, had refuted the reality of time and downgraded it to being a mere subjective illusion. This miscomprehension is widespread not only amongst philosophers but, above all, amongst scientists.

The following quote from the British philosopher, and founder of analytical philosophy of time, John M.E. McTaggart provides a significant example of the insistence with which this miscomprehension had established itself within philosophy. In his famous essay *The Unreality of Time* he writes, "In philosophy, again, time is treated as unreal by Spinoza, by Kant, by Hegel and by Schopenhauer" (McTaggart, 1908, p. 31). Scientists such as Albert Einstein or Kurt Gödel also went along with this prejudice. Gödel, whose view was that time had lost its 'objective meaning' (Gödel, 1970, p. 557) through the 'relativity of simultaneity' (Gödel, 1970, p. 557) which Einstein had proven, writes: "In short, it seems that one obtains an unequivocal proof for the view of those philosophers who, like Parmenides, Kant, and the modern idealists, deny the objectivity of change and consider change as an illusion or an appearance due to our special mode of perception" (Gödel, 1970, p. 557). Just as Gödel praises Einstein's work as being the physical evidence of the unreality of time, McTaggart commends his own work as being an analytic variant of the proof of time's unreality, as is allegedly demanded by Kant. In this context McTaggart writes, "I believe that time is unreal. But I do so for reasons which are not, I think, employed by any of the philosophers whom I have mentioned (. . .)" (McTaggart, 1908, p. 31).

At this point it would be going too far to expand on McTaggart's proof in detail. In summary, however, it may be said that what McTaggart proves is nothing other than what Kant had shown long ago: namely not – as McTaggart believed – that time is absolutely unreal, but rather that time has no reality which is independent of the subject. This is an important difference. If time has no subject-independent reality, then it lacks only a certain kind of reality – and not reality altogether. Thus it is not the case that time is unreal in an indiscriminate sense, and just a mere illusion. Further, to have no subject-independent reality is by no means a deficit which devalues time's being in contrast to other things. As Michael Dummett highlighted in his essay *McTaggart's Proof of the Unreality of Time: A Defence*, (Dummett, 1960), the idea of time as a subject-independent, fully describable reality is in itself a fiction. A fiction which presupposes that we have access to a world through which – detached from our finite perceptory conditions – we connect, in a quasi-devine sense, with entities" inner-being. It is this fiction to which Kant had put an end.

In contrast to the assumptions made by McTaggart and Gödel in the quotes mentioned here, it was by no means Kant's aim to question the objectiveness of time by reducing it to the level of 'illusion' or 'mere appearance'. The Kantian coupling of time, traditionally conceptualized as a structure of the world in itself by Newton and Leibniz, with the transcendental subject is far more an attempt to base time's objectivity on a new, transcendental, plane whilst considering the justified doubts expressed by Hume about the traditional Leibnizian-Newtonian view. The point of Kant's reasoning is that time can be ineluctable and a priori – i.e. generally valid and necessary – only when it is proven to be an intersubjective condition for the possibility of knowledge in general.

Kant accentuates sensible intuition as the fundament of all human knowledge – this in contrast to the traditional, Platonic idea of knowledge, prevalent until the time of Leibniz and Newton, according to which only the intelligible can be a true object of knowledge. This basic proposition is contained in the first sentence of the *Critique of Pure Reason*, which reads, "In whatever manner and by whatever means a mode of knowledge may relate to objects, intuition is that through which it is in immediate relation to them, and to which all thought as a means is directed" (Kant, 1985, p. 65, [B33]). It is this basic proposition, of the primacy of intuition as the main condition in enabling all human knowledge, which one must consider in order to understand how far Kant's proof that time is "a pure form of sensible intuition" (Kant, 1985, p. 75 [B 47]) simultaneously assures its empirical objectivity and transcendental quasi-universality.

Kant's simple proposition, which Gödel fails to consider along with most other scientists who have come up against the Kantian conception of time, is that all knowledge accessible to humankind – and that includes humankind in our pursuing science (e.g. in physics) – is sensible, that is temporal intuition. Thus Kant tries to assert the empirical objectivity of time through its transcendental subjectivism. This connection is expressed in the following, much cited, excerpt from the *Transcendental Aesthetic*. First it appears that Kant really does want to deny time all reality. He writes, "Time is therefore a purely subjective condition of our (human) intuition (which is always sensible, that is, so far as we are affected by objects), and in itself, apart from the subject, is nothing" (Kant, 1985, p. 77 [B 51]). However, the next sentence, which is mostly omitted in citation, is decisive. This states, "Nevertheless, in respect of all appearances, and therefore of all the things which can enter into our experience, it is necessarily objective" (Kant, 1985, p. 78 [B 51]). With this in mind, Kant then speaks of the 'empirical reality' of time, that is to say of its "objective validity in respect of all objects, which allow of ever being given to our senses" (Kant, 1985, p. 78 [B 52]).

Along with the misconception, mentioned above, about the unreality of time, an influential second shortfall distinguishes the reception of Kant's philosophy of time. Strictly speaking, this shortfall is less a misunderstanding than a failure to understand, that is to say, perceiving the theory with a narrowed outlook. Decisive aspects of Kant's thinking about time have long been eclipsed, in the mould of Schopenhauer and Hegel, by its being equated with the treatment of time in the *Transcendental Aesthetic*. This equation was addressed, with recourse to insights gained from Heidegger's book *Kant and the Problem of Metaphysics* (1929), by the German philosopher Klaus Düsing in his *Examination of Kant's Theory of Time and its Critical Modern Reception* (Düsing, 1980). As Düsing emphasizes at the beginning of his examination, "Kant's theory of time is of course contained only incompletely in the *Transcendental Aethetic* of the *Critique of Pure Reason*; essential details of this theory are found in the following sections (. . .)" (Düsing, 1980, p. 2). Similarly in Section 10 of Heidegger's *Kant and the Problem of Metaphysics* we read: "The following interpretation will reveal how time in the course of the development of the several stages of the foundation of metaphysics comes more and more to the fore and thereby reveals its proper essence in a more original way than is possible by means of the provisional characterization in the *Transcendental Aesthetic*" (Heidegger, 1962, p. 52).

The obscuring of Kant's thoughts on time which go beyond the *Transcendental Aesthetic* is based on another, profoundly narrowed outlook, upon which Düsing does not expand. This narrowing of outlook consists in the failure to have apprehended the fact that Kant himself had not only implicitly but also explicitly relativized his own transcendental universalization of time. Whilst the transcendental universality of time in modern philosophy after Kant is retained as a dimension which constitutes the subjectivity of the subject by both the finite, intentional subject (Husserl) and the living self of pure duration (Bergson), Kant himself had already questioned the universality of time which he had initially presupposed. In so doing he opened up a field of discussion which was further set out by Heidegger and is currently addressed by Rorty, Derrida, Lyotard and others.

Kant's relativization of the time, whose transcendental universalization was completed in the *Transcendental Aesthetic*, is not found in the *Transcendental Aesthetic* itself, but is developed contiguously within the context of his *Transcendental Logic*. The distinction which Kant makes, in a footnote in the B edition of the *Transcendental Deduction*, between time as a "form of intuition" and as "formal intuition" (Kant, 1985, p. 170 [B 160]) is central here. The distinction to which this footnote relates is already introduced in the main text. The main text reads, "In the representations of space and time we have a priori *forms* of outer and inner sensible intuition; and to these the synthesis of apprehension of the manifold of appearance must always conform, because in no other way can the synthesis take place at all. But space and time are represented a priori not merely as *forms* of sensible intuition, but as themselves *intuitions* which contain a manifold [of their own], and therefore are represented with the determination of the *unity* of this manifold (*vide* the Transcendental Aesthetic)" (Kant, 1985, p. 170 [B 160]). Thus, according to Kant's own understanding, the subject of the *Transcendental Aesthetic* is not the form of intuition as such, but a quasi-objective construction: time as formal intuition.

This is made quite explicit in Kant's annotation. This reads, "Space, represented as *object* (as we are required to do in geometry), contains more than mere form of intuition; it also contains *combination* of the manifold, given according to the form of sensibility, in an *intuitive* representation, so that the *form of intuition* gives only a manifold, the *formal intuition* gives unity of representation" (Kant, 1985, p. 170 (B 160f). With an eye to the *Transcendental Aesthetic* the annotation continues, "In the Aesthetic I have treated this unity as belonging merely to sensibility, simply in order to emphasize that it precedes any concept, although, as a matter of fact, it presupposes a synthesis which does not belong to the senses but through which all concepts of space and time become possible. For since by its means (in that the understanding determines the sensibility) space and time are first *given* as intuitions, the unity of this a priori intuition belongs to space and time, and not to the concept of the understanding" (Kant, 1985, p. 170 [B 161]). The conceived unity of time in the *Transcendental Aesthetic*, which in the "Transcendental Exposition of the Concept of Time" (Kant, 1985, p. 76 [B 48]) at the same time serves as a fundament of the "general doctrine of motion" (Kant, 1985, p. 76 [B 49]), in itself preconceives time as an objectivized, and thus conceptual and categorial synthesis. It is this unifying, linear conception of time, which may be introduced 'by analogies' (Kant, 1985, p. 77 [B 50]) in describing "a line progressing to infinity" (Kant, 1985, p. 77 [B 50]), which Kant universalizes in its 'empirical reality' (Kant, 1985, p. 78 [B 52]) aiming at a new epistemological foundation of Newton's physics.

At the same time, however, the second representation of time on which the concept of objectivized time is based evades transcendental philosophical explanation. For time, as a form of intuition in the strict sense, forms the horizon which Kant fails to illuminate, in which time can first be dealt with as formal intuition. The universality of the notion of objective time in the *Transcendental Aesthetic* is decentralized and at the same time methodically relativized by this horizon's irrevocable transcendental philosophical dimension. In this context Heidegger underlines in Section 9 of his *Phenomenological Interpretation of Kant's Critique of Pure Reason* (Heidegger, 1977), that, from Kant's perspective, "formal intuition is not a primordial, but a derived conception" (Heidegger, 1977, p. 132). How Heidegger's analysis of temporality presents itself against the background of this insight is to be elaborated in what follows.

2.4. The reflexive temporalization of time in modern philosophy by Heidegger

Heidegger developed his analysis of temporality in the second division of the first part of his *Being and Time* (1927). In this, Heidegger's early, still fragmentary master work one must differentiate between two things: the non-realized, but suggested undertaking of a fundamental ontology, and the factually accomplished analysis of *Dasein.* In the following I shall concentrate on the analysis of temporality developed in the second division of *Being and Time*, which bears the title *Dasein and Temporality.* The work's broader perspective of a fundamental ontology is drawn upon only insofar as it affects the analysis of temporality.

Unlike Husserl and Bergson, who did not directly relate their theories of time to Kant's, Heidegger's early thinking takes issue directly with Kant. This was clearly expressed in his lecture *Phenomenological Interpretation of Kant's Critique of Pure Reason*, in the year *Being and Time* was published, as well as in the book *Kant and the Problem of Metaphysics*, published in 1929, and in references to Kant found in *Being and Time* itself. In directly contesting Kant, Heidegger broke through the theoretical means of approach to the problem of time, as established by the *Critique of Pure Reason* and retained by Bergson and Husserl.[2] With Heidegger, the question of time as a pure form of sensible intuition – which is left open by Kant and reformulated by Bergson and Husserl as a question of the intrinsic temporality in subjectivity – becomes a question of the genuinely practical means of temporal self-projection in human existence.

'*Dasein*' is Heidegger's alternative term for what is called 'subject' or 'I think' in Kant. Heidegger is of the opinion that Kant reduces the transcendental subject to being an aspect of theoretical knowledge. According to Heidegger, man is not a creature which aims first and foremost to cognize the present-at-hand (*das Vorhandene*). As *Dasein*, he is far more a being which has always been cast amidst its 'there' (*Da*), and thus did not first begin artificially and retrospectively to construct an epistemological relationship to the outside world, but rather one which had always found itself practically related to a concrete environment – to the 'ready-to-hand' (*das Zuhandene*) (Heidegger, 1992a, p. 98).[3] Here Heidegger highlights: "The 'I' is not just an 'I think', but an 'I think something' " (Heidegger, 1992a, p. 367). And he explains: "Kant has indeed avoided cutting the 'I' adrift from thinking; but he has done so without starting with the 'I think' itself in its full essential content as an 'I think something', and above all, without seeing what is ontologically 'presupposed' in taking the 'I think something' as a basic characteristic of the self" (Heidegger, 1992a, p. 367). This postulate is the 'Being-in-the-world' of *Dasein.* Since Kant "did not see the phenomenon of the world"

[2] See my reconstruction of Bergson's and Husserl's philosophies of time in Sandbothe, 1998, 82–98.
[3] For an analysis of 'Heidegger's Pragmatism' in *Being and Time*, see Okrent, 1988.

(Heidegger, 1992a, p. 368), the basic Heideggerian insight must continue to obstruct him: "In saying 'I', *Dasein* expresses itself as Being-in-the-world" (Heidegger, 1992a, p. 368).

Like Kant, Heidegger also asks about the conditions of possibility. For him, however, it is not an abstract enquiry about the possible conditions of knowledge, but quite concretely about the possible conditions of our Being-in-the-world. In the second division of *Being and Time* Heidegger reveals that 'temporality' (Heidegger, 1992a, p. 274 and elsewhere) is the basic existential structure forming the fundamental dimension underlying Dasein's structure of care (*Sorge*), upon which he had expanded in the first division of *Being and Time*. With recourse to Kierkegaard he describes the 'double-movement' (Kierkegaard, 1987, pp. 36, 119; cf. Kierkegaard, 1980) which effects the *Da* ('there') in Dasein, and which opens the world, as a doubly temporal occurrence. The first partial movement in this occurring exists in the anticipation of the future; the second partial movement in coming back to the present as an openness for the encountered world determined by the past – or as Heidegger put it, the 'having been' (Heidegger, 1992a, p. 373). In summary Heidegger writes: "Coming back to itself futurally, resoluteness brings itself into the Situation by making present. The character of 'having been' arises from the future, and in such a way that the future 'has been' (or better, which is 'in the process of having been') releases from itself the present. This phenomenon has the unity of a future which makes present in the process of having been; we designate it as '*temporality*'" (Heidegger, 1992a, p. 374).

Here, on the existential level of conditions of possibility, the concern is not the concrete future, determined by certain substantive aims, but the future in general, of which is written: "By the term 'futural', we do not here have in view a 'now' which has *not yet* become 'actual' and which sometime *will be* for the first time. We have in view the coming [*Kunft*] in which Dasein, in its ownmost potentiality-for-Being, comes towards itself" (Heidegger, 1992a, p. 373). Heidegger's designation of this basic structure of Dasein as 'transcendence' (Heidegger, 1992a, p. 62, 414ff) has also given cause to infer theological implications here. Heidegger defended himself against such a reading of his work from an early stage. Even in his early lecture *The Concept of Time* to theologians in Marburg in 1924, in which he had just formulated the general ideas behind his analysis of temporality, he emphasized: "The philosopher does not believe. If the philosopher asks about time, then he has resolved *to understand time in terms of time* (. . .)" (Heidegger, 1992b, 1f). To understand time in terms of time means to think about time temporally, or to be in favour of a temporalization of time. Such is the through and through secular Heideggerian programme, and one must also understand his designation of 'futurality' as being the "coming [*Kunft*] in which *Dasein*, in its ownmost potentiality-for-Being, comes towards itself" (Heidegger, 1992a, p. 373) against this backdrop.

Unlike Kierkegaard, for whom the double-movement of human existence only fails to lead us to desperation when it happens in the consciousness of belief in God, Heidegger considers successful temporal self-fulfilment to be possible in the absence of divine transcendence. Heidegger describes the anticipation of one's own future as a 'Being-towards-death' (Heidegger, 1992a, 278ff), but he means that this anticipation of the "possibility of the measureless impossibility of existence" (Heidegger, 1992a, p. 307), which represents death, allows a kind of 'authentic' existence. A kind of existence in which the experience of a radical finiteness does not occasion Kierkegaardian desperation, but which instead opens up a new horizon of manifold possibilities, one within which our everyday *Dasein* was always organized, without the essential characteristics of its possibilities having entered our consciousness. This radical view of 'the future as coming towards' (*Zu-kunft*) in anticipating one's own death as the "ownmost, non-relational possibility, which is not to be outstripped" (Heidegger, 1992a, 310) is also understood by Heidegger to be the self's own "resoluteness" to itself: as authentic "potentiality-for-Being-one's-Self" (*Selbstseinkönnen*) (Heidegger, 1992a, p. 312).

Heidegger contrasts this distinguished basic form of human temporality with its opposite, or what he calls our "everyday understanding of time" (Heidegger, 1992a, p. 278). He attempts to show how the everyday understanding of time arose as a derivative of the original temporality of human *Dasein*. Or to put it another way: Heidegger's goal is to show why and how the objectivized time which we read off our clocks and calendars arises from the temporal processes of our self-constitution, that is, from the authentic temporality of the double-movement of human existence. Heidegger's perception is that we can only hold ourselves temporarily – in distinguished moments of our *Dasein* – in the authentic temporality, or the resolute anticipation of death. In the normal run of things we anticipate a future whose content is determined by our concrete needs and plans, and whose final horizon, death, is excluded. This reduced, practical everyday form of double-movement is what Heidegger calls 'inauthentic temporality' (Heidegger, 1992a, p. 378). Inauthentic temporality is again different from what Heidegger in Section 81 of *Being and Time* calls '*vulgäres Zeitverständnis*', a vulgar[4] or "ordinary conception of time" (Heidegger, 1992a, 472 ff). Whilst in the inauthentic, practical everyday temporality a "reflect[ion of] the ecstatical constitution of temporality" (Heidegger, 1992a, p. 461) can still be sensed, the temporal origin of time from the temporality of human *Dasein* is totally obscured in the vulgar conception of time.

Heidegger makes this distinction quite clear through our use of clocks. He refers here to a paradox which economists and managers of time have yet to

[4] Macquarrie and Robinson translate the German 'vulgär' as 'ordinary' at this and other points. In the following I prefer to translate this as 'vulgar' in the interest of clarity of the distinction between this and inauthentic temporality.

overcome. This paradox is that "precisely that *Dasein* which reckons with time and lives with its watch in its hands (. . .) constantly says 'I have no time'" (Heidegger, 1992b, p. 15). How is it that the greatest strategist of time at once suffers the greatest stress due to time? Heidegger's answer is: because to the professional manager of time, time has congealed into a series of nows of exchangable seconds, minutes, hours, days, weeks, months and years, into an objectivized external authority of time, that is just as an infinitely divisible, endless line lying before him and which he can never really succeed in filling. Objectivized time slips through his fingers. Any time he saves through skilful time management immediately imposes itself on him again as empty and in need of being filled with work. It is no longer concrete concerns and needs which dictate planning of time, rather it is the emptiness itself, which awakens new needs and forces its own capitalization. Whilst this form of dealing with time has become the norm in the second half of the twentieth century (cf. Rinderspacher, 1985), Heidegger was still able to view this vulgar conception of time as being an extreme case, from which the inauthentic temporality could still be clearly delineated. In the practical context of everyday concerns time appears to be not exactly an external, nor still the physically determined power of the clock or 'nature-time' (Heidegger, 1982, p. 262), but to be built into our everyday concerns and determined by these as 'world-time' (Heidegger, 1982, p. 262).

Heidegger identifies the aspects of datability, tension and publicness as being the three central characteristics which distinguish inauthentic temporality from the vulgar conception of time. The Heideggerian standpoint can be shown particularly clearly by taking datability as an example. Whereas in the vulgar conception of time the 'now-point' (*Jetztpunkt*) (Heidegger, 1992a, p. 482) is defined solely through its imminent relation to other now-points, or through the abstract relationship earlier/later, time perceived with respect to everyday concerns is always integrated with concrete regard to daily business, whose datability is provided by: there is a "now that . . ." (*Jetzt, da . . .*) (Heidegger, 1992a, p. 461). In this context Heidegger notes: "When we look at the clock and say 'now' we are not directed toward the now as such but toward that *wherefore* and *whereto* there is still time now; we are directed toward what occupies us, what presses hard upon us, what it is time for, what we want to have time for" (Heidegger, 1982, p. 259). He concludes from this: "The fact that the structure of datability belongs essentially to what has been interpreted with the 'now', 'then' and 'on that former occasion', becomes the most elemental proof that what has thus been interpreted has originated in the temporality which interprets itself. When we say 'now', we always understand a 'now *that* so and so . . .' though we do not say all of this. Why? Because the 'now' interprets a *making-present* of entities. In the 'now that . . .' lies the ecstatical character of the Present. The *datability* of the 'now', the 'then' and the 'on that former occasion', reflects the *ecstatical* constitution of temporality, and is *therefore* essential for the time itself which has been expressed" (Heidegger, 1992a, 460f).

In summary it can be said: In Heidegger's differentiation between authentic temporality, inauthentic temporality and the vulgar conception of time there is a continuation of the relativization of objective time, which Kant began with his distinction between time as 'formal intuition' and time as a 'form of intuition'. Heidegger radicalizes it under the concrete conditions of human Being-in-the-world. This continuation has a dual aspect. On the one hand, Heidegger relativizes the objectiveness on which the vulgar conception of time is based with recourse to the pragmatic, inauthentic temporality which pervades our dealing with time in relation to daily concerns. On the other hand, Heidegger relativizes not only the objectiveness on which the vulgar conception of time is based, but also the pragmatic understanding of time, from which inauthentic temporality arises. This he does with recourse to the superior and, in his view, fundamental form of authentic temporality. From this fundamental form of temporality Heidegger believed he could effect the transition from an analysis of *Dasein* to a fundamental ontology. It simultaneously marks the inner turning point at which Heidegger's phenomenology of the temporality of human *Dasein* is subsumed within and forcibly enshrouded by the fundamental ontological perspective in *Being and Time*.

This last aspect in Heidegger's thinking – the fundamental ontological lapse into a new time theoretical universalism – was brought to the fore by Rorty in his Heidegger critique. In his book *Contingency, Irony and Solidarity* (Rorty, 1989) the American pragmatist writes: "Heidegger seems seriously to have thought, when he was writing *Being and Time*, that he was carrying out a transcendental project, namely, giving an accurate list of the 'ontological' conditions of possibility of merely 'ontic' states. (. . .). Just as Kant seems never to have asked himself how, given the restrictions on human cognition the *Critique of Pure Reason* had discerned, it was possible to assume the 'transcendental standpoint' from which that book was purportedly written, so the Heidegger of this period never looks into the question of methodological self-reference. He never asks himself how 'ontology' of the sort he was busy producing was, given its own conclusions, possible" (Rorty, 1989, 109f). Rorty adds: "In remarking on this early unselfconsciousness, I am not trying to denigrate Heidegger's early (internally inconsistent, hastily written, brilliantly original) book. Heidegger was, after all, not the first philosopher to have taken his own idiosyncratic spiritual situation for the essence of what it was to be a human being" (Rorty, 1989, p. 110).

That the Kantian theory of time should only be partially affected by Rorty's criticism has already been made clear above in reference to the relativization of time by Kant in the *Transcendental Logic*. In closing, it can be shown to be similar for Heidegger. To this end, Rorty's initially positive reading of Heidegger's incipient intentions is quoted. In connection with this, Rorty summarizes the basis of *Being and Time* as follows: "Heidegger would like to recapture a sense of what time was like before it fell under the spell of eternity, what we were before we became obsessed by the need for an overarching context which would subsume and explain us (. . .). To put it in another way: he would like to recapture a sense of *contingency*, of the fragility and riskiness of any human project (. . .)" (Rorty, 1991, p. 34). This productive intention, continues Rorty, was undermined by Heidegger's absolution of authentic temporality and its fundamental ontological elucidation.

Nonetheless, an objection to this essentially justified criticism is that the subsumption of Heidegger's analysis of temporality into a fundamental ontology, although projected, was not carried out in reality. At the same time rudimentary attempts to relativize the model of temporal double-movement, i.e. to understand this temporally in itself, are found in Heidegger's early work on time. This step, on which Heidegger's work borders in several places, marks the principal feature of a radical temporalization of time in its full consequence. This radical temporalization of time, is implicitly anticipated above all in Heidegger's addressing of "the ideas of Count Yorck" (Heidegger, 1992a, 349) which is found in Section 77 of *Time and Being*. Here Heidegger ascertains positively: "And Yorck (. . .) did not hesitate to draw the final conclusion from his insight into the historicality of *Dasein*" (Heidegger, 1992a, p. 453). As evidence Heidegger approvingly quotes from correspondence between Yorck and Dilthey: "Behaviour and historicality are like breathing and atmospheric pressure; and – this may sound rather paradoxical – it seems to me methodologically like a residue of metaphysics to not historicize one's philosophizing" (Yorck, quoted by Heidegger, 1992a, p. 453). Heidegger above all explicitly demands the reflexive temporalization of time in his early lecture *The Concept of Time*. Here Heidegger notes: "(. . .) we must talk temporally about time. Time is the 'how'. If we inquire into what time is, then one may not cling prematurely to an answer (time is such and such), for this always means a 'what'" (Heidegger, 1992b, p. 22). And Heidegger concludes: "*Time itself* is meaningless; time is temporal" (Heidegger, 1992b, p. 21).

If one takes Heidegger literally at this point, a description of *Dasein's* temporality may be given free of the fundamental ontological implications engendered by the singling out of one specific temporal structure as the 'authentic temporality'. In this way the bonding of pragmatic temporality with the formal structure of the temporal double-movement can be retained, without requiring that the hierarchy of temporal structures which Heidegger constructs be adopted. This modification is tantamount to a radical pluralization of Heidegger's analysis of temporality. A plurality is meant here which goes beyond the simple pluralization which typifies the second basic tendency in contemporary philosophy of time. This is the case insofar that it no longer attempts to defuse the plurality of time through the speculative evidence of a negatively conceived unity of time. Rather, in the language of Musil's metaphor introduced above, the banks of the river of time are swept along by the radical historization and relativization of time.

The radical pluralization of time seen in Heidegger has two aspects. First it leads to an internal pluralization insofar as the Heideggerian time structures are no longer to be understood as being founded in a hierarchical context furnished with normative implications (authentic/inauthentic). With this background, the coupling of pragmatic temporality, temporality based on certain projections of the future, within *Dasein's* temporal double-movement is to be understood as being boundness within a horizon from which future first appears to be concretely experienceable and compellingly comprehensible in its contingency. The modification in the apprehension of the temporal double-movement is, secondly, bound to an external pluralization. It no longer concerns only the temporal structures which Heidegger describes, but also incorporates alternative forms of subjectivity and temporality which can no longer be understood under the conditions presupposed by Heidegger's "priority of the future" (Heidegger, 1992a, p. 378). The range of diverse types of temporality is to be considered here, reaching from Kant's 'reflective faculty of judgement', Freud's 'free association', via Proust's '*mémoire involontaire*', Benjamin's '*Jetztzeit*' and Newman's 'now', through to Lyotard's '*passage*' or Derrida's '*écriture*'. Against the background of the differentiations described here the task for philosophically founded transdisciplinary research on time might be to retrace the intertwinements existing between the plurality of time concepts which play varying pragmatic roles both in academic disciplines and in our everyday understanding of self and the world.

Translated by Andrew lukpin

References

Baumgartner, H.-M. (Hrsg.), 1993, Das Rätsel der Zeit. Philosophische Analysen, Freiburg/Munich, Alber.

Dummett, M., 1960, A Defense of McTaggart's Proof of the Unreality of Time. In: Philosophical Review, **69**, 497–504.

Düsing, K., 1980, Objektive und subjektive Zeit. Untersuchungen zu Kants Zeittheorie und zu ihrer modernen kritischen Rezeption. In: Kant-Studien, Bd. 71, 1–34.

Gimmler, A./Sandbothe, M./Zimmerli, W.Ch. (Hrsg.), 1997, Die Wiederentdeckung der Zeit. Reflexionen-Analysen-Konzepte, Darmstadt, Wissenschaftliche Buchgesellschaft and Primus-Verlag.

Gödel, K., 1970, A Remark About the Relationship Between Relativity Theory and Idealistic Philosophy. In: Albert Einstein. Philosopher – Scientist, ed. by Paul Arthur Schilpp, The Library of Living Philosophers, **7**, La Salle (Illinois), Open Court, 557–562 (first published 1949).

Griffin, D.R. (ed.), 1986, Physics and the Ultimate Significance of Time. Bohm, Prigogine, and Process Philosophy, New York, State University of New York Press.

Heidegger, M., 1962, Kant and the Problem of Metaphysics. Translated by James S. Churchill, Bloomington, Indiana University Press (German original first published: Bonn 1929).

Heidegger, M., 1977, Phänomenologische Interpretation von Kants Kritik der reinen Vernunft. In: ders., Gesamtausgabe. II. Abteilung: Vorlesungen 1923–1944, Bd. 25, Frankfurt a.M., Klostermann.

Heidegger, M., 1982, The Basic Problems of Phenomenology. Translated by Albert Hofstadter, Bloomington, Indiana University Press, 1982 (German original first published in: Martin Heidegger, Gesamtausgabe, II. Abteilung: Vorlesungen 1923–1944, Bd. 24, Frankfurt a.M., Klostermann, 1975)

Heidegger, M., 1992a, Being and Time. Translated by John Macquarrie and Edward Robinson, Oxford and Cambridge (Mass.), Blackwell (first published 1962; German original first published: Tübingen 1927).

Heidegger, M., 1992b, The Concept of Time. Translated by William McNeill, Oxford and Cambridge (Mass.), Blackwell (German original first published: Tübingen 1989).

Kant, I., 1985, Critique of Pure Reason. Translated by Norman Kemp Smith, London, Macmillan (first published: London 1929; German original first published: Riga 1781[A]; 1787[B]).

Kierkegaard, S., 1987, Fear and Trembling. Edited and translated with introduction and notes by Howard V. Hong and Edna H. Hong, Princeton, Princeton University Press (Danish original first published: Kopenhagen 1843).

Kierkegaard, S., 1980, The Sickness Unto Death: A Christian Psychological Exposition for Upbuilding and Awakening. Edited and translated with introduction by Howard V. Hong and Edna H. Hong, Princeton, Princeton University Press (Danish original first published: Kopenhagen 1849).

Krohn, W./Küppers, G./Nowotny, H. (ed.), 1990, Selforganization. Portrait of a Scientific Revolution, Dordrecht/Boston/London, Kluwer Academic Publishers.

Le Poidevin, R./McBeath, M., 1993, The Philosophy of Time, Oxford, Oxford University Press.

Levinas, E., 1979, Le Temps et l'Autre, Montpellier, fata morgana.

Lübbe, H., 1992, Im Zug der Zeit. Verkürzter Aufenthalt in der Gegenwart, Berlin/Heidelberg/New York, Springer.

Macey, S.L., 1991, Time. A Bibliographic Guide, New York/London, Garland.

McTaggart, J.M.E., 1908, The Unreality of Time. In: Mind, **17**, 457–474.

Musil, R., 1954, The Man without Qualities, **II**(2), London, Secker and Warburg (German original first published: Berlin, 1930–1933).

Prigogine, I., 1973, Time, Irreversibility and Structure. In: Jagdish Mehra (ed.), Physicist's Conception of Nature, Dordrecht/Boston, D. Reidel Pub, 561–593..

Prigogine, I., 1980, From Being to Becoming: Time and Complexity in the Physical Sciences, San Francisco, W. H. Freeman.

Prigogine, I., 1997, The End of Certainty. Time, Chaos, and the New Laws of Nature, New York, Free Press.

Prigogine, I./Stengers, I., 1984, Order out of Chaos: Man's New Dialogue with Nature, New York, Bantam Books.

Prigogine, I./Stengers, I., 1988, Entre le temps et l'éternité, Paris, Fayard.

Ricoeur, P., 1984, Time and Narrative, **1**. Translated by Kathleen McLaughlin and David Pellauer, Chicago, University of Chicago Press (French original first published: Paris 1983).

Ricoeur, P., 1988, Time and Narrative, **3**. Translated by Kathleen Blamey and David Pellauer, Chicago, University of Chicago Press (French original first published: Paris 1985)

Rinderspacher, J.P., 1985, Gesellschaft ohne Zeit. Individuelle Zeitverwendung und soziale Organisation der Arbeit, Frankfurt a.M. and New York, Campus.

Rorty, R., 1989, Contingency, Irony, and Solidarity, Cambridge, Cambridge University Press, 1989.

Rorty, R., 1991, Essays on Heidegger and

others, Philosophical Papers, **2**, Cambridge, Cambridge University Press.
Rorty, R., 1995, Philosophy and the Future. In: Herman J. Saatkamp (ed.), Rorty and Pragmatism, Nashville and London, Vanderbild University Press.
Sandbothe, M., 1998, Die Verzeitlichung der Zeit. Grundtendenzen der modernen Zeitdebatte in Philosophie und Wissenschaft, Darmstadt, Wissenschaftliche Buchgesellschaft (engl. translation: The Temporalization of Time. Basic Tendencies in Modern Debate on Time in Philosophy and Science, forthcoming)
Theunissen, M., 1991, Negative Theologie der Zeit, Frankfurt a.M, Suhrkamp.
Wood, D., 1989, The Deconstruction of Time, Atlantic Highlands (NJ), Humanities Press.
Zimmerli, W.Ch./Sandbothe, M. (Hrsg.), 1993, Klassiker der modernen Zeitphilosophie, Darmstadt, Wissenschaftliche Buchgesellschaft.

chapter 3

UNDERSTANDINGS OF TIME IN COMPLEMENTARISTIC LANGUAGE

LARS LÖFGREN
University of Lund

Time in Contemporary Intellectual Thought
Patrick Baert (Editor)

CONTENTS

3.1. Time and language

Although we usually think of time as a *dynamic* phenomenon, a little reflection reveals, perhaps surprisingly, that, after all, we do write and communicate our ideas of time in terms of *static* sentences, like those in a paper or book. We do so by using a temporal vocabulary. Obviously, the dynamic/static opposition suggests difficulties for explanations of the temporal vocabulary.

Should we try to describe temporality, and aim at completeness, we encounter a self-referential situation. Time cannot be completely described but with some reference to time itself. In writing about time, we have to use temporal words that are already understood as such.

Notice how this self-referential situation is characteristically tied with attempts at *describing* time. This does not prevent us from communicating intended ideas of time, which we try to write down in sentences, provided that the remaining non-described parts of the intended ideas can be regenerated by the interpreter. This is the case if the interpreter and the writer have in common a *shared language*, *L*. A language, that is, *as a whole of description and interpretation processes*. Fundamental interpretations of *L* may then be shared (on evolutionary grounds), even if beyond complete description in the language *L* itself. The communicating parts may well understand each other's temporal vocabulary even if they cannot describe it completely in *L*. We will later explain this as a *complementaristic understanding*.

A complementaristic understanding of time is well compatible with an awareness like that in St. Augustine's *Confessiones*:

> "What, then, is time? If no one asks me, I know: if I wish to explain it to one that asketh, I know not."

In outline, time is a complex phenomenon going beyond description in the sense that there can be no complete descriptive *theory* of time. A descriptive theory is theory, with proper axioms as well as logical, in a language. Also, *logics* for time turn out to be incomplete. A *language* of time, however, becomes a proper frame of reference – provided that language is comprehended as a *systemic whole*. Beyond that, we cannot go as long as we require our knowledge of time to be *communicable* – holistic language is the ultimate prerequisite for communication.

Both time and language need complementaristic comprehension. Otherwise, we can only describe *fragmented* or *partial* aspects, and the question arises whether the involved fragmentation into parts is justifiable or not. We touch here upon a very basic *part-whole* issue, sometimes discussed as a general *fragmentation problem for descriptions*.

In the next section, we will give some *historical* highlights, referring to this century, of the fragmentation problem for time and for language itself.

In a later section, after having presented the holistic conception of language, and of time, we shall argue that holistic language, which at first may seem similar to language as conceived in the received view of *semiotics*, in fact is more radical, with consequences for the conception of time.

In our account of language, we choose semiotics, rather than linguistics, as a first natural starting point towards holistic language. The reason is simply that linguistics is much too fragmented, whereas semiotics aims at exploring the more global features of language.

3.2. Historical glimpses of the fragmentation problem for time and language

Every description, even a whole descriptive theory, is a description of something, not everything. Were it not for the remarkable property of nature that it seemingly allows fragmentation, as in our becoming conscious of a particular phenomenon as target for description, or as in the isolation of a particular physical phenomenon in an experimental set-up for measuring an observable, every attempt at describing nature would fail. We face a fragmentation problem.

> **The fragmentation problem** concerns the question if whether nature in itself is fragmentable, and thereby non-distortively describable; or if it is our linguistic description processes which make nature appear fragmentable.

In his boot-strap philosophy, Chew explains fragmentation as follows:

> Chew (1968). "A key discovery of Western culture has been the discovery that different aspects of nature can be individually 'understood' in an approximate sense without everything's being understood at once. All phenomena ultimately are interconnected, so an attempt to understand only a part necessarily leads to some error, but the error is sufficiently small for the partial approach to be meaningful. Save for this remarkable and far from obvious property of nature, scientific progress would be impossible."

Difficulties in describing time can be viewed as difficulties in isolating time, as concept as well as physical phenomenon. How well can time be isolated as target for description? Similarly, how well can language be isolated as a target for conceptualization?

3.2.1. Bergson and Russell on disciplinary versus holistic comprehensions of time

In Bergson's work on evolution, Bergson (1911), he introduces the concept of complementarity (in a weak sense, yet somewhat similar to Bohr's later use in quantum physics, and to ours). In his discussions of time, he considers complementary types of knowledge. One is the physical knowledge, namely with time described in a physical theory in terms of "moments of time, which are only arrests of our attention". The other refers to the flow of time, "the very flux of the real".

> Bergson (1911, p. 342). "The first kind of knowledge has the advantage of enabling us to foresee the future and of making us in some measure masters of events; in return, it retains of the moving reality only eventual immobilities, that is to say, views taken of it by our mind. It symbolizes the real and transposes it into the human rather than expresses it."

> Bergson (1911, p. 343). "The other knowledge, if it is possible, is practically useless, it will not extend our empire over nature, it will even go against certain natural aspirations of the intellect; but, if it succeeds, it is reality itself that it will hold in a firm and final embrace. Not only may we thus complete the intellect and its knowledge of matter by accustoming it to install itself within the moving, but by developing also another faculty, complementary to the intellect, we may open a perspective on the other half of the real. . . .
> To intellect, in short, there will be added intuition."

> Bergson (1911, p. 344). "The flux of time is the reality itself, and the things which we study are the things which flow. It is true that of this flowing reality we are limited to taking instantaneous views. But, just because of this, scientific knowledge must appeal to another knowledge to complete it.
> "In our hypothesis, on the contrary [to ancient science and metaphysics], science and metaphysics are two opposed although complementary ways of knowing, the first retaining only moments, that is to say, that which does not endure, the second bearing on duration itself."

We understand Bergson's saying "that of this flowing reality we are limited to taking instantaneous views", as referring to scientific limitations, as in scientific measurements (measurements always transform even dynamic phenomena into static, readable, measurement results). The conclusion "just because of this, scientific knowledge must appeal to another knowledge to complete it", with reference to metaphysics as a complementary way of knowing, is, no doubt, an early insight compatible with the linguistic complementarity to be introduced later.

The insight that science needs a complementary way of knowing, in order to be able to cope with time (without distorting it by fragmentation) was of course a very bold thought at the time when analytic thinking was dominant – which it has also continued to be. Only recently, notably in quantum physics, non-local phenomena have been discovered which cannot be approximated with classical fragmentations into local parts.

Russell, although expressing misgivings, considered the problem seriously:

Russell (1914, p. 157). "The view urged explicitly by Bergson, and implied in the doctrines of many philosophers, is, that a motion is something indivisible, not validly analysable into a series of states. This is part of a much more general doctrine, which holds that analysis always falsifies, because ... [of the following part-hole doctrine]"

Part-whole doctrine. Russell (1914, p. 157). "the parts of a complex whole are different, as combined in that whole, from what they would otherwise be."

We find Russell's formulation of this part-whole doctrine interesting in its close connection with the fragmentation problem, of whether nature can be fragmented in parts, allowing a non-distortive description. In this connection, the doctrine implies that nature is not non-distortively describable.

Russell rejects the doctrine and insists on analysis.

3.2.2. Disciplinary versus holistic comprehensions of language; linguistics versus holistic semiology

In a disciplinary account of logic, as in mathematical logic, the concept of language is either not defined at all, or is considered as partly outside the domain of the discipline. Compare Shoenfield's book on mathematical logic:

Shoenfield (1967, p. 4). "We consider a language to be completely specified when its symbols and formulas are specified. This makes a language a purely syntactical object. Of course, most of our languages will have a meaning (or several meanings); but the meaning is not considered to be part of the language."

Shoenfield's honest account of his disciplinary approach indicates how the fragmentation into mathematical logic 'makes' language devoid of meaning. This is a clearly distortive approach, or a high price to be paid for mathematical clarity.

Also in a wider, philosophical linguistic context, admitting meaning as part of language, the fragmentation problem is apparent:

Putnam (1975, p. 215). "Analysis of the deep structure of linguistic forms gives us an incomparably more powerful description of the *syntax* of natural languages than we have ever had before. But the dimension of language associated with the word 'meaning' is, in spite of the usual spate of heroic if misguided attempts, as much in the dark as it ever was." . . .

In my opinion, the reason that so-called semantics is in so much worse condition than syntactic theory is that the *prescientific* concept on which semantics is based – the prescientific concept of meaning – is itself in much worse shape than the prescientific concept of syntax."

In semiotics, sometimes referred to as the science of language, there is an explicit recognition of syntax and semantics, as well as pragmatics. This emphasis on very central parts of language, is no doubt a step towards a general understanding of language – taken without explicitly recognizing a lurking fragmentation problem. In our opinion, the following quote from Carnap, where he clearly proposes that the whole science of language can be fragmented into three individually understood parts, is to be conceived as a *fragmentation hypothesis* for language.

Carnap's fragmentation thesis. Carnap (1968, p. 9). "If we are analyzing a language, then we are concerned, of course, with expressions. But we need not necessarily also deal with speakers and designata. Although these factors are present whenever language is used, we may abstract from one or both of them in what we intend to say about the language in question. Accordingly, we distinguish three fields of investigation of languages. If in an investigation explicit reference is made to the speaker, or, to put it in more general terms, to the user of the language, then we assign it to the field of **pragmatics**. (Whether in this case reference to designata [what the expressions refer to] is made or not makes no difference for this classification.) If we abstract from the user of the language and analyze only the expressions and their designata, we are in the field of **semantics**. And if, finally, we abstract from the designata also and analyze only the relations between the expressions, we are in (logical) **syntax**. The whole science of language, consisting of the three parts mentioned, is called **semiotic**."

Carnap here seems to take it for granted that the three parts mentioned can be individually understood – at the same time that they, as individually understood, constitute the whole science of language (recall Russell's part-whole doctrine and the fragmentation problem).

We want to mention that semiotics itself has not escaped the scientific drifts towards fragmentation. For example, in Nowakowska (1982) there is a distinction between global semiotics and local, or formal, semiotics.

After having outlined the holistic conception of language, we will at the end of the next section return with a critical comparison between this general understanding of language and carnapian semiotics.

3.3. The complementaristic comprehension of language

Phenomena of language are at the bottom of all human activity and are, indeed, at the roots of all forms of life as genetic processes. The phenomena are extremely rich, and exceedingly difficult to conceptualize without distorting them in the act. Yet, at the same time, our communication languages are so natural and easy for us to use that we hardly notice them. It is as if they were universal, as if what we are saying had an absolute meaning which were independent of the language in use. As if the language could be detached from the ideas we are talking about. Such impressions fade away, however, when we try to objectify, or conceptualize, language.

On the basis of several earlier investigations of language phenomena, from genetic language, through programming language, formal language, observation language, inner cerebral language, to external communication language, we have come to the conclusion that there is a common concept of language, of which all these phenomena are species. That is language as a *whole of complementary description and interpretation processes* (cf Löfgren (1979, 1981, 1984, 1988, 1992, 1993, 1994, 1998)).

Involved is a *linguistic complementarity*, to be formulated below. In Löfgren (1992) we have argued the validity of the linguistic complementarity separated from the functional role of any language, namely to admit communication or control. This requires that the descriptions are *finitely representable*, as well as *locally independent of time* (static). It is the contrast between these conditions and the fact that descriptions in a language, L, can be interpreted in L as non-finite phenomena, as well as dynamic, that are behind the non-describability in L of its interpretation processes (which is one view of the complementarity).

More detailed arguments for the complementarity of specific languages, like programming and formal languages, are naturally developed in terms of the available metamathematics for these languages. Compare arguments in Löfgren (1992, 1998).

The linguistic complementarity. In general, complementarity refers to holistic situations where (a classical) fragmentation into parts does not succeed. In its complementaristic understanding, the phenomenon of language is such a whole of description and interpretation processes, yet a whole which has no such parts fully expressible within the language itself. Instead, within the language, the parts are complementary or tensioned (rather than classically contradictory). There are various related ways of looking at the complementarity:

(i) as descriptional incompleteness: in no language can its interpretation process be completely described in the language itself;
(ii) as a tension between describability and interpretability within a language: increased describability implies decreased interpretability, and conversely;
(iii) as degrees of partiality of self–reference (introspection) within a language: complete self–reference within a language is impossible;
(iv) as a principle of "nondetachability of language".

Languages may change and evolve, and with them their capacities for describing and interpreting. Yet, at each time that we want to communicate our actual knowledge, even on the evolution of language, we are in a *linguistic predicament*, namely to be confined to a language with its inescapable complementarity.

The linguistic closure. Our thinking abilities are usually looked upon as free and unbounded. But when it comes to *communicable thought*, we are confined to some shared communication language. The systemic wholeness, or the complementaristic nature, of this language implies a closure, or circumscription, of our linguistic abilities – be they 'pure thoughts' communicable in a formal mathematical language, or constructive directions for an experimental interpretation-domain of a physics language. The nature of this closure is not that of a classical boundary of a capacity, like describability, or interpretability. It is a *tensioned and hereditary* boundary of the systemic capacity of *describability-and-interpretability* admitting *potentialities* in two directions:

(a) The closure is **tensioned**. Within the language there is a tension between describability and interpretability (view (ii) of the linguistic complementarity), whereby it may be possible to increase the describability at the cost of a lowered interpretability, and conversely. In other words, what the closure bounds off is neither describability, nor interpretability, but their interactive whole as a linguistic unit of *describability-and-interpretability*.

(b) The closure is **hereditary**. Languages may evolve, and at a later time we can have access to another shared communication language of greater capacity for communication. However, we are then back to the *linguistic predicament*: at each time that we try to communicate thoughts – even introspective thoughts about language and their evolution – we are confined to a shared language, however evolved, and the linguistic complementarity of that language restricts our communicability in the tensioned way according to (a).

In order to emphasise, and further clarify, the concept of complementarity in the holistic conception of language, let us compare with *semiotics*. Holistic language consists of description and interpretation processes as complementaristic parts. Semiotics ("the whole science of language"), is fragmented into parts, namely syntax, semantics, and pragmatics. In a comparative understanding, the static descriptions are the objects for syntax, the interpretations the objects for semantics, and the processual nature (of the description and interpretation processes) the objects for pragmatics. There is, however, in this comparison *a fundamental difference*, namely concerning fragmentability.

Incompatibility between complementaristic language and Carnapian semiotics. According to the linguistic complementarity, classified fragmentation of wholistic language does not succeed. According to Carnap's fragmentation thesis, semiotics allows classical fragmentation,.

By the way of further comments, let us recall the previous quote of Carnap (1968, p. 9). Here Carnap suggests that the whole science of language, semiotics, really can be fragmented into three parts, syntax, semantics, and pragmatics, which are abstracted in the analysis of language as individually understandable. Now, consider the language in which this whole science of language occurs (cf. the linguistic predicament in the linguistic closure). By the linguistic complementarity for that language, it consists of description and interpretation processes that are complementary in the language. This contradicts the assumption that syntax, semantics, and pragmatics can be isolated as individually understood disciplines – if their union shall constitute the whole science of language. *What is missing is the complementaristic interaction between the suggested parts.* In other words, if Carnapian semiotics is considered the whole science of language, it is distortive of language.

Now, with the holistic concept of language delineated, we are in a position to explain, or comment on, conceptions of time which, as outlined, needs holistic language as a frame of reference. This is what we will do in the following sections.

3.4. McTaggart's time-concept as germ for a time-language fusion

McTaggart (1908) reveals two sides of our conception of time with such a clarity that a similarity with the interpretation and description sides of language becomes transparent. It has been suggested that McTaggart's clarification of the two sides of time is a predecessor of what we today call semantics and syntax (in a semiotic sense) for temporal language. In our view, McTaggart's conception is, furthermore, compatible with a development from (Carnapian) semiotics to complementaristic language. It is suggestive for a time-language fusion in complementarity.

The two sides of time, that we refer to, are McTaggart's A and B series of time, outlined as follows.

Positions in time, McTaggart argues, are distinguished in two fundamentally different ways. On the one hand we conceive of time in a dynamic way, where we look for each position either as past, present, or future. This dynamic conception is associated with an *A series*. On the other hand we also conceive of time in a static or tenseless way, where each position in time is earlier than some and later than some other position. This static conception is associated with a *B series*.

Notice that if an event, for a given observer, is conceived as earlier than another event, it is always conceived as earlier, which indicates the static nature of the B series. But an event which is now conceived as present, was future, and will be past, which indicates the dynamic nature of the A series.

McTaggart argues at length that it is essential to the reality of time that its events form an A series as well as a B series. We never observe time except as forming both series.

It is the dynamic A series that offers considerable problems, and which makes McTaggart conclude the "unreality of time".

In Löfgren (1984) we have analyzed McTaggart's reasoning from the perspective of our complementaristic conception of language. In distinguishing between descriptions and models (interpretations) of time, we find that the static B series qualifies as a proper description of one of the properties of time. In contrast, the intrinsic dynamic nature of the A series reveals that it is not a description but an interpretation. A description is always, like a sentence, something static, timeless, something that does not change as long as it is a description. This does not prevent a description from describing a dynamic phenomenon in a shared language. In this way the dynamic A series is a model, or interpretation, of time.

In our comparison, with the A series an interpretation, and the B series a description of time in a time language, view (i) of the linguistic complementarity for this language suggests that the A series cannot be fully described in the language.

In Löfgren (1984) we have sharpened this suggestion into a general time-description incompleteness theorem.

Time incompleteness theorem. In no time-language L can there be a sound complete time-theory T, i.e., a T in which all true temporal L-sentences are provable.

The proof is based on Gödel's revised conception of a formal system, where the concept of Turing machine replaces his earlier "finite procedure". Accordingly, a rule of inference, even within pure syntax, has to be interpreted (whereby syntax cannot be understood as isolated from semantics – in opposition to Carnap's thesis). This admits a temporal interpretation of the proofs in the formal system, which is a key point in our proof of the incompleteness theorem.

At the time of McTaggart (1908), however, the metamathematical understandings of incompleteness phenomena were in the future, and his conclusion about the unreality of time may have been understandable. Today, our conclusion is that time is to be complementaristically conceived, beyond complete describability.

3.5. The Bergson-Russell time discussion revisited

Let us recall from subsection 3.2.1 how Bergson argues that time needs to be understood beyond a mere scientific account, and that scientific knowledge of time must appeal to another knowledge to complete it. Russell, on the other hand, insists on a pure scientific account of time, and provides in his arguments an interesting formulation of a philosophical part-whole doctrine. Russell rejects the doctrine and insisists on analysis.

Although Bergson never discusses his term 'complementarity' in relation to language, but rather with intuitive appeal to different kinds of knowledge, as in science and in metaphysics, we have in Löfgren (1992) found Bergson's complementarity compatible with the linguistic complementarity. We accept Bergson's complementaristic view on time even though his arguments lack support in an explicit recognition of knowledge to be communicable.

We do not accept Russell's rejection of the part-whole doctrine. The doctrine, namely that "the parts of a complex whole are different, as combined in that whole, from what they would otherwise be", is in fact affirmed by the wholistic concept of language. Its constituent interacting description and interpretation processes, producing descriptions and interpretations, are intuitively clear in a classical global perspective (with descriptions as finitely representable, static, objects – with meanings, or interpretations, beyond any such restrictions; cf. Section 3.3). However, when these parts are made objects for investigation in the language itself, they become non-classical tensioned objects (cf. the linguistic closure).

If Russell's 'analysis' is viewed as a description procedure within a language, aiming at complete description, we see from the linguistic complementarity how analysis either fails, or has to be paid for by non-interpretability. It should be noticed that this latter possibility has been hinted at by Russell himself in his witticism: a mathematician is a person who does not know what he is talking about, and does not know if what he says is true.

We find Russell's formulation of the part-whole doctrine interesting in its close connection with the fragmentation problem, namely whether nature can be fragmented into parts, allowing a non-distortive description. Our affirmation of the doctrine will mean that if 'a complete description of time' is generally accepted, it will be distortive of a holistic conception of time, like Bergsonian time.

3.6. Time for reference, and for unfolding of self-reference; time and type

We will here consider a partial property of time, namely the property that our clocks produce. Such a property is usually extended to a noun, clock-time, and we can then both say that our clocks measure clock-time, and that clock-time is what our clocks produce. The latter aspect comes from our linguistic view of a measurement process as a constructive kind of interpretation process, explained in Löfgren (1993).

Clock-time is what is used *for reference* in coordinating events in virtually every natural description. We do describe – and measure – in terms of space and time. Compare, for example, our symbolism for a physical state, $\psi(r,t)$, parametrized as it is in terms of a space vector r and a clock-time t.

In going from a measurement of a quantity, *in terms of time*, to a measurement of time itself, we face a *self-referential* situation, namely of measuring time in terms of time. (In the previous sections on complementaristic language, we have explained a more general self-referential situation, that of conceiving language in terms of language.)

At instances, such self-reference may be explained by an *unfolding* in terms of *levels* of reference. By way of an elementary example, consider the unfolding of the behaviour of a circular feed-back circuit (a possible clock) in terms of levels of reference – which here are themselves times. Instead of saying that a state s at a node in the circuit is the cause of itself, we say that $s(t)$ is the cause of $s(t+T)$, where T is the cycle time of the circuit.

By way of another, but similar, example, consider Russell's unfolding of self-membered sets, $S \in S$, in terms of levels, or types, t. If a set S is a member of a set R, i.e., $S \in R$, the type t of R is one unit higher than the type of S. Hence, for typed sets, S^t, we cannot have $S^t \in S^t$ but can have $S^t \in S^{t+1}$. If the hierarchical domain of typed sets is projected down on a non-typed domain, self-membership problems arise – which are unfolded in the hierarchical domain. In this example the types are tenseless – unless we consider the formation of $\in$-chains as explanation *processes* for sets.

In typed explanations of biological evolution processes, which may become self-referential in a non-typed perspective, the types are associable with times.

These examples support the idea that time has a coordinating function, not only on a primary descriptive level. Time can also have a coordinating function over types in an unfolding hierarchical description process. Time then obtains a *direction* from the hierarchy.

3.7. On Gödel's space-time model for Einstein's general relativity theory

Gödel (1949a, 1949b, 1995) develops a space-time model for Einstein's general relativity theory with remarkable properties. The interpretation allows closed time-like lines, whereby it is possible to send a light signal along a line running back into itself. This means that the light signal will come back at exactly the same moment at which it is sent. Moreover, the model allows lines such that a light signal, sent along such a line, will return earlier than it was sent. Accordingly, it is possible to send light signals into the past. What is more, since a light path can be approximated as closely as one wishes by a path of a material particle, one can travel into the past on a rocket ship of sufficiently high velocity.

These illustrations of Gödel proves that general relativity theory has unexpected interpretations, and the question is, as Einstein suggests, "whether these are not to be excluded on physical grounds".

In commenting on Gödel's interpretation, Einstein writes as follows after having recalled that, in the sense of thermodynamics, the sending of a signal is an irreversible process whereas "according to our present knowledge, all elementary processes are reversible":

> Einstein (1949): "If, therefore, *B* and *A* are two, sufficiently neighbouring, world-points, which can be connected by a time-like line, then the assertion: "*B* is before *A*", makes physical sense. But does this assertion still make sense, if the points, which are connected by a time-like line, are arbitrarily far separated from each other? Certainly not, if there exist point-series connectable by time-like lines in such a way that each point precedes temporarily the preceding one, *and if this series is closed in itself*. In that case the distinction "earlier-later" is abandoned for world-points which lie far apart in a cosmological sense, and those paradoxes, regarding the *direction* of the causal connection, arise, of which Mr. Gödel has spoken.
>
> Such cosmological solutions of the gravitation equations (with not vanishing Λ-constant) have been found by Mr. Gödel. It will be interesting to weigh whether these are not to be excluded on physical grounds."

We find this an interesting illustration of the fragmentation problem. Even if time is not fragmented from space, and instead a whole of space-time is considered according to Einstein's theory, we seem to face a further problem of fragmentation. If no physical ground is found for excluding Gödel's interpretation, then a still more holistic view needs to be taken for space-time. Perhaps a linguistic view, where description and interpretation are not separated as in the above example (cf. Löfgren (1994) for a similar holistic view on quantum mechanics).

We have a somewhat comparable situation in the Skolem-Löwenheim theorem of metamathematics for first-order predicate languages. Accordingly, a theory, if it has a denumerable model, must also have a model of any higher cardinality. It is not possible to prevent these further models by some describable restriction. Instead, the usual procedure in a case where one wants to describe, say, *precisely* the denumerable set of natural numbers, is to talk of this set as a "standard" model of the description. That is, with some intuitive appeal to an undescribed shared language.

In the Gödel-Einstein case we do not know whether a comparable principally clear incompletability situation exists. But already at this stage of development, it seems plausible that the above quest for a describable *physical* exclusion of unwanted models will turn into a quest for an exclusion which is describable in a broadened framework. A framework of a more introspective physics that objectifies also *linguistic* processes of certain basic levels of constructivity.

3.8. On interdisciplinary approaches and 'strangification'

As we have argued, time needs to be conceived in a framework of holistic language. Attempts at relativizing time in specific disciplinary contexts, although quite natural for increasing our understandings of time, are not without problems. Since there can be no complete descriptive theory of time, in spite of all elaborate theoretical achievements, there is always the fear that some such theory will be taken as definitions of time. Fear, that is, because it is bound to be distortive with respect to the prescientific holistic nature of time which needs complementaristic comprehension.

In this situation, an interdisciplinary approach, in the form of a collection of disciplinary accounts of time, may be illuminating. It may allow the reader to find out contextual assumptions, even if not fully understood within the individual approaches. Actually, there is a general procedure, called "strangification" (*Verfremdung*) to this effect. It is proposed in Wallner (1992, p. 95):

> "**Strangification** (*Verfremdung*). Strangification is a central concept (or even method) in Constructive Realism. Simply speaking it means to transfer a certain system of (scientific) sentences from one context (of scientific theories, paradigms, etc.) into another system of sentences. The aim is to "detect" and reveal "hidden" structures and implicit presuppositions in the system of scientific sentences; i.e., to make explicit the implicit assumptions of the set of sentences, of theories, or even paradigms. These implicit assumptions can be seen quite well, if the set of sentences is taken out of its original context into a completely different field, discipline, etc., because these structures become explicit, if one uses them in a different context."

Our own contribution, which can be related to the field of global semiotics, has suggested views of time, in particular its sensitivity to fragmentation, that may help reveal fragmentability presuppositions for studies of time in other fields as well.

References

Bergson, H., 1911, Creative Evolution. New York: Henry Holt.

Carnap, R., 1968, Introduction to Semantics and Formalization of Logic. Cambridge, Massachusetts: Harward University Press.

Chew, G.F., 1968, 'Bootstrap: a Scientific Idea?' Science, **161**, 762–765.

Einstein, A., 1949, 'Reply to Criticisms.' In: Schilpp, A, (ed.), Albert Einstein: Philosopher–Scientist. New York: Harper and Brothers Publishers. 687–688.

Gödel, K., 1949a, 'A Remark about the Relationship Between Relativity Theory and Idealistic Philosophy.' In: Schilpp, A., (ed.), Albert Einstein: Philosopher–Scientist. New York: Harper and Brothers Publishers. 557–562.

Gödel, K., 1949b, 'An Example of a New Type of Cosmological Solutions of Einstein's Field Equations of Gravitation.' Reviews of Modern Physics, **21**, no 3, 447–450.

Gödel, K., 1995, 'Lecture on Rotating Universes.' In: Feferman et al, (eds), K. Gödel Collected Works, Volume III; Unpublished Essays and Lectures. New York, Oxford: Oxford University Press. 269–287.

Löfgren, L., 1979, 'Goals for Human Planning.' In: Ericson R., (ed.), Proc Silver Anniv Int Meeting of the Society for General Systems Research. Berlin, Heidelberg, New York: Springer Verlag. 460–467.

Löfgren, L., 1981, 'Knowledge of Evolution and Evolution of Knowledge.' In: Jantsch, E., (ed.), The Evolutionary Vision : AAAS selected symposium 61. Boulder: Westview Press, 129–151.

Löfgren, L., 1984, 'Autology of Time.' Internat J Gen Systems, **10**, 5–14.

Löfgren, L., 1988, 'Towards System: From Computation to the Phenomenon of Language.' In: Carvallo, (ed.), Nature, Cognition, and System I. Dordrecht: Kluwer. 129–155.

Löfgren, L., 1992, 'Complementarity in Language: Toward a General Understanding.' In: Carvallo, (ed.), Nature, Cognition, and System II. Dordrecht: Kluwer. 113–153.

Löfgren, L., 1993, 'Linguistic Realism and Issues in Quantum Philosophy.' In: Laurikainen et al, (ed.), Symposia on the Foundations of Modern Physics 1992: the Copenhagen Interpretation and Wolfgang Pauli. Singapore: World Scientific. 297–318.

Löfgren, L., 1994, 'General Complementarity and the Double–Prism Experiment.' In: Laurikainen et al, (ed.), Foundations of Modern Physics 1994: 70 Years of Matter Waves. Paris: Editions Frontieres. 155–166.

Löfgren, L., 1998, 'Nonseparability of Inferribility and Measurability in Quantum Mechanics as a Systema Magnum.' In: Trappl, Robert, (ed.), Cybernetics and Systems '98, Vol. I. Austrian Society for Cybernetic Studies. 113–118.

McTaggart, E., 1908, 'The Unreality of Time.' Mind, **17**, 457–474.

Nowakowska, M., 1982, 'Formal Semiotics and Multidimensional Semiotic Systems.' In: Trappl, R., Findler, N., Horn, W., (ed.), Progress in Cybernetics and Systems Research, Vol XI. Washington: Hemisphere Publ. Corp. 211–226.

Putnam, H., 1975, "The Meaning of 'Meaning'." In: Putnam, H.: Mind, Language and Reality; Philosophical Papers, Volume 2. Cambridge: Cambridge University Press. 215–271.

Russell, B., 1914, Our Knowledge of the External World : As a Field for Scientific Method in Philosophy. London: Allen & Unwin.

Shoenfield, J.R., 1967, Mathematical Logic. Reading, Mass.: Addison–Wesley.

Wallner, F., 1992, Acht Vorlesungen über den Konstruktiven Realismus. Wien: WUV–Universitätverlag.

chapter 4

THE ORIGIN OF THE UNIVERSE

(or: whatever happened to metaphysics?)

WH NEWTON-SMITH

University of Oxford

Time in Contemporary Intellectual Thought
Patrick Baert (Editor)

CONTENTS

4.1. Scientific metaphysics

Once upon a time it was to philosophy that one turned for answers to the ultimate questions. Philosophers, perhaps aided by theologians, would tell us, if anyone could, what the origin of the universe was, for example. But philosophers, if not their theological colleagues, have given up all this with a considerable air of satisfaction. The questions persist; only those who give answers are different. In this era they are to be found in the departments of physics, astrophysics and cosmology. Thereby hangs a curious tale which raises the pressing question: is the scientist now succeeding where the philosopher failed? But first the curious tale.

Define metaphysics as unconstrained speculation which goes beyond the empirical data in the hope of providing fundamental explanations. The empirical data is what we might observe with our senses, perhaps with a little help from our instruments. Unconstrained speculation is speculation that goes so far beyond this data that there is no realistic hope that the data will decide between rival speculations. By fundamental explanations I mean explanations which purport to account for the ultimate constitution of the physical world and/or about the ultimate origin of the universe.

Much of the history of philosophy consisted of metaphysics under this definition. It is a case of 'consisted' and not 'consists', for at least in the English-world in the Twentieth Century, there has been a concerted attempt to eradicate metaphysics. The attack on metaphysics started modestly enough in the early years of this century. It began with philosophers of science who sought to remove metaphysics from physics. Pierre Duhem, noting the irresolvable character of metaphysical disputes, proposed constituting the goal of physics so as to make it an 'autonomous' subject. As long as metaphysical speculation was part of physics, the physicist would depend on the metaphysician; and the sins of the latter would be visited on the former. Referring to the metaphysicians of the nineteenth century, Duhem said that "the noise of their battles and the fracas of their collapse have wearied physicists". To avoid metaphysical contamination, he advocated 'instrumentalism'. The goal of science was not to be explanatory truth but empirical success. Theories were to be judged exclusively in terms of their abilities to make correct predictions. Duhem was fond of the image of the scientist as a do-it-yourselfer. This scientist had a utility cabinet with partitions grouping tools for different purposes. He took out a tool, a theory, and, as for the handyman, the only question of interest was the usefulness of that theory in dealing with the empirical world, the world of things we can observe.

Anti-metaphysics was not exactly news in philosophical circles. Hume, after all, had condemned all "ideas not derived from impressions". In more modern terminology, we would say that for Hume no word had a meaning unless it stood for a type of experience or could be defined in terms of words which did stand for a type of experience. Among the ideas condemned on Hume's edict were Newton's conceptions of absolute space and absolute time. For absolute space, space considered apart from its contents, was not something which could be experienced or even defined in terms of things which could be experienced. While the anti-metaphysical perspective was not new, what was new was the take-up of this idea in scientific circles. Einstein himself played an important role in the conversion of the scientific community. For in sweeping away the Newtonian conceptions of space and time, Einstein gave Hume credit for his insights. It is in fact rather doubtful that reading Hume had anything to do with his generation of the Special Theory of Relativity. But the fact that he said this had considerable impact. Contemporary cosmologists take themselves to be working without metaphysics. Barrow and Silk claim it to be one of the great achievements of modern cosmological theory that "it has transformed the study of the universe from metaphysics into physics" (Barrow and Silk, 1983, p. 226).

The Vienna Circle, the Logical Positivists of the twenties and thirties, never tired of citing this as an example of what could be achieved in science once metaphysics was banished. Reichenbach roundly condemned the greatest scientist of all time, Newton, for being a dogmatic metaphysician. The Positivists argued that science had made great strides through being anti-metaphysical. And they sought to legitimize their anti-metaphysical philosophy of science on the grounds that it described what it was about science that made science the great success it has been this century; science, they said, was controlled by the empirical data; science had progressed by banishing what was not so controlled, such as the Newtonian conceptions of space and time.

If science could achieve this by being anti-metaphysical, there was hope for philosophy if only it would emulate science. And so there was to be a new, scientific philosophy, which would be ruthlessly anti-metaphysical. This meant that there was little left of traditional philosophy. Any sentence was condemned as meaningless if there was no way of testing that sentence in experience. The only role left for philosophy in future was the analysis of the language of science. The glorious future was not long-lived. Someone was ungracious enough to point out that the fundamental principle of positivism – "untestable sentences are meaningless" – was itself an untestable sentence.

If positivism as such was in turn to be rejected, it left a legacy: a collective nervousness in the philosophical community about departing from the realm of the empirical. Some philosophers took the data to be ordinary language and the task of philosophy that of determining how words for notions which traditionally interested philosophers such as truth, causality or knowledge functioned in the vernacular. For another philosopher, the later Wittgenstein, the very asking of the traditional questions stemmed from an objectionable impulse to seek general, speculative explanatory theories. His therapeutic philosophy sought to remove those misunderstandings of ordinary language which had promoted this impulse to theorize illegitimately. In so far as anything legitimate lay behind these questions, the answers were a matter for science.

Perhaps things have softened a bit recently. It is fashionable, for instance, to talk in philosophy departments of the need for an explanatory theory of language or meaning. Such a theory might explain how it is that we can understand the meaning of a sentence of our language which we have never encountered before; but these theories do not address the ultimate questions. Metaphysics in the grand style is not having a resurgence among those employed as philosophers.

By and large, contemporary Anglo-Saxon philosophers have either sought to distance themselves from their metaphysical past or to excuse it. The standard excuse involves representing past metaphysical activity as having contributed to science. Descartes' claim that the essence of matter was extension, for instance, is represented as an idea which played a role in the development of scientific theories of matter. Popper in particular has stressed the historical importance of metaphysics in providing ideas which were refined at the hand of the scientist. To my knowledge, however, this is at best an historical thesis. No respectable philosopher indulges in speculative metaphysics now in the hope that it will be a source of future scientific ideas.

What has this history to do with the origins of the universe? It may have nothing to do with the origins: it has a lot to do with speculations about the origin. The impulse to speculate in a metaphysical vein goes deep in our culture. It has not gone away just because philosophers are hesitant to get involved. One of my themes is that it is, in fact, alive and well; and that it is being nurtured within cosmology. And this is very ironic in the light of my story of the fate of metaphysics this century. A modest move to get the metaphysics out of physics ends in metaphysics having been removed from philosophy, yet it is back in physics; and, to add insult to injury, Stephen Hawking ends his *A Brief History of Time* with a castigation of philosophers for not doing metaphysics! Cosmologists are indulging in the sort of metaphysical speculation that would prevent a young philosopher from getting tenure. And the reception of this, as evidenced by the position of Stephen Hawking's book on the best seller lists for over two years shows that interest in metaphysics is as strong as ever within our culture.

This is metaphysics, but it is metaphysics with a difference. Philosophical metaphysics lacked the kind of following that scientific metaphysics has. We have such a positive image of science that if an idea can be represented as scientifically respectable it is given a credibility that no idea receives if it is represented as being philosophically respectable. To a large extent, this arises because of the impressive success of modern science. One might claim success for philosophy but that claim is essentially contestable. In the case of science, even the most severe critics of science, such as Feyerabend, are forced to recognise the success of science. These critics have no hesitation in using the technological spinoffs of science. They fly to conferences, use word-processors and appear on television.

The respect we accord to science is not restricted to wonder at the obvious ways in which it has achieved Bacon's aim for science of giving us power over nature. A trivial but telling illustration of our collective respect for the scientific can be found in the notorious Ben Johnson fiasco. One evening in the autumn of 1988, 26 million Canadians wanted to believe that Johnson had won the gold without the use of steroids. A couple of scientists were reported as having conducted a test which showed Johnson to have traces of the steroid in his urine. Probably fewer than a hundred of those 26 million would have understood theories assumed in the testing procedure; and not even those 100 would have reviewed the details in this particular use of the test. But contrary to what the millions wanted to believe, they came to believe on faith what they were told. If it had been announced that two officials of Reverend Moon's Unification Theological Church had come to the same conclusion using a procedure understood only by them, it would have had no credibility. That it was represented as scientific meant that it had immediate credibility. I am not for the moment suggesting that this was wrong or irrational. I am not part of the "bring back Ben Johnson" movement. My point is simply to illustrate how we give credibility to claims which are represented as being scientifically respectable by persons whom we are told are experts. Hawking reports that scientists now believe that time had a beginning and that is what the public thinks that they should believe about the matter.

One might think that my example is not very telling. For we can ascertain whether these sorts of procedures work and whether the scientists in question are adept at using them. If the same scientists cannot tell from the urine of an obviously drunk person that he has been drinking, suspicion is cast on their techniques. Or, if we find that scientists who claim to show that sugar substitutes are harmful receive generous funding from sugar companies, we might wonder about them. But in the case of cosmological speculations about the origins of the universe, there is nothing much you and I can do to check the techniques being used. The theories involved go so far beyond the empirical data that the possibilities of checking are remote. And, as for the scientists themselves, there is unlikely to be any commercial interests who care whether or not there was a Big Bang or a little whimper. Still, it may be that this is rash. For we will need to consider in due course whether ideological inter tests, specifically theological interests, may be playing a role in the scientists' choice of cosmological theories.

4.2. Big Bang metaphysics

To this juncture my story has had two themes. The first drew attention to the irony of a situation in which metaphysical speculation has passed from philosophy to science: it is a case of irony because philosophy jettisoned metaphysics in order to be more like science. The second theme sounded a word of caution. We do tend to attach credibility to the deliverances of the members of the scientific community. If that community is now offering us metaphysical speculation, it may be in the interests of rationality to be a little cautious. There is one more theme which needs attention before the detailed question of the origin of the universe is taken up. This too is a cautionary tale. It will further serve to suggest that I am some kind of scientific Luddite. That suggestion is misplaced; for I am a fan of science. It is a good thing; but one of the sad facts of life is that one can have too much of a good thing.

Reference was made above to the impact of the Einsteinian revolution in science. The particular description of how that revolution came about has been used to legitimate anti- metaphysics. The very fact that we have come to see the growth of science in terms of the notion of revolutions grounds my other cautionary tale. Whewell, writing in the early part of the Nineteenth Century said there were never genuine revolutions in science. Change, no matter how radical, amounted to adding to the corpus of scientific knowledge. Whewell's building block model of science did not seem outlandish in a period in which the Newtonian programme was still being successfully extended after 150 or so fruitful years. We now know differently. Einstein did not simply extend Newton; the basic Newtonian laws were rejected. In consequence, we no longer think of space and time as absolute; in their place scientists have put a four- dimensional relativistic spacetime. Endless books have been written describing the change in our entire world view that Einstein is said to have brought about.

One can overdo this use of the image of revolution. In a political revolution, the agreed procedures for making political choices are replaced, if only temporarily, by force. This image suggested to some that in scientific revolutions everything is swept aside. Kuhn has been misinterpreted as saying that the normal procedures for scientific theory choice are suspended in favour of rhetoric, propaganda and the influence of personalities. Kuhn did appear to say at one stage that the concepts changed so radically that the theories of Einstein and Newton were incommensurable; that is, expressed in such different languages that no translation between them was possible. But that is absurd; much is preserved in a revolution including a commitment to the basic procedures for making scientific decisions. What is also preserved is the empirical successes of the old theory. Einstein did not deny that the Newtonian worked well for most cases; problems arose in cases involving great distance or rapid motion. In fact, part of the success of the Einsteinian theory derives from the fact that Einstein's theory explains just why

Newton's theory worked when and where it did work. Newton's laws need a correcting factor, a factor which drops out for low velocities; it is this correcting factor which is responsible for the great transformations in our thinking about space and time. And what we should learn from this 'revolution' is that it is highly likely that in due course some correction will be required in, say, Einstein's general theory of relativity. A slight correction, a correction that does not matter most of the time, can well lead to quite radical transformations in the general picture of things, the world view, provided by the theory. This is the cautionary tale. Be minded that the metaphysical speculations supported by a currently fashionable theory are highly likely to be swept away in due course. The next theory will preserve the empirical successes of Einstein but may well not preserve the associated world-views.

At this juncture, I turn to the metaphysics in physics. It derives from a particular feature of current cosmological theories; namely, they represent the universe as having had an origin. A brief history of cosmological speculations will bring this out. At one stage in the not too distant past, it was commonly held that one explained the current large scale structure of the cosmos by reference to previous states of the universe. An account of how the universe once was, together with the causal laws governing the universe, explained why the universe is the way it now is. It was thought that there were no limits to this process of backwards extrapolation to earlier and earlier states of the universe. A scientific account of the universe would simply trace the history of the universe endlessly back. In a scientific context, no question of the ultimate origin of the universe arose. The universe had always been there and in principle one could reconstruct its history as far back as one liked. Once upon a time, popular science books sketched this picture as one which displaced any need for theological accounts of the origin of the universe. For quite simply, it had no origin. Something of this view survives among the very diminished band of supporters for the Steady State Model of the Universe.

Even when this picture held sway, some were not satisfied and sought to ask why there was this universe with an endless past rather than nothing at all. Those who had an antecedent commitment to a conception of a divine creator sought to invoke him or her at this juncture. But, interestingly, the scientific picture did not invite or encourage a theological move. And it took some ingenuity to provide any account of how anything like a Christian God could have exercised his role as a creator in the face of a universe which had no origin. The days of this 'origin-less' universe appeared to be numbered in this century. There was evidence produced by Hubble in 1929 that the universe was expanding; a model of an expanding universe, run backwards in time, packed all the matter tightly together. Going far enough back, it was packed too densely. The prevailing theories of matter precluded such dense states and this fuelled speculation that the universe came into existence at some point in time just after the time at which it would have been otherwise impossibly dense. The Abbe Georges Lemaitre, familiar for other reasons with the idea of a created universe, had suggested in 1927 that it all began with a 'primeval atom' some 30 times larger than the Sun. Lemaitre's atom underwent a 'Big Bang' generating by repeated fission an expanding universe. With the discovery of background cosmic blackbody radiation by Arno Penzias and Robert Wilson in 1965, the Big Bang model became the orthodoxy.

Viewed in scientific terms, however, the Big Bang model did not convincingly establish that the universe had an origin. That is, the model itself did not require the assumption that the Big Bang was the first event. One could regard the Big Bang as a first event which was uncaused. Or one could regard it as the first physical event having a non-physical cause in the actions of a creator or creators. Or one could regard it as an important event the causal origins of which we did not know. That is, it might represent not an absolute beginning but a limit to our abilities to extrapolate backwards. In terms of scientific respectability at the time, this latter option was the best. For it was taken as known that all macroscopic physical events had physical causes and that the Big Bang was a macroscopic event. Given that there was no reason at the time to doubt these physical assumptions, from the scientific point of view the Big Bang appeared as the most important event in the history of the universe tell; an event beyond which we could not currently extrapolate. That being so, this cosmological model did not particularly prompt speculation as to the origin of the entire universe. At the same time, unlike models which could be extrapolated endlessly backwards, it did give comfort to those who shared the Abbe Lemaitre's religious commitments.

In 1974, Roger Penrose and Stephen Hawking established an intriguing theorem which was taken as negating the import of the these reflections concerning the Big Bang. For they showed that any universe satisfying certain conditions would have a singularity in its past; a singularity is a point at which both the curvature of spacetime and the density of matter is infinite. The theorem applies given that the General Theory of Relativity holds of the universe together with the following conditions which were deemed 'at the time' to be entirely reasonable: no time travel, no repulsive graviation, a closed spacetime and a distribution of matter which was not too symmetric; as it was held that these conditions were met in our universe, the assumption of a limit to the density of matter fell by the wayside. The Big Bang was deemed to have arisen from the singularity, the Big Bang Singularity (hereafter cited as BBS), in the past of our universe.

The singularity theorems created the greatest excitement. It became commonplace in cosmological circles to assume that the only scientifically respectable option was to regard the BBS as being the first physical event, as being the ultimate beginning of the universe. As the BBS was a microscopic, not a macroscopic, event, and as quantum mechanics had taught us that microscopic events could lack causes, there was no longer any possibility of applying to some general background assumption of universal causation in arguing that the Big Bang must have had a physical cause. Furthermore, at the singularity various parameters took on infinite values and this meant that the laws of nature, including the laws which lead to the posit of the singularity, broke down. If the laws actually broke down at this point, it seemed that the difficulties in any further backward extrapolation arose as a matter of principle and not merely as a result of our ignorance.

The very positing of the BBS was metaphysical. The BBS involved various parameters taking on infinite values. Just what it is for an actual physical quantity to take on an infinite value is quite obscure. Once upon a time it was held to be a serious defect in a theory if it permitted this. The charge of metaphysics derives from the fact that there could be no direct empirical evidence that a physical parameter took on an infinite value rather than some enormously large finite value, as large as you like. It is no answer to say that an infinite value is to be preferred to a very large finite value on the grounds that a nice model involves that value. For what is at issue is the viability of that particular model. Mach and Duhem, who set physics on its anti-metaphysical course, would have firmly resisted any theory giving infinite values to parameters.

If we describe the posit of the BBS as metaphysical, we will have to describe the additional assumption that the BBS was the ultimate physical beginning of the universe as 'metaphysics squared'. This description is adapted from Alan Guth who aptly described one of his ideas as being 'speculation squared'. For this assumption of a beginning went so far beyond the empirical data, that alternative hypotheses which fitted equally well with the data were easily available. One might suppose, for instance, that the universe had existed before the BBS, governed by quite different laws and culminating in the BBS. Indeed, John Wheeler evinced considerable enthusiasm for this alternative. It is not something I would advocate. For such metaphysical speculation strikes me as singularly unrewarding. But, on the same grounds, I would not advocate the thesis that the BBS was the ultimate physical beginning of the universe.

The reasonable, if unexciting thing to say, given that one was even prepared to countenance the existence of singularities, was that we could say nothing about what went before or did not go before the BBS. This plea for a candid acknowledgement of ignorance did not find favour; it was said that it failed to attend to the defining characteristic of a singularity; a singularity is a boundary beyond which the spacetime manifold cannot be extended. Certainly it is a property of singularities that they are points where the manifold is infinitely curved and matter is infinity dense. But their most important property is that of being a terminus to the further temporal extrapolation backwards of the spacetime. That being so, it is said that the singularity theorems show the time of the singularity to be the first time. Consequently, there was a first event.

The above argument commits what will be called the Stephen Leacock fallacy. When in school we were regularly tormented in algebra class by situations in which it was said, for instance, that A can lift the square of the number of bricks that B can. B in turn lifts the square of the number of bricks that C can, C the square of D and so on. How many men is it going to take to build the house in so many days? The teacher heaped scorn on the results of my careful calculation; I calculated that it would be 2 or root minus 1 of them. This was said to be ridiculous; extraneous roots were to be thrown out. Throwing out extraneous roots is what you do when some feature of a useful mathematical system is just a feature of the mathematical system, having no counterpart in reality. The Stephen Leacock Fallacy (so named in honour of his insightful essay on algebra) amounts to assuming that any solution of any system of mathematical equations set up to model some process describes a physical possibility.

The mere fact that a complex system of mathematical equations generates a solution containing a singularity is not on its own anything like an adequate ground for saying that that solution represents a physical possibility. It would be entirely reasonable for someone to have regarded the singularity theorems as extraneous roots; one can push this scientific Ludditism too far. It has been on occasion a successful ploy in science to take seriously what appeared to be an extraneous root; sometimes it has turned out that these can be given a physical interpretation. Sometimes a theory may be so overwhelmingly successful that we school our intuitions about the absurdity of the solutions. But at the time the singularity theorems were fashionable we were not faced with such a theory. The General Theory of Relativity does not have the vast body of observational successes of a theory like quantum mechanics, nor has it generated impressive technological spin-off. Given, in particular, that the singularity theorems said that the General Theory broke down at the BBS, the empirical data did not favour holding the BBS to be an absolute beginning over holding the BBS to be a limit to our abilities to know. One can imagine it being objected that if we cannot extend the manifold backwards, there is no content to the supposition that there might have been anything before the BBS. But this verificationist move would be utterly unconvincing if urged by the proponents of the singularity theorems. For anyone tainted by verificationism could not possibly attach any credibility to the idea of singularities in the first place.

One might well have regarded the singularity theorems as providing a mere model; that is, with Duhem one could treat the theory which generates these results instrumentally. At one point this is how Hawking suggests all cosmological theories should be treated:

> I shall take the simple-minded view that a scientific theory is just a model of the universe, or a restricted part of it, and a set of rules that relate quantities in the model to observations that we make. It exists only in our minds and does not have any other reality (whatever that might mean). A theory is a good theory if it satisfies two requirements: It must accurately describe

> a large class of observations on the basis of a model that contains only a few arbitrary elements, and it must make definite predictions about the results of future observations [Hawking, 1988, p. 9].

On this construal, we can investigate the properties of the model and we can use it as a tool to make empirical predictions, but we do not take the model as a true or even plausible representation of how the universe is. The problems involved in thinking of an actual physical quantity having an infinite value lapse. But we would no longer be able to use the model to derive conclusions (other than low level observational conclusions) about the universe. The singularity theorems could not be said to show, for instance, that the universe had a beginning; that would simply be a feature of the model. In passing, one should note that Hawking is quite inconsistent on this point. In spite of this official statement of how he views theories in general, he proceeds throughout his book to treat cosmological theories as providing representations of how the world actually is; witness in this regard his assertions that time had a beginning [Hawking, 1988, p. 34].

It is also worth remarking that even if we do not treat the singularity theorems instrumentalistically, the BBS was not much of an explanation of the large scale structure of our universe today. For quantum indeterminacy applies to singularities and there is no possibility of determining what might come out of a singularity. What comes out of a singularity is not deterministically related to the singularity. As Hawking has put it, a television set or a volume of Shakespearean sonnets is as likely to come out as what did in fact come out. Hence, the existence of a singularity is not going to explain the character of a subsequent universe. It might be said that it at least explains the existence of a universe; but even this seems dubious; for one of the possibilities is that nothing comes out. In which case, the singularity cannot even be represented as an explanation of the fact that something or other came out. Those who seek an understanding of the origin of the universe are unlikely to be satisfied with the kind of understanding that a story in terms of an original singularity provides.

To treat the BBS as really existing was to introduce metaphysics into physics. To treat it as the beginning of the universe was *metaphysics squared.* For some cosmologists, the metaphysics did not stop even there. Being used to seeking explanations, they asked what could explain the occurrence of the BBS. *Ex hypothesis*, it was the first physical event or state, so no possibility existed of providing a causal explanation of its occurrence in terms of prior states of the universe. In what sense could it be explained? To some the only possible answer was a theological one; if there was an omnipotent, omniscient creator the occurrence of the BBS could be explained by reference to his or her actions. As it is standard scientific practice to infer the existence of something if that something provides the best possible explanation of some phenomenon (in this case the BBS), to posit the existence of such a creator was scientifically respectable.

This attempt to use an explanatory strategy to legitimate the notion of a creator is evaluated below. But first it is necessary to distinguish two ways in which the notion of a creator might be introduced. Consider someone who thinks of themselves as having an argument for the existence of a creator quite independently of reflections on cosmology. Such a person, on learning of the putative BBS, might think it quite natural to explain the BBS by reference to the creator. If one believed in the existence of a creator who created the universe it would be reasonable to suppose the creator to have started the universe with the BBS. When else could he or she have acted? To have acted after the BBS would render his or her activities otiose. To have acted before the BBS would have given an empty spacetime, making it somewhat mysterious just why the BBS should have occurred sometime later. This position will not be further considered.

In what follows I am interested in the thought that seeking an explanation of the BBS could take one in a scientifically respectable way to the notion of a creator. For convenience the putative creator will be called X; assume for the sake of argument that up to sexuality, X has whatever properties the god of the Christians is supposed to have.

Would it be methodologically reasonable from a scientific point of view to invoke X as the explanation of the BBS? The answer is quite clearly 'no' for at least three reasons. First, it is a principle of scientific methodology that all things being equal, the weakest of the hypotheses which will explain the data is to be preferred. Clearly there are weaker hypotheses available. One could invoke, for instance, the existence of Y who is really powerful without being omnipotent (powerful enough to bring this world into existence but not powerful enough to bring certain other universes into existence) and really quite knowledgeable without being omniscient (Y could not quite see how things would work out in detail). All things being equal, Y is to be preferred to X. Indeed, if we imagine also that Y is not perfect, Y is much to be preferred. For even a cursory glance around the universe makes it highly unlikely its creator, if any, was perfect. For that case, one would expect British trains to run more to schedule; and one would have expected humans to be slightly more disposed to exercise their freewill in ways beneficial to their fellows if the creator had been all good.

A second methodological failure of the hypothesis of X (and Y for that matter) is that it does not generate novel predictions capable of falsification. Some methodologists, most notably Popper and his disciples, hold that no hypothesis counts as scientific unless it generates testable predictions. Even those methodologists who do not take such a strong line would hold that a hypothesis not generating novel falsifiable predictions was for that very reason suspect and not capable of playing an explanatory role.

All things being equal, it is methodologically undesirable to introduce what might be called 'terminal hypotheses' in science. A terminal hypothesis is one which precludes the possibility of seeking further explanations. If one were to ask in turn what explains the existence of X, one is likely to be told either that X is self-explanatory or that the request for a further explanation is inappropriate. To see how methodologically suspect such a response is, imagine someone in the early days of the atomic hypothesis, offering a version of that hypothesis which said either that atoms were self-explanatory or that a request for the explanation of the behaviour of atoms was inappropriate. This would not have had any credibility in comparison with an atomic theory which invited further explanatory investigations into the structure of atoms. On the basis of this and of the other two methodological considerations cited above, one has to conclude that X is methodologically suspect. It was not at all scientifically respectable to argue from the hypothetical BBS to the existence of a creator. Those scientists so inclined move from 'metaphysics squared' to 'metaphysics cubed'.

4.3. Inflationary Big Bang and anthropic principles

Enough has now been said amply to illustrate one of the three themes of my opening remarks; namely, that metaphysics, driven out of philosophy, has found a home in cosmology and this is so quite apart from the theological speculations considered above. And so far we have only considered the more restrained of the metaphysical cosmologists; of the others, more below. Another theme remarked upon concerned the fact that theories in science have a finite life time; and a mere relatively small change in a theory, a change introduced perhaps to cope with very special circumstances, can lead to a radical change in the general world-view or metaphysical picture which the theory supports or which it might be thought to support. It was also noted that we tend to give great credibility to claims which are represented as being scientifically respectable. It should be clear by now that when those claims are baroque metaphysical constructions built on speculative theories, our caution should be great. And if the theories are drawn from an area of science which is itself in great turbulence, caution should give way to complete agnosticism. That is the situation in contemporary cosmology; great changes in cosmological theorising have been noted above and further changes are to be considered below.

Paul Davies, cited above as someone inclined to invoke God to explain the Big Bang, has more recently had a change of heart concerning the precise role that God might have played:

> Whenever I give a lecture on cosmology one question never fails to be asked: What caused the big bang? A few years ago I had no real answer. Today, I believe we know what caused the big bang [Davies, 1984, p. 183].

If we are now in the position of 'knowing' the cause of the Big Bang (and it is not a theological cause which Davies has in mind), something very significant must have happened in cosmology. In rough terms, this potentially exciting story runs something like the following. The singularity theorems did not take account of quantum mechanics. Using only the General Theory of Relativity and extrapolating back to a singularity, a point with infinite density, produced a situation in which the very laws used in the extrapolation broke down. Something had to be done about that and in view of the fact that we were now dealing with the tiniest of events, it was appropriate to consider a quantum mechanical treatment of the Big Bang.

Furthermore, there were observed features of the universe which could not be explained by reference to the character of the Big Bang as traditionally conceived (call this the 'cold Big Bang'); for instance, the distant galaxies are very like each other. This suggests that the matter from those differing galaxies was once in close causal interaction. This homogeneity could be explained if things were held tightly together and then expanded much faster, at a rate truly dramatic 10^{50} times faster than in the cold Big Bang. This is the inflationary hypothesis of Alan Guth advanced in 1980. To make it work, the assumption that gravity was always attractive had to be dropped. Once repulsive gravity is allowed (technically, negative pressures are introduced), the singularity theorems no longer apply and we have a Big Bang without a point at which matter is infinitely dense. The inflationary expansion is short lived and after some very short time, 10^{-30} seconds, inflation ceases and the universe expands as described by the traditional Big Bang model. This model with an inflationary Big Bang but no singularity will be cited as IBB.

The IBB depends crucially on the fact that quantum mechanics allows for a continual ferment where pairs of particles come into existence, *ex nihilo*, and after the briefest of existences, cease to be. The IBB treats spacetime as being what is called a 'false vacuum'. Such a vacuum involves a particular vehement ferment and that is what launches the inflationary phase. Thus, the proponents of the IBB see the universe as coming into being as a result of quantum mechanical processes occurring in an empty spacetime. It is for this reason that Guth has described the universe as a 'free lunch'. Quantum mechanical fluctuations of the empty spacetime lead to an inflationary Big Bang.

It appears that this takes us one step further back. We no longer stop at the Abbe Lemaitre's primeval atom or Hawking and Penrose's BBS. We are back to nothing, to empty spacetime. But appearances are here as elsewhere somewhat misleading. For on the one hand, modern cosmology treats spacetime as a substantial item; it has a definite structure and causally affects matter; and so it cannot be thought as literally nothing. And this spacetime region of inflation, is a very active region in which seas of virtual particles are being created and which issue in the creation of real particles.

There is no doubt that this is an intriguing model. But no one could legitimately claim to 'know' on the basis of this model the cause of the Big Bang. To treat the model as representing how things actually were, is to indulge in utterly inflated speculation. But inflated speculation, like monetary inflation, goes on relentlessly with every prospect of becoming run-away. Having concluded that there is a respectable scientific account of how the Big Bang came to be out of empty spacetime, the question arises as to where that empty spacetime came from.

To this juncture, the non-theological explanations we have considered involved two ingredients; initial conditions and laws. In the case of the first Big Bang model, the initial condition was the explosion of densely packed matter. In the case of the Big Bang Singularity, the existence of that singularity constituted the initial conditions. In the case of the Inflationary Big Bang, the initial conditions is an 'empty' spacetime subject to quantum mechanical fluctuations. In all cases, the application of laws to the conditions gave a model for the evolution of the universe. No explanation invoking initial conditions can be ultimate. For the question arises as to what produced those initial conditions.

Davies and others are attracted by the possibility of extending the approach of the IBB to remove any reference to initial conditions; the story goes as follows. Quantum mechanics has made us familiar with the idea that particles can come into existence *ex nihilo*. It is then speculated that the fluxing empty spacetime might be pictured as coming into existence in the same way (whatever that might mean):

> In the beginning the universe erupted spontaneously out of nothing. From a featureless ferment, bubbles of empty space began to inflate, bootstrapping colossal reserves of energy into existence. The new physics holds out a tantalising promise that we might explain from science how 'all' things came to exist. No longer do we need to 'put them in by hand' at the beginning. We can see how all the fundamental features of the physical world could have arisen 'automatically', purely as a consequence of the laws of physics, without the need to assume that the universe was set up in a very special state initially [Davies, 1984, p. 204].

This passes from the realm of merely outlandish speculation to that of entirely unintelligible gibberish; this is not the place to enter into a discussion of the nature of laws. Suffice it to say that on any plausible account of the laws of nature, laws of nature describe regularities in something or other. What are the laws would depend on what came into existence, not the other way around. Outside a theological framework, one just cannot make sense of the laws existing prior to anything to which they might apply.

This is not the end of the matter for Davies. Having reduced the unexplained to simply the laws of nature, we are to seek an explanation of the laws; and that, for Davies, would appear to bring God or something purposeful back into the story once again:

> The new physics and the new cosmology hold out a tantalising promise: that we might be able to explain how all the physical structures in the universe have come to exist, automatically, as a result of natural processes. We should then no longer have need for a Creator in the traditional sense. Nevertheless, though science may explain the world, we still have to explain science. The laws which enable the universe to come into being spontaneously seem themselves to be the product of exceedingly ingenious design. If physics is the product of design, the universe must have a purpose, and the evidence of modern physics suggests strongly to me that the purpose includes us [Davies, 1984, p. 243].

The methodological objections detailed above to any attempt to argue to the existence of God as the explanation of the BBS would apply equally to any move to invoke God as the explanation of the existence of the laws of nature; and, in any event, it is far from clear that it is intelligible to suppose that the laws of nature existed (existed ?) before (before?) the coming-into-be (?) of spacetime; and if this is not intelligible, the question of how to explain it simply does not arise.

Science is not, in general, an area of human activity much subject to the vissicitudes of fashion. Cosmology is the exception; the half-life of theories is exceedingly short. In their day the theories are enthused about, their consequences explored and writ large. But quickly the cold Big Bang gives way to the BBS which in turn gives way to the IBB. This is an intellectual space in flux. And in the changing fashions, ideological factors, factors not properly scientific, play a role uncommonly large for a physical science. The reason for the fashions and the ideology is not hard to discern. In virtue of operating at such removes from the hard empirical data, in virtue of involving *speculation squared*, the speculator is not easily caught out by the data. The speculator speculates and if, say, his (sadly in this case 'his' is correct) religious convictions play a role in what speculations he pursues, that is not going to run him into the sort of trouble with the data that would plague one whose Christianity of a fundamentalist sort was playing a role in his or her selection of evolutionary (or non-evolutionary) theories. The scope for extraneous factors to play a role in cosmology means that one has to take with a grain of salt the *ex cathedra* pronouncements of cosmologists like Davies.

Many cosmologists have evinced an interest, implicitly or explicitly, in having ultimate explanations [Hawking 1988, p. 167]. And it would indeed be very nice to have an explanation that did not simply prompt the request for yet further explanations. We have seen that Davies' attempts at the ultimate are of dubious intelligibility. Where he and others have sought to introduce an other-worldly dimension in this quest, others have opted for a terminus in us and not in a creator. This is what lies behind much of the discussion of the so-called 'anthropic principles' in cosmology. Bizarre as it may seem, the suggestion is that the initial conditions of the universe can be explained by reference to us. These conditions had to be what they were in order that we should exist. Just how some have thought that such a conclusion could be scientifically respectable requires patient and imaginative reconstruction of their thought processes, a task to which I turn below.

The story begins innocuously enough with what has been called the 'weak anthropic principle' or WAP:

> The observed values of all physical and cosmological quantities are not equally probable but they take on values restricted by the requirement that there exist sites where carbon-based life can evolve and by the requirement that the Universe be old enough for it to have already done so [Barrow and Tipler, 1986, p. 16].

This hardly merits elevation into the status of being a grand principle of cosmology. It is simply the application to cosmology of a standard theorem of probability; Bayes' theorem. This states that the probability of a hypothesis, h, on a given piece of evidence, e, is equal to the product of the probability of the hypothesis and the probability of the evidence given the hypothesis divided by the probability of the evidence: $p(h/e) = p(h) \cdot p(e/h)/p(e)$. To see how Bayes' theorem might be applied let e be the existence of carbon. In the use of WAP nothing turns on the existence of carbon-based life. The existence of carbon itself is all the evidence that is needed. We are as confident as can be that there is carbon and thus we set $p(e) = 1$. Consider two cosmological hypotheses, h' and h'', which initially we are inclined to regard as equally likely to be true; $p(h') = p(h'') = c$. Thus we have:

$$p(h'/e) = c \cdot p(e/h')$$

$$p(h''/\mathrm{e}) = \mathrm{c} \cdot \mathrm{p}(\mathrm{e}/h'')$$

Suppose that h'' makes the existence of carbon less likely than does h'. It might be the h'' gives a lower value to the age of the universe than does h', making it less likely that carbon would have evolved. In that case $p(h'/e)$ is greater than the $p(h''/e)$ because $p(e/h')$ exceeds $p(e/h'')$. In this case the existence of carbon favours the adoption of h' over h''. Barrow says of WAP that

> [it] should not be viewed as a falsifiable theory or theorem. It is a methodological principle which one ignores at one's peril [Barrow, 1988, p. 357].

It is not a methodological principle, it is a theorem of the mathematics of probability. In application to cosmology it amounts to saying that as we exist (or, better, as carbon exist), the theory to be adopted must be compatible with that fact. Or, more accurately, theories are to be favoured to the extent that they increase the probability of what is obviously the case. As such WAP is an evidential principle, it has nothing directly to do with explanation.

It is thus most misleading of Hawking to remark:

> One example of the use of the weak anthropic principle is to 'explain' why the big bang occurred about ten thousand million years ago – it takes about that long for intelligent beings to evolve [Hawking, 1988, p. 124].

WAP may well favour a cosmological model which gives the universe this age rather than a lesser age. Our existence is evidence for the Big Bang having occurred then. It does not explain why it occurred then. My existence now is very good evidence for the existence of grandparents in the past. My existence does not explain their existence. But perhaps Hawking is at least dimly aware that this is not really explanatory for he regularly places 'explain' in these contexts within quotation marks.

WAP could not be used in explaining why the universe had the initial conditions it did. In the face of this the response of the proponents of the anthropic approach has been simply to offer stronger principles. Thus we have:

The Strong Anthropic Principle or SAP

> The Universe must have those properties which allow life to develop within it at some stage in its history [Barrow and Tipler, 1986, p. 21].

The Participatory Anthropic Principle or PAP

> Observers are necessary to bring the universe into being [Barrow and Tipler, 1986, p. 22].

The Final Anthropic Principle or FAP

> Intelligent information-processing must come into existence in the universe, and, once it comes into existence, it will never die out [Barrow and Tipler, 1986, p. 23].

If we cast plausibility and probably also intelligibility to the winds and assume that life 'had' to come into existence, then if certain initial conditions were necessary for life to come into existence, one might have an explanation of sorts of the initial conditions. One way of explaining something is to show that what happened had to happen given some other conditions. If our existence is necessary, then, so the argument runs, the initial conditions had to be the way they were. But the premises are completely crazy. There is no reason whatsoever to hold that life has to exist, that our existence is necessary. Necessary existence is the preserve of gods and numbers. The entire explanatory structure offered by the proponents of SAP runs backwards. It is the initial conditions of the universe and the laws of the universe that explain our existence. Our existence does not bring about (backwards in time) the initial conditions!

To turn from SAP to PAP and FAP is to turn from bad to worse. This is metaphysics with a vengeance. We are necessary to bring the universe into being? We (or something like us) can never go out of existence? It is not surprising that one writer (Martin Gardiner) was driven to introduce yet another anthropic principle, CRAP or the Completely Ridiculous Anthropic Principle.

4.4. Quantum physics and anthropic principles

What could drive otherwise intelligent persons with a training in the physical sciences to re-vitalize a metaphysics at least as outlandish, if much less polished, than that of Leibniz'. There is an answer and that is to be found in another area of physics where metaphysics is alive and well. This is quantum mechanics. In this context no more can be done than to indicate a controversy and a response to that controversy which in part prompts the excesses of FAP and PAP.

Quantum indeterminacy sets limits to how accurately one can measure certain associated quantities. For example, if one determines precisely the position of a particle, one cannot determine the momentum. In classical physics one would have assumed that the particle had a definite position and a definite momentum even though one could not ascertain both of these definite values. To assume, as some do, that quantum quantities have definite but unknowable values (the hidden variables interpretation) generates serious problems. Indeed, it was argued that this assumption is incompatible with the basic axioms of quantum mechanics. Suppose I set up a system which will enable me to measure either a definite momentum or a definite position of a particle. Prior to the measurement, I cannot assume that the particle has, say, a definite position. For the particle does not know which measurement I will perform. Some interpreters of quantum mechanics (the Copenhagen school) suggest that prior to measurement the particle did not have a momentum and did not have a position. The act of measurement brought into existence a definite position or a definite momentum, depending on which measurement was performed; apart from measurement, quantum systems do not have definite states.

Given the importance that the notion of measurement assumed in the Copenhagen interpretation, the question as to what constituted a measurement became particularly pressing. A quantum measurement looks like an interaction between a micro-system, the quantum system, and a macro-system, the measuring system. But the measuring system itself is ultimately composed of items to which quantum mechanics applies. What then guarantees that the macro-system is itself in a determinant state? If measurement gave a determinate state to the micro-system, what gives this to the measuring apparatus itself? An apparent answer can be found if measurement is replaced by observation and if observation is tried as a conscious act; in this case, quantum quantities take on determinant values when they are observed. The observation in this case consists in someone having such and such a state of consciousness. Macro objects do not suffer from noticeable quantum indeterminacy, for our observing them guarantees determinacy. This interpretation will be called the Wigner interpretation of the theory (or WIT for short).

On such an interpretation of quantum mechanics, nothing would have a determinant value if it were not for us. While this takes us someway to the extreme views embedded in SAP and FAP, it does not go quite all the way. For it would not follow that nothing existed apart from us; only that what existed lacked determinant properties. That might not seem much. Perhaps the thinness of what would exist on WIT given the absence of conscious agents is what inclined Barrow, Tipler and Co to say that nothing would exist in the absence of such agents.

One source of WIT was Schrodinger's cat. Schrodinger had a sick thing about cats. He liked to imagine a cat in a box with a vial of prussic acid. An electron is to be fired at a half silvered mirror. If the electron passes through the mirror, a mechanism releases the prussic acid and the cat dies in great distress. If the electron is reflected, the cat survives for the moment. The electron, on hitting the mirror goes into one of two definite states (reflection, passing through). If it goes through the cat is in the state: DEAD. If it bounces off, the cat is in the state: ALIVE. If we treat the cat as a quantum mechanic system, we can only say prior to measurement that the cat is in the state DEAD with a probability of 50% and in the state ALIVE with a probability of 50%. The cat takes on a definite state on being observed; and if observed means seen by a conscious mind to be in a certain state, then the cat does not become definitely DEAD or definitely ALIVE until someone looks into the box; and if this takes some days, the cat may well be dead in either case. Indeed, the cat may become DEAD at a time at which it is very badly decomposed.

The only reasonable reaction to all of this is to go back to the drawing board; WIT will not do. Some other interpretation will have to be found; and if we are told simply that there is no other interpretation which is any better, we should not opt for WIT as a case of *faute de meilleur.* For there is always the option of treating the theory instrumentalistically pending the articulation of a sensible interpretation.

There is another outlandish interpretation of quantum mechanics which has been deployed in an attempt to render some strong anthropic principle plausible. Within the confines of this paper neither this version of the principle nor the link with quantum mechanics can be explored. The interpretation concerned has it that when a quantum event occurs which could go one way or the other, it actually goes both ways. The universe conveniently splits at the time into two universes to accommodate the two outcomes. Thus, when the electron strikes the mirror there comes into existence a branch of the universe with cat DEAD and a branch with cat ALIVE. Given the millions, no billions, of quantum events occurring per second in your vicinity as you read this, you see that appearances notwithstanding there are really an awful lot of branches in the universe.

The strengthening of anthropic principles to reach the level of CRAP (SAP, PAP and FAP) presupposes an utterly implausible and totally metaphysical interpretation of quantum mechanics. Apply this interpretation to the already metaphysical position in cosmology goes far beyond metaphysics squared or cubed; it is metaphysics raised to the power of metaphysics.

All of this makes one want to bring back Cardinal Bellarmine. Bellarmine persecuted Galileo on behalf of the church, apparently with some reluctance. One of Bellarmine's tasks was the calculation of the date of Easter, something which he found it easier to do using a Copernican system rather than a Ptolemaic system; and consequently Bellarmine attempted to persuade Galileo to become an instrumentalist. If Galileo offered his version of the Copernican system as a mere tool, as calculating device generating correct predictions, there would be no clash with the Scriptures. The Church in the person of Cardinal Bellarmine could use Galileo's device and Galileo could be encouraged to work on it further. This is not a very good reason for being an instrumentalist and Galileo declined to follow Bellarmine. Galileo claimed his theory to be true. Remarking that the Scriptures were also true and that the truth cannot conflict, he recommended that Bellarmine find an interpretation of the Scriptures which would render them consistent with his theory – a remark which did Galileo no good at all.

There had been good reasons, however, for being an instrumentalist with regard to the theory of Copernicus. For at the time the theories of Ptolemy and Copernicus fitted the data equally well. Either could be used to get the same, correct, predictions (even if Bellarmine found Copernicus computationally simpler). In the end, observations came to light which rendered Copernicus a better bet than Ptolemy; the situation in cosmology at present is somewhat similar. There is a range of theories that go so far beyond the available empirical data, that there is no reason to select one rather than the other with any degree of confidence. In this situation, the reasonable thing to do is to regard these theories as devices, one of which may some day warrant treatment as something more than a mere device, a mere model. Agnosticism should be the order of the day.

To take these models as true representations of the universe is bad science. To offer to the public these models as being the truth ("I believe we now know the cause of the Big Bang . . ."), is entirely irresponsible. It is to play on the credulity of the public with regard to what is represented to them as being scientifically respectable. Having taken the models as literally true descriptions of the universe and then to seek and derive grand metaphysical conclusions is to indulge in the Abstruse Philosophy that Hume wished to see ended. Hawking castigates the philosophers for not doing metaphysics. No doubt the philosophers need castigation (see below) but in this case the castigation should properly be directed at the cosmologists who pass beyond their data into the realm of utter fancy. The fruits of their labours will not render metaphysics respectable. Once the nonsense being offered is exposed, it is likely to bring cosmology into disrepute.

Philosophers need castigating; they adopted the wrong attitude in the face of the collapse of the 'scientific philosophy' of the logical positivists. Positivism sought to restrict the scope of philosophy; the philosophers in rejecting positivism nonetheless accepted the emasculation of their discipline. It is as if they blamed science for the excesses of positivism and subsequently tended to take a studied disinterest in science. Throughout the fifties and sixties, few philosophers took an interest in the details of scientific theories; and even now this is the prerogative of only certain specialists in philosophy. Philosophers may not be good at much but one thing they are good at is deflating Abstruse Metaphysics. Their failure to make the interpretation of physical theories a matter of major concern is one reason why the grandiose metaphysical schemes considered in this paper have flourished. It will of course be said that the specialists have been doing just this; and to some extent that is so. That it is so only points to the other reason philosophers need castigating. Hawking and Co are communicating with the general public. The philosophers are not. Philosophy has lost sight of what ought to be one of its predominant goals: the dissemination of its reflections to the general public. The general public may not have adopted Hume's anti-metaphysics but at least it had the opportunity to follow his arguments. Hume put his reflections on the science of his day to the public in a lucid, lively and accessible fashion. That is what is particularly needed now. Unless philosophy as an institution provides this, the case of its survival as a discipline is questionable.

This chapter has been concerned with the metaphysical excesses indulged in and encouraged by certain cosmologists and physicists. These excesses prompt a metaphysical conjecture in turn, to be called "Newton-Smith's Conjecture" or NUTS for "No Ultimate Theory Succeeds". Cosmology provides us with models. These models have fascinating structures which it is intriguing to explore. They may well provide now or in the future useful instruments for predicting observable phenomenon; and certain of their theoretical conjectures are sufficiently well-grounded to merit belief; for instance, that some cataclysmic explosive event of some character or other occurred a long time ago and was responsible for some of the large scale features of the universe. But it is touchingly naive to assume that any currently fashionable model correctly describes the world in detail. This would be a two-dimensional naivety; naivety about the history of science and philosophy (a failure to appreciate the fate of theories) and naivety about scientific method (a failure to see how remote these theories are from anything like convincing empirical support). But mere naive credibility does not violate NUTS. One violates that when one decides what to believe now or how to interpret current theories on the assumption that some ultimate theory is in fact possible. To assume that it is possible to have a theory which answers all questions about the origin of the universe without giving rise to any further questions, is empty metaphysical posturing. To let that implausible metaphysical assumption influence what one makes of particular cosmological conjectures is unacceptable.

In the face of cosmological models which give a certain representation to the origin of the universe, models which go vastly beyond the data, we have always the option of silence. Silence is a much under-valued virtue in both science and philosophy. One can enjoy the study of such models and one can glean what one can of the world from such empirical successes as the models have, without passing judgment on the success of those aspects of their representation which transcend the evidence. Such silence is much preferred to the metaphysical noise (of which Duhem complained) generated by those who violate NUTS in assuming that there is some ultimate theory.

References

Barrow, J., 1988, The World Within the World, Oxford Clarendon Press.

Barrow, J.D. and Tipler, F.J., 1986, The Anthropic Cosmological Principle, Oxford University Press.

Barrow, J.D. and Silk, J., 1983, The Left Hand of Creation, New York, Basic Books Publishers.

Barrow, J.D., 1991, Theories of Everything, Oxford Clarendon Press.

Davies, P., 1984, Superforce, New York, Simon and Schuster.

Duhem, P., 1981, The Aim and Structure of Physical Theory, New York, Atheneum.

Duhem, P., 1969, To Save the Phenomena, Chicago University Press.

Guth, A., 1981, The Inflationary Universe: A Possible Solution to the Horizon and Flatness Problems, Physical Review.

Hawking, S., 1988, A Brief History of Time, New York: Bantum.

Hawking, S. and Israel, W., 1987, Three Hundred Years of Gravitation, Cambridge University Press.

Kuhn, T.S., 1962, The Structure of Scientific Revolutions, Chicago University Press.

Leslie, J., Physical Cosmology and Philosophy.

Newton-Smith, R.K., Supernovae: What are they? (unpublished manuscript).

Newton-Smith, W.H., 1980, The Structure of Time, London, Routledge.

Weinberg, S., 1993, Dreams of a Final Theory, London, Hutchinson Radius.

chapter 5

A CLASH OF DOCTRINES: THE ARROW OF TIME IN MODERN PHYSICS

PETER COVENEY

Queen Mary and Westfield College
University of London

Time in Contemporary Intellectual Thought
Patrick Baert (Editor)

CONTENTS

5.1. Introduction

There are two worlds, it would seem: on the one hand there is an objective world, devoid of emotion, with which science is concerned, aiming to describe reality-in-itself. And on the other hand, there is the world which the artists and poets describe, a subjective world which is the fruit of our human experience. Different concepts of time correspond to these two worlds. Whilst in literature, time is usually characterised by its irreversibility, Albert Einstein and a majority of twentieth century scientists have advocated the contrary point of view: "for we convinced physicists, the distinction between the past, the present and the future is only an illusion, however persistent . . ." (Speziali, 1972).

The apparent disagreement between human experience and science finds its roots in the three centuries which have followed the scientific revolution initiated by Newton when he proposed a mechanical description of motion. From Newton's equations of motion, which describe the movements and forces existing between bodies in motion, one can predict completely the behaviour of a system at any instant in time. The only essential prerequisite is that the initial conditions are given, that is one must know the initial co-ordinates – the positions – of the bodies, as well as their initial velocities.

Newton's equations exhibit a feature which is essential to the subject at hand: they impose no privileged direction to the parameter time t, because the latter only occurs as its square (t^2). In other words, the two senses of evolution, towards the future ($+t$) or towards the past ($-t$) are equally permissible (Fig. 5.1). Einstein's remark thus seems well founded.

Let us take the case of simple systems, where the equations can be solved exactly, and which are periodic, as for example is the case of the motion of the Earth around the Sun (neglecting the presence of other bodies). Contrary to our own experience, there can be no arrow of time (to use Eddington's expression), for such systems always return arbitrarily close to their starting point. But one important characteristic is also missing from such a mechanical picture: this is *dissipation,* which is the irreversible loss of energy from a system, for example, its kinetic energy under the action of friction.

Even though the Sun rises every day, there are many events associated with its cycle, including thermonuclear reactions, and others taking place on Earth which

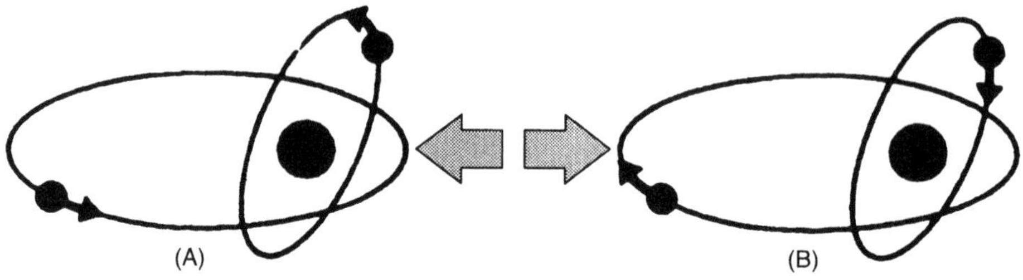

Fig. 5.1. The reversibility of Newton's mechanics. There is no distinction between time running forwards or backwards. Consider two planets orbiting the Sun as in A. If we film them and run the film in reverse, we obtain B. But how are we to distinguish which of the two states is really going forward in time?

affect us directly: for example, biological evolution and the processes associated with life, such as birth, ageing and death. Can one describe such events, which appear to have a temporal directionality associated with them, in a mechanistic way?

The two scientific revolutions which have occurred in the twentieth century – relativity and quantum mechanics – afford no place for irreversibility.[1] Indeed, the rigorously deterministic theory of relativity no longer describes time as an absolute parameter, since the duration of an event depends on the relative speed of an observer. Space and time are not independent entities; instead the relevant concept has become that of space-time. Neither does relativity admit the existence of an arrow of time; in other words, all the phenomena it describes are reversible.

Quantum mechanics, on the other hand, stipulates that one can give only a probabilistic description of phenomena by virtue of Heisenberg's uncertainty principle. In fact, all that can be known about a system is contained in the wave function. The evolution of the wave function in time is described by a differential equation – Schrödinger's equation (although the wave function is not itself an observable property of a system) – which is both deterministic and reversible in relation to time: but, at a given moment, one can only calculate the *probability* of a specific event, for example, the emission of a photon by an excited atom. Time plays no new role in this theory, or at least almost none. For there exists a phenomenon called CP violation (C for charge conjugation and P for parity) which is associated with the existence of dynamical laws that break the symmetry between past and future. In studying the disintegration of certain particles called kaons, Christenson et al. found in 1964 that a quantum mechanical quantity – the product of the parity P with charge conjugation C – was not conserved (Christenson et al., 1964). In accordance with the CPT theorem, the product of parity, charge and time reversal is invariant; CP violation shows that these decays

[1] Except in a way that we shall discuss later.

is not time symmetric. The explanation and the implications of this observation remain an open problem today; we shall not deliberate further on it.

The main problem is that on the microscopic level, the laws of physics are all symmetric with respect to time, yet time asymmetry is manifest on the macroscopic scale (Penrose, 1979). Indeed, it is necessary to return to the last century to find a physical theory where irreversibility occupies a central position. Classical thermodynamics describes the behaviour of macroscopic systems without reference to the microscopic world of atoms and molecules. And yet today the microscopic world is described by the equations of mechanics which are reversible. Thus we have a paradox to surmount: how to reconcile microscopic reversibility with macroscopic irreversibility? This is the subject of the present chapter.

Alone among the laws of thermodynamics, the second provides a criterion for the evolution of systems by stipulating their irreversible tendency towards randomness, to the suppression of inhomegenetics and consequently to complete disorganisation (Denbigh, 1981). In one of its main formulations, the second law states that a certain function of the state of the system, the entropy, increases monotonically during all irreversible processes. The German physicist Rudolph Clausius formalised the second law in 1854 by defining the entropy as a function of state, that is a function whose value depends only on the instantaneous parameters of the system – the volume, temperature for example – and not on the particular path by which the state is reached.

The second law admits a very simple interpretation in the particular case of isolated systems which do not exchange heat with their surroundings: the entropy of these systems increases monotonically and reaches a maximum at the equilibrium. Thus the idea arises naturally to identify the increase of the entropy with the arrow of time. Once it has reached the final state for which the entropy cannot increase further, such a system is then in thermodynamic equilibrium; it no longer has the capacity to change. At equilibrium, the system can evolve no more, it is in a maximally degenerate state completely characterised by the value of the associated thermodynamic parameters: temperature, volume, etc. This is the reason why the entropy, which is then maximal, is interpreted at the microscopic level as a measure of randomness. In fact, it is only for such equilibrium states that the entropy is properly defined. Away from equilibrium, one knows only that it increases with time.

Many scientists have been tempted to apply the second law directly to the entire universe by treating it as an isolated system, and thus (by omitting the important role played by gravity) erroneously claiming to deduce its heat death. Isolated systems are very special cases. In general, most real systems are closed – they exchange energy, but not matter, with the environment – or open, if they can exchange both energy and matter (Fig. 5.2); indeed, later we shall discuss the universe as an open system. Let us take the familiar example of the expansion of

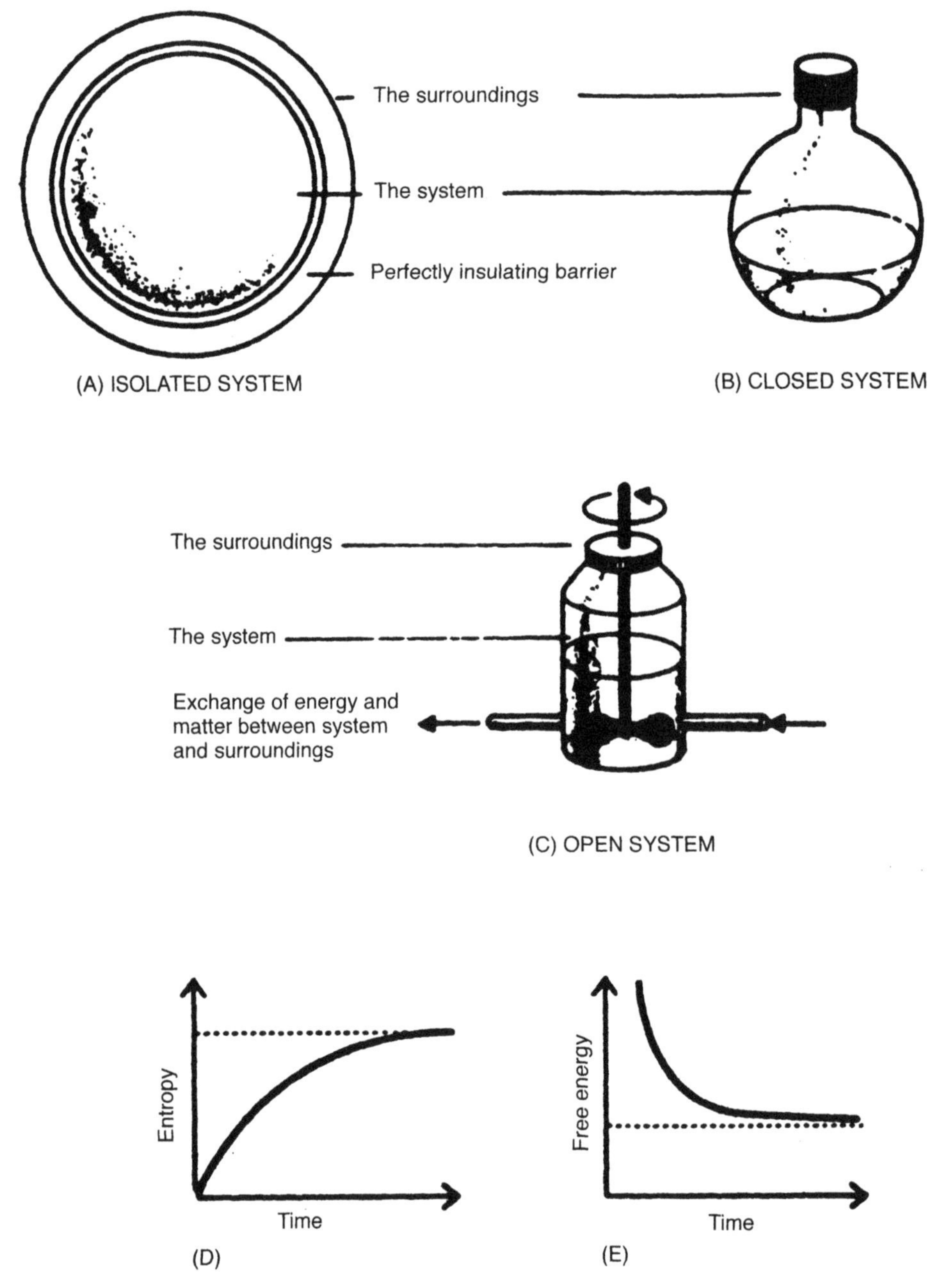

Fig. 5.2. The three types of system encountered in thermodynamics: isolated (A), closed (B) and open (C). The two graphs show increasing entropy (D) and decreasing free energy (E).

a gas by means of a piston. As the latter is not isolated, one must take into account the flow of the entropy due to the exchange of energy with the surroundings and not only the internal production of the entropy by the gas. The total entropy does not change if the increase on the internal entropy of the gas is compensated at each instant by the reduction in the entropy of the environment; in other words the system and the environment stay continuously in equilibrium.

Nothing prevents one imagining an equivalent process to this, which would be reversible: it is sufficient that at each instant the difference between the gas pressure and that of the surroundings stay infinitesimal so as to preserve the state of equilibrium. However, to stay continuously in equilibrium, such a reversible process must last indefinitely; in reality all processes are of finite duration and therefore take place out of equilibrium, thus being globally irreversible. In fact, the concept of total entropy does not allow us to describe the irreversible evolution of a closed or open system, because we must consider in this case the sum of the system's and the environment's entropies. Thus effectively we remain with a new isolated system – the original system together with its surroundings.

Instead, we must search for another extremum principle which guides the thermodynamic evolution of a system, which might be isolated or not isolated, in equilibrium or away from equilibrium. For example, we have seen that isolated systems evolve by maximising their entropy, but this principle cannot be maintained in the case of non-isolated systems, nor indeed for systems far from the equilibrium. Do there exist one or more functions of state analogous to the entropy, which can play the role of a thermodynamic potential in these cases? The study of a system's evolution would then amount to the search for the extrema (maxima or minima) of this potential. As in mechanics, the potential should indicate under which conditions a system is stable with respect to the random fluctuations to which it will be subjected, thus giving a criterion for predicting its evolution in time.

5.2. From equilibrium to non-equilibrium thermodynamics

For about a century, thermodynamics was primarily concerned with a very special and very simple case, that of equilibrium, where the total entropy stays constant and where there is thus no real evolution in time. Then the total entropy remains invariant, giving no information on the thermodynamic state of the system itself. The two 'free' energies of Helmholtz and of Gibbs (usually written A and G respectively), were introduced for closed and open systems in equilibrium, each indicating the position of the equilibrium for the relevant conditions (Fig. 5.2) in terms of properties of the system alone.

Equilibrium thermodynamics is very useful in various fields of physics, chemistry and biochemistry, but it does not give information on how systems, such as chemical reactions or biological organisms, evolve in the course of time.

Life itself is a non-equilibrium process wherein ageing is a manifestation of its irreversibility. Developed about fifty years ago, non-equilibrium thermodynamics studies systems of this type which are maintained away from equilibrium by the imposition of constraints. An example is provided by the Bénard convection cells which appear when a layer of liquid is subjected to a temperature gradient owing to the presence of a heat source beneath the fluid. As soon as the temperature difference between the upper and lower parts of the cell is increased beyond a certain threshold, the liquid begins to move against gravity and a convective structure is established. If one removes the heat, the convection cell becomes blurred and the liquid eventually returns to the state of equilibrium, becoming homogenous once more. If one maintains these non-equilibrium constraints, the convection cells take on a macroscopically organised form indefinitely. It is clearly a macroscopically ordered state, because the resulting structure is characterised by length scales of the order of centimetres and not of the order inter-molecule separations. Such irreversible phenomena seem at first sight to contradict the second law of thermodynamics, since their time evolution is not accompanied by a progressive randomisation of the state of the system, but rather by an increase in the degree of organisation. How can one account for this type of evolution in thermodynamic terms?

To address this question, it is helpful to single out two cases, depending upon whether the system considered is close to or truly distant from equilibrium. Lars Onsager accounted for the first case in the 1930s by showing that the non-equilibrium fluxes are proportional to the thermodynamic forces imposed on the system; the linear thermodynamics which results merges smoothly into equilibrium thermodynamics as the forces are reduced to zero. The relevant idea to consider in this case however is no longer the entropy itself, since the latter increases with time without reaching its maximum in a non-equilibrium state. Therefore one may ask: what is the correct thermodynamic potential which governs these systems? Prigogine demonstrated in 1947 that such systems evolve to a stationary state for which the entropy *production* – the rate at which entropy is produced within the interior of the system – reaches its minimum value (Prigogine, 1947). In other words, the relevant potential is no longer the entropy itself but instead the *system's* entropy production, a result enshrined in the so-called theorem of minimum entropy production.

Thermodynamicists, in particular Prigogine and Glansdorff, have tried to extend this theorem to systems far from equilibrium (Glansdorff and Prigogine, 1971). To achieve this these authors attempted to formulate an appropriate thermodynamic potential that describes the behaviour of a system in a deterministic manner. In fact, Keizer and Fox were among the first to raise qualms that Prigogine's theorem has no equivalent far from the equilibrium, because the evolution of these systems, in general, no longer follows a simple extremum principle. In short, there is no useful purely thermodynamic criterion of evolution far from equilibrium (Keizer and Fox, 1974).

The absence of such a potential is an important result, because it shows that the evolution of these systems depends in large measure on their microscopic details and not merely on macroscopic parameters. In the vicinity of equilibrium, thermodynamic systems are stable with respect to small fluctuations: the existence of a potential ensures their stability. By contrast, when such a system is far from the equilibrium, the slightest fluctuation – often of thermal origin but possibly also due to random external perturbations – can lead to the onset of fundamentally new behaviour, such as the emergence of macroscopic organisation. The 'thermodynamic branch', which follows the evolution of the system in keeping with the minimal production of entropy, becomes unstable. Instead of the existence of a single stable state, as the distance from equilibrium is increased a myriad of bifurcations emerge, which can change the state of the system in a non-deterministic manner into qualitatively distinct states in which matter becomes structured in time and/or space (Fig. 5.3). These are the *self-organised* states of matter that Prigogine called 'dissipative structures', and of which the Bénard convection cells are an example.

The existence of these structures was quite unexpected in thermodynamics. People were used to the (actually erroneous) idea of the heat death of the universe, the idea that the universe was running down, but today it is clear that irreversible non-equilibrium processes can be harbingers of organisation. But how can we describe the evolution of these irreversible processes, which, as we have seen, are no longer controlled by an extremum principle indicating the exact state to which the system will evolve?

If one wishes to follow in detail the temporal evolution in these cases, it is necessary to adopt another strategy, based on the analysis of the differential equations which govern the behaviour of the system in time. Thermodynamics thus merges into dissipative non-linear dynamics, which studies the evolution of the solutions of such equations in the course of time. The only – but fundamental – difference with Newtonian mechanics is that the dynamical equations in question are *irreversible*. Before analysing the connections between dynamics and irreversibility, let us see how we can model, through the use of differential equations, a self-organised system by giving the example of a non-linear chemical reaction.

5.3. Self-organised systems and non-linear dynamics

As we have seen, self-organised structures – or dissipative structures – appear if one maintains a system far from the equilibrium by exchanging energy and matter with its environment. Dissipative structures however do not occur in all systems far from the equilibrium; their appearance depends crucially on detailed features

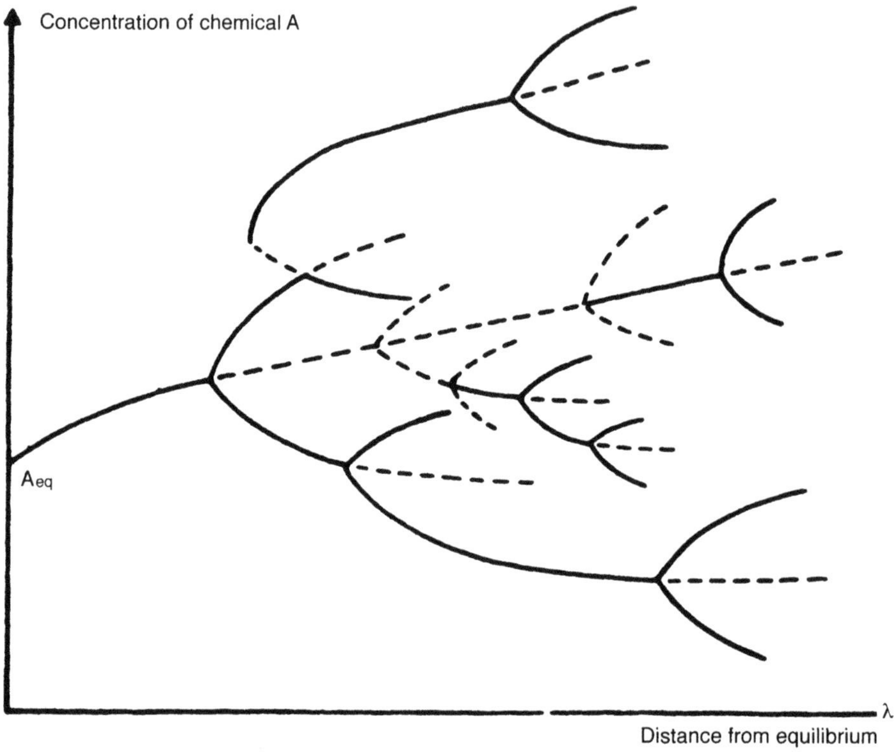

Fig. 5.3. Bifurcation diagram for a chemical reaction far from equilibrium. The concentration of a species A is plotted against distance from equilibrium; at equilibrium, the concentration of A is denoted A_{eq}. As the system is driven further from equilibrium, a whole range of different possible states arise through a sequence of bifurcations.

of the system, details which determine the kinetic equations which describe the process. The equations which determine the *dynamics* of the reaction may be non-linear in the reagent concentrations and it is this non-linearity which underpins their special self-organising characteristics. If one is interested in inhomogoneous systems, where the concentrations of chemical species also vary from one point to another in space, it is equally necessary to take into account the effects of diffusion.

Let us take the example of an *autocatalytic* chemical reaction – that is to say one that produces its own catalyst as the reaction proceeds. The British mathematician, Alan Turing (Turing, 1952) was the first to envisage such a situation for chemical reactions, in the framework of his research on the chemical basis of morphogenesis in biology. A more complicated scheme, named the 'Brusselator', was introduced in 1968 by Prigogine and Lefever. They discussed four simple coupled steps, where two products A and B gave birth to two others C and D, after some intermediate stages involving species X and Y:

$$A \rightarrow X$$

$$B + X \rightarrow Y + C$$

$$2X + Y \rightarrow 3X$$

$$X \rightarrow D$$

The third reaction is autocatalytic, trimolecular – it effectively represents a collision between three molecules – and non-linear. Because of the law of mass action, which expresses the rate of change of concentrations of chemicals in terms of their concentrations, one obtains a system of differential equations in which the non-linearity is due to the fact that three molecules are involved in the course of the third step. This model, consisting of only four steps, is too elementary to describe any actual experiments, but it gives considerable insight into the processes of self-organisation far from the equilibrium. The 'Oregonator' of Field and Noyes (Noyes and Field, 1974; Nicolis and Prigogine, 1977) which is more complex still, explains in large measure many of the self-organised structures that one can see in the celebrated Belousov-Zhabotinsky reaction (Nicolis and Prigogine, 1977).

Today, the study of complex non-linear chemical reaction dynamics has grown enormously, the annual rate of publication on oscillating reactions having doubled every three years since the late 1960s, five times the doubling rate for chemistry as a whole (Winfree, 1987). One particularly fascinating model system has recently been proposed by Gray, Scott and coworkers and has since been studied in some detail (Gray and Scott 1990; Scott 1991); called the crosscatalator, it is arguably the simplest (three variable) chemical model which can display chaotic behaviour, but it also supports a truly vast range of other complex temporal states (Chaudry et al., 1994).

Self-organisation is a ubiquitous phenomenon in biology. For example, the glycolytic oscillations observed during the glycolytic cycle, itself the source of adenosine triphosphate – that is to say of chemical energy – in living cells, can be interpreted as dissipative structures. Even the organised macroscopic structures associated with the origins of life most likely were created in the prebiotic soup by molecules which reproduce in an autocatalytic manner, as do DNA and RNA today (Coveney and Highfield, 1990 and 1995).

Although far from equilibrium processes are perfectly compatible with the second law of thermodynamics, this law is rather vague and does not supply much information relevant for a given system, because non-linear systems possess peculiar characteristics which are often sensitive to the slightest fluctuations which they are subjected to once they are beyond the thermodynamic branch describing equilibrium and linear thermodynamics. It is thus necessary to analyse separately the differential equations which govern each case; nevertheless, the non-linear dynamics of dissipative systems supplies the qualitative tools which describe in general terms the long time behaviour of these systems and reveals the extreme complexity that their evolution can produce. Among the relevant concepts, which we shall not dwell on here, are most notably those of fixed points, periodic and strange attractors, deterministic chaos and so on. (Ruelle, 1989; Coveney and Highfield, 1990 and 1995).

A basic question nevertheless remains unanswered by all of this: how can we reconcile the irreversibility which is manifest at the macroscopic scale with the 'fundamental' reversible laws of mechanics at the microscopic level?

5.4. Statistical mechanics

Statistical mechanics is the field of physics which builds a bridge between the microscopic and macroscopic levels of description used by scientists. It studies the behaviour of macroscopic systems in so far as these are described by the behaviour of truly vast numbers of atoms (of the order of 10^{24}). Whereas at the macroscopic level the description of such a system is made in terms of a very small number of variables, such as temperature, pressure, hydrodynamic variables and so on, at the microscopic level the number of degrees of freedom is immense. For example, in classical mechanics, these degrees of freedom correspond to all the spatial co-ordinates and the momenta of each particle (Balescu, 1975). If there are N particles, the dimensionality of the resulting 'phase space' is 6N.

In fact, such microscopic details are only ever known in a probabilistic manner and so it is quite impossible to determine the trajectories used in classical mechanics; this applies *a fortiori* to the wave functions of quantum mechanics. One must replace these concepts with the probabilistic concepts of the distribution function (also known as the ensemble density) and of the density matrix respectively, the latter assigning probabilities to various possible quantum states of a system, each of which is described by a wave function. The two cases, classical and quantum, formally obey the same Liouville equation with various changes being made in the interpretation of symbols. Unfortunately, the Liouville equation is also invariant under time reversal; in other words, it is also reversible. There is thus no hope of extracting directly from it a function possessing the properties of the macroscopic entropy. This result applies moreover to all purely mechanical models of time evolution because these induce unitary mathematical transformations in the relevant phase space, that is to say transformations which do not alter the probabilities describing the various states of the system. As long as the time evolution is unitary, irreversibility cannot appear in an explicit form.

During the entire history of statistical mechanics, born about 120 years ago, physicists have done their best to bypass this difficulty. Of course, their work is first and foremost concentrated on the case of the state of equilibrium where the *dynamical* problem no longer exists. The basic tool of equilibrium statistical mechanics is a very complicated function, called the *partition function*, which does not depend on time but on the temperature, the volume and the macroscopic energies permitted to the system. All the equilibrium thermodynamic properties, including entropy, can be expressed in terms of this partition function, which connects the microscopic and macroscopic levels together in a very elegant and intellectually satisfying way (Balescu, 1975). Unfortunately it is very difficult to calculate exactly in most situations. Worse still, the partition function does not exist out of equilibrium, so that one can no longer define a non-equilibrium entropy simply by analogy with the equilibrium case.

5.5. Boltzmann's contribution

More than a century ago, Boltzmann envisaged in a highly original way a much more effective description of non-equilibrium systems (Boltzmann, 1872). His approach was based on the atomic theory of the matter, a theory which was not fashionable at the time; in applying the Liouville equation to a macroscopic quantity of a (classical) gas composed of a vast number of particles, he derived an *irreversible* equation, now known as the Boltzmann equation, for the single particle distribution function. From this equation he constructed a new function, which he called H, that decreased monotonically in the course of time: this result is known as the 'H theorem'. Boltzmann identified this function H with the negative of entropy, $H = -S$, thus ensuring his entropy increased monotonically with time, as demanded by the second law. From this brief account, it might seem that he had succeeded in generalising the definition of entropy to non-equilibrium systems.

In fact, to reach this result, Boltzmann made a dynamical approximation – the hypothesis of 'molecular chaos' – which consists of not taking account of correlations which are created between the molecules during collisions. It follows that Boltzmann's equation describes correctly the time evolution of dilute gases, but it does not apply in more general situations. Despite the considerable progress that Boltzmann's work has led to, most notably in the study of the molecular basis of transport properties such as diffusion and viscosity in rarefied gases, it has proved impossible to generalise his kinetic equation in any simple manner, and hence to find a general definition of non-equilibrium entropy. These difficulties have paradoxically dissuaded many physicists from giving any significance to the problem of irreversibility. Since the time-reversible laws of mechanics, be they classical or quantal, seem to work perfectly in their own (restricted) fields of application, these people find it preferrable to dismiss irreversible phenomena as being the result of 'inevitable' approximations which arise due to the limited character of the human observations. This is the basis for the idea of 'coarse-graining', which is often claimed to explain the second law of thermodynamics.

The advocates of 'coarse-graining' maintain that, even in classical mechanics, an observer can never resolve the details of microscopic motions beyond a certain length scale. To describe events which occur on shorter length scales, it is then necessary to calculate dynamical averages for these cells in phase space – which is what is meant by 'coarse-graining'. At this level, one can effectively demonstrate the existence of a generalised H theorem and thus 'deduce' irreversibility, provided one assumes that at each instant Boltzmann's hypothesis of 'molecular chaos' holds, which is of course in general incompatible with dynamics. Even ignoring this problem, the difficulties have not really disappeared because nothing indicates rigorously on which length scale one should introduce coarse-graining, thus making coarse-grained entropies rather arbitrary. Indeed, if coarse-graining is done in an ad hoc manner, there is no guarantee that the entropy will increase with time – it may even decrease. If the presence of an observer could change our knowledge of phenomena in this manner, it would follow that all macroscopic processes – which are manifestly irreversible – would merely be the reflection of our own approximations. This point of view has led to some very important work which finds its place in the subject known as 'information theory', building on the influential ideas of Shannon and Jaynes (Brillouin 1962) for the treatment of statistical problems based on the selection of appropriate probability distributions when only incomplete knowledge is available. Information theoretic formulations of statistics mechanics do not furnish an objective description of an independent objective physical reality, but rather provide a means of making statistical inferences. In short, information theory alone is incapable of making genuine predictions about the independent reality which science aims to describe.

According to this school of thought, the entropy is a measure of 'missing information' (the uncertainty) in our description, which is why it is only natural that the entropy depends on how the macroscopic description is formulated. The concept of entropy formulated in this way has led to the 'maximum entropy' formalism, which has many practical applications in data processing and image analysis. But entropy here remains a measure of *our* ignorance of the details of the underlying microscopic processes. As Jaynes put it, "it is not the physical process that is reversible, but rather our ability to follow it" (Jaynes, 1957). Nonetheless, many people think that the analogy between entropy and information is a purely formal one, in a mathematical sense (Wehrl, 1978; Penrose, 1979).

Despite the indisputable fruitfulness of these information-theoretic ideas, they do not admit an objective definition of irreversibility on the microscopic scale. There are, however, many people whose basic point of view is very different. Convinced that irreversibility has no privileged scale on which it emerges – it must be formulated on the microscopic level as well as on the macroscopic level

– their research has focused on an attempt to endow irreversibility with a rigorous microscopic status. This work, while not yet providing a definitive resolution to the problem, throws interesting light on the subtle relationship between dynamics and thermodynamics.

According to this school of thought, it is the notion that irreversibility is intimately linked to dynamical instability which brings the macroscopic and the microscopic fields together. The study of unstable, more popularly known as 'chaotic', dynamical systems has shown that it is essential to use probability ensembles (or density matrices in the quantum case) rather than trajectories (or wave functions) for their description. The criterion for the presence of irreversibility is thus dependent on intrinsic, objective, properties of a microscopic system – its dynamical instability – and not on subjective criteria: it is a consequence of such objective properties that one can no longer describe a mechanical system's dynamical state and time evolution exactly in terms of trajectories.

5.6. In search of a microscopic entropy

Let us consider the example of classical mechanics. If a dynamical system is stable, this means that a slight alteration in the initial conditions selected in any experiment does not alter its evolution in the long term, but as the instability increases, initially neighbouring states lead to increasingly different long time evolution (Fig. 5.4). For sufficiently unstable systems, including those called mixing and K flows (K is the initial of the famous Russian mathematician

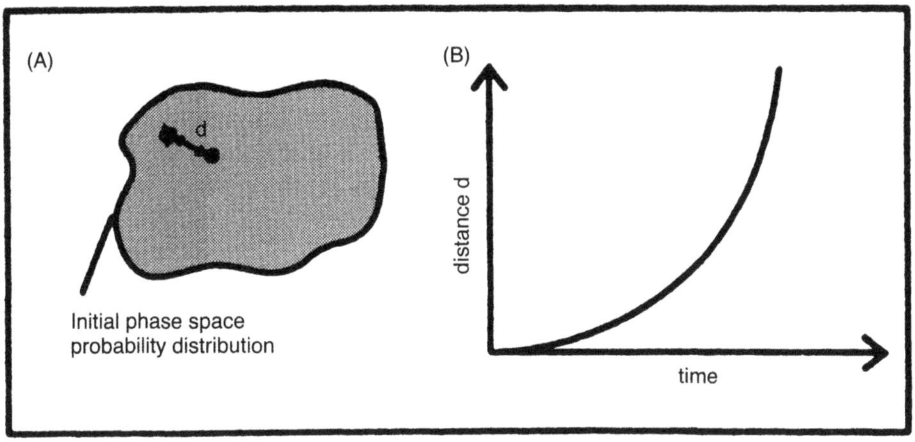

Fig. 5.4. Dynamical instability in chaotic classical systems. (A) The initial probability distribution showing two initial trajectories a distance *d* apart. No matter how close together these points are initially, they diverge exponentially fast as time goes on, as shown in (B).

Kolmogorov) in ergodic theory[2] – one example being a system as simple as that comprising two or three billiard balls placed on a frictionless table – neighbouring trajectories diverge exponentially fast as time goes by no matter how close these are initially. In physical terms, the reason for this is due to the effect of collisions on the dynamics: small changes in trajectories can cause collisions where before there were none, and conversely. Since one can never measure initial conditions with infinite precision – this process would require the accumulation of an *infinite* quantity of information – it becomes essential for intrinsic and objective reasons to invoke the probability distribution function to describe the dynamical behaviour of such chaotic systems. Given that the overwhelming majority of macroscopic systems are at once complex and dynamically unstable, it becomes clear that, even in classical mechanics, the probabilistic description is essential at both the microscopic and macroscopic scales.

The concept of dynamical instability has an impact on Poincaré's recurrence theorem, which shows that the trajectory of a classical system passes arbitrarily close to its initial dynamical state after a finite time, albeit in most cases an extremely long one. Although, in accordance with this theorem, irreversibility would appear to be prescribed for finite classical dynamical systems within a framework in which the concept of trajectories remains meaningful, it does exist in those unstable systems where it is necessary to use probability ensembles.

Moreover, to associate with a dynamical system a microscopic entropy which grows in the course of time, Baidayanath Misra established in 1978 that it was necessary that the system be sufficiently unstable, in essence at least a K flow (Misra, 1978). In 1987 Misra published a proof that even some classical fields are K flows and hence admit an intrinsically irreversible representation (Misra, 1987). In quantum mechanics, however, the situation is very different, because all finite quantum systems are in fact almost periodic, thereby admitting a stronger form of Poincaré's recurrence theorem, for which not only wave functions but also the density matrix must return arbitrarily close to the initial state. This version of the theorem can only be violated for *infinite* systems, which are realised in the so-called thermodynamic limit of finite systems which is frequently used in statistical mechanics to describe macroscopic behaviour (the thermodynamic limit is one in which the number of particles and the volume containing them both tend to infinity in such a way that the density stays constant).

---

[2] Ergodic systems are ones for which there is only one constant of the motion – the total energy; thus they sample all possible dynamical states compatible with their energy. The foundations of ergodic theory were laid down by von Neumann, Birkhoff, Hopf and Halmos, and more recently developed by Kolmogorov, Anosov, Arnold and Sinai. Their work has revealed a whole hierarchy of ergodic properties in these systems. From the standpoint of statistical mechanics, the most important ergodic property is 'mixing', a form of dynamical chaos which ensures that the associated probability density approaches an equilibrium state for long times. K flows are even more chaotic than mixing flows. See Coveney and Highfield (1990) for more details.

The foregoing results emanate from a rigorous mathematical analysis of the properties of the Liouville equation for unstable dynamical systems, from which it follows that if a *microscopic* entropy S exists (with the property of monotonic increase), it has a positive definite value and thus is equal to the square of a new quantity Λ acting on the distribution function or the density matrix. This transformation Λ has one essential property: it is non-unitary, although it does preserve the average values of dynamical observables. Its action on the distribution function or the density matrix is such as to break the symmetry between the forward and backward directions of time. Thus, for the case of sufficiently unstable dynamical systems, the existence of a non-unitary transformation enables one to explicitly describe an irreversible temporal evolution towards the state of equilibrium.[3] Indeed, Misra et al. (1979) defined a new quantity, called the 'internal time', which is supposed to represent the age of a dynamical system. Loosely speaking, the age reflects the system's irreversible thermodynamic aspects, while the Liouville equation for the same system portrays its purely reversible dynamical features. In fact, there is a kind of uncertainty principle linking these features of chaotic systems together: complete knowledge of the 'age' renders the reversible dynamical description meaningless, whilst complete certainty in the dynamical description similarly disables the thermodynamic view. More general situations represent a blurred picture of both aspects of the system.

5.7. The measurement problem in quantum mechanics

Thus the existence of S and Λ effectively rule out dynamical descriptions based on trajectories or wave functions, the latter having important consequences for the problem of measurement in quantum mechanics. As we have seen, quantum mechanics describes the behaviour of a system with the help of a deterministic and reversible equation, the Schrödinger equation, where the unknown function is the wave function. In fact, the wave function can be expressed in terms of a linear combination of all the possible results that can be obtained from a given measurement, the square of the coefficients in this expansion being the probability of the respective outcomes, a representation which is reminiscent of that used to describe localised waves in terms of their individual Fourier components, characterised by their wave lengths or frequencies. In the case of the wave function, each term in the linear superposition represents the result of a measurement, after which the system is 'reduced' to a single eigenstate. The ultimate description of the system is given by a density matrix which groups the various results that one can obtain, modulated by their probabilities of occurrence.

[3] This work is actually incomplete: while such representations have been shown to exist, an explicit development remains to be carried out. It is likely to be connected with some of the measure-theoretic ideas of Penrose and Coveney (1994); see Section 5.9 below.

The awkward problem that the process of reduction of the wave function poses is that it cannot be explained through the use of reversible unitary transformations acting on the wave function, although these are the only kind of transformations available from the Schrödinger equation. For the process of reduction is irreversible, non-unitary and non-deterministic. The overwhelming majority of physicists accept this state of affairs, which they deal with quite happily on the basis of von Neumann's projection postulate (von Neumann, 1955). But there is nothing in the formalism to indicate how, from such a linear superposition, this irreversible reduction of the wave function leads to a well-defined measurement result (Jammer, 1974).

One possible way of dealing with this problem turns on the macroscopic and thus irreversible nature of any experimental measurement device, a point repeatedly stressed by Niels Bohr in his attempts to interpret quantum theory (Wheeler and Zurek, 1983). As described above, it is first necessary to move from a description based on the wave function to one based on the density matrix. Due to the irreversibility of the measurement process, 'pure' states, described by wave functions, are converted into mixtures, which can only be described by density matrices. The existence of these irreversible representations of dynamics incorporating non-unitary transformations into the dynamical description, means that it is no longer necessary to resort to ad hoc hypotheses, such as von Neumann's, resting on no physical foundations.

It is interesting to note here a rather close similarity between this resolution of the measurement problem and one advocated by the mathematical physicist Roger Penrose. He too is persuaded of the objective and time asymmetric nature of wavefunction collapse.[4] However, coming from a background in general relativity, he believes that the source of this asymmetry should originate within an as yet unknown mathematically consistent unification of quantum theory and relativity; he thus maintains that wave function collapse is caused by gravitational interactions (Penrose, 1989) although to most people this seems implausible.

5.8. The kinetic equations of statistical mechanics

In statistical physics, several kinetic equations have been derived which govern the irreversible behaviour of certain systems far from equilibrium, including most notably the Boltzmann equation, the diffusion equations, as well as the equations of Vlasov (strictly, this equation does not describe truly irreversible processes), Landau, and Balescu-Lenard in plasma physics (Balescu, 1975). These differential equations describe the evolution of single particle distribution functions and their solutions describe the *transport phenomena* of mass, momentum, and energy, the meat and drink of non-equilibrium statistical mechanics.

[4] He also believes that the non-unitary transformations controlling wave function collapse are of a non-algorithmic nature and are hence not compatible (Penrose, 1989).

A central problem in non-equilibrium statistical mechanics concerns the derivation of such equations. Most attempts to derive such equations rigorously are valid only in limiting cases where a coupling parameter representing the strength of interactions becomes vanishingly small (De Masi and Presutti, 1991). The best known example of this kind is Lanford's derivation of the Boltzmann equation for a system of hard spheres, in which the small parameter is the diameter of a sphere (Lanford, 1975).

In the late 1960s and early 1970s, a group working in Brussels (Prigogine et al., 1969, 1973; George et al., 1972) pioneered an approach whose aim was to derive kinetic equations which were valid not just in a limiting case, but for some finite range of values of the coupling constant. The so-called Brussels formalism, which has attracted considerable interest, has also been applied in fields outside statistical mechanics, such as in dynamical systems theory (Hasegawa and Saphir, 1992) and in the theory of decaying states in quantum mechanics (Petrosky et al., 1991). Using formal mathematical arguments, these authors showed that the evolution of certain unspecified dissipative – and thus irreversible – systems divides in general into two independent parts or 'subdynamics': a kinetic part which describes the asymptotic evolution of the system towards the equilibrium, together with a non-kinetic part which is present initially, but disappears during the course of evolution. Coveney and Penrose (1992) recently provided a set of sufficient conditions under which at least a part of the formalism is valid. Under these conditions, it is found that the kinetic component of the distribution function obeys a general kinetic equation at all times. As a result of this work, it seems likely that the formalism holds only for certain classes of chaotic dynamical systems, but not in general.[5]

5.9. Canonical non-equilibrium ensembles

Coveney and Penrose (1994) and Coveney and Evans (1994) have recently shown that it is possible to reconcile the apparent conflict between dynamics and thermodynamics in terms of the properties of the probability distributions or ensembles used in non-equilibrium statistical mechanics. They have proposed a rigorous measure-theoretic method for constructing canonical non-equilibrium ensembles; the word 'canonical' here is intended to convey the idea that the probability ensemble furnishes a simple, standard, reduced description that contains the most important features of the approach to equilibrium. For simple chaotic dynamical systems their work shows explicitly that the probability ensembles which describe a system's approach to equilibrium are not symmetric under time reversal. At the same time, they have established direct connections

[5] In technical terms, the norm of the collision operator must be bounded above by an exponentially decaying function of time.

with the Brussels formalism and Ruelle's dynamical theory of resonances (Ruelle, 1986, 1989).[6]

One interesting feature which emerges from this work is that it is necessary to place a restriction on the class of dynamical variables that can be observed, even in classical mechanics. Such a restriction is necessary, since if any dynamical variable whatsoever could be observed, it would be possible to predict or retrodict infinitely far into the past or the future, and even the concept of equilibrium itself would break down.[7]

5.10. Cosmology

The importance of entropy is not only limited to problems arising at the microscopic and macroscopic levels. It also has a cosmological significance. The universe possesses an abundant reservoir of entropy by virtue of the presence of the photons within the black body microwave radiation background which pervades all of space. However, even this entropy resource is, to quote Roger Penrose (1989), 'utter chicken feed' by black hole standards. For, by the Bekenstein-Hawking formula (Penrose, 1989), black holes – assuming that they do indeed exist – contribute in overwhelming terms to the total entropy of the universe. And yet according to modern cosmology, the structure of the universe is described by the Einstein equations of general relativity. As these equations are time symmetric, they cannot explain the origin of this entropy in a consistent manner. In 1987, Gunzig et al. (1987) suggested applying the thermodynamics of open systems to cosmology itself, making use of the connections between entropy and dynamical instability. According to these authors, the creation of matter in the universe is an irreversible process, directly linked to the instability of the vacuum.

Their evidently speculative model[8] is based on some earlier work by Brout et al., which constituted an alternative to the Big Bang scenario (Brout et al. 1978, 1979). Whereas according to the latter picture, the universe was born from an exploding singularity, the model of Brout et al. suggested that it appeared *ex*

---

[6] It should be emphasised that most of the rigorous results that we have discussed can only be established for systems with specified ergodic properties. It follows that there are two limitations to their applicability. Firstly, by no means all systems are ergodic: simple systems such as the three body problem considered by Poincaré are non-integrable but they are not ergodic. Moreover, establishing the ergodic properties of systems in general is a notoriously difficult task, and so only a small number of realistic systems have been shown to have these properties. In a hand-waving manner, most people would argue that sufficiently large systems do display appropriate ergodic properties: since the approach to equilibrium is such a widespread phenomenon, this seems to provide an experimental proof of the mixing property of these dynamical systems.

[7] This restriction is quite weak: in the model described by Penrose and Coveney (1994), any function with a bounded derivative satisfies it.

[8] It has been said that there is speculation, more speculation, and cosmology (see Coveney and Highfield, 1990).

nihilo, as a result of an instability[9] of the quantum Minkowskian vacuum field which describes space-time devoid of matter, which would have been, in geometrical terms, flat (Brout et al., 1978; Gunzig and Nardone 1987). Various other quantum-field-theoretic models for the spontaneous creation of the universe have since been published, each amounting to a modification or a development of this model (Hawking 1988).

Heisenberg's uncertainty principle allowed energy to be lent at no cost over a short period of time to create the universe. According to Einstein's mass-energy relationship, this energy produces matter (in the form of mini black holes) which causes curvature of spacetime, which is more familiar to us as gravity. From nothing we get a very substantial something; the overall energy cost of creating the universe is nevertheless zero, since the energy of all the gravitational forces in the universe is negative and exactly cancels out the positive energy of the mass produced.

In the model of Gunzig et al. (1987), the process of matter formation is taken to be irreversible on a cosmological scale, and it is this primeval process which produces the entropy which resides in the black body radiation. After its creation, the universe is considered to undergo a period of inflationary expansion (during which the black holes evaporate) until it switches over to a universe composed of a mixture of matter and radiation of the kind familiar today. It is interesting to note that, were the universe open (in the sense that there were insufficient matter to drag it towards a big crunch – no clear answer to this is known today), then this model suggests that, as it continued to expand, eventually the universe would possess matter in a highly diluted form. This state of affairs would correspond to an essentially flat spacetime again, whereupon the whole show would be repeated, albeit on a vastly greater scale.

5.11. Conclusion

Evidently, the relationship between time-symmetric mechanics and irreversible processes is difficult to unravel in general. The search for a consistent formulation for irreversibility on the microscopic scale poses a whole series of problems of an essentially mathematical nature, which remain under active investigation today. However, the irreversibility paradox of statistical mechanics can be resolved on the basis of the fact that the non-equilibrium probability ensembles which it

[9] There is some dispute about whether the Minkowski vaccum is actually unstable in this sense. For this reason, Gunzig et al. (1987) avoided the problem by applying phenomenological thermodynamic arguments to describe the situation.

employs for the description of complex – typically macroscopic – processes are not symmetric under time reversal. This is in agreement with the fact that such systems are seen to approach equilibrium states, but not to leave them. In favourable cases, we can even formulate criteria for selecting such canonical ensembles (Penrose and Coveney, 1994). Why nature should be such as to prefer the existence of only those ensembles which show an approach to equilibrium is less clear.

Despite the many attempts that have been made by reductionists to relegate irreversibility to the realms of the illusory, this does not seem to be a tenable position today at a time when the time-asymmetric nature of complex systems is widely recognised as being one of their central features. The attempt to resolve the irreversibility paradox along microscopic and objective lines not only does much to enhance the unity of physics and science more generally; it also provides a lesson in the dangers of ad hoc reductionist arguments in general, as well as encouraging an intellectual reconciliation between our scientific and wider human experiences.

References

Balescu, R., 1975, Equilibrium and Non-Equilibrium Statistical Mechanics (Wiley, New York).

Boltzmann, L., 1872, Sitzungsber. K. Akad, Wiss, Wien, **66**, 275.

Brillouin, L., 1962, Science and Information Theory (Academic Press, New York).

Brout, R. et al., 1978, Ann. Phys. **115**, 78.

Brout, R. et al., 1979, Phys. Rev. Lett. **43**, 417.

Chaudry, A., Coveney, P. and Billingham, J. 1994, J. Chem. Phys. **100**, 1921.

Christenson, J.H. et al., 1964, Phys. Rev. Lett. **13**, 138.

Coveney, P. and Evans, A., 1994, J. Stat. Phys., **77**, 7939.

Coveney, P. and Highfield, R., 1990, The Arrow of Time (W.H. Allen, London).

Conveney, P. and Highfield, R., 1995, Frontiers of Complexity (Faber, London).

Coveney, P. and Penrose, O., 1992, J Phys A: Math Gen. **25** 4947.

De Masi, A. and Presutti, E., 1991, Mathematical Methods for Hydrodynamic Limits Lecture Notes in Mathematics 1501 (Springer, Berlin).

Denbigh, K., 1981, The Principles of Chemical Equilibrium, Cambridge University Press 4th ed.

George, C., Prigogine, I. and Rosenfeld, L., 1972, K Danske Vidensk Sesl Mat-fys Medd **43** 1.

Glansdorff, P. and Prigogine, I., 1971, Structure, stabilité et fluctuations (Masson, Paris).

Gray, P. and Scott, S., 1990, Chemical Oscillations and Instabilities (Oxford University Press).

Gunzig, E. and Nardonne, P., 1987, Fund. Cosmic Phys. **11**, 311.

Gunzig, E. et al., 1987 Nature **330** 621.

Hasagawa, H. and Saphir, W., 1992, Phys. Lett. **161A** 477.

Hawking, S., 1988, A Brief History of Time (Bantam, London).

Jammer, M., 1974, The Philosophy of Quantum Mechanics (Wiley, New York).

Jaynes, E., 1957, Phys. Rev. **108**, 171.

Keizer, J. and Fox, R., 1974, Proc. Natl. Acad. Sci. USA, **71**, 192.

Lanford, O.E. III, 1975, Time Evolution of Large Classical Systems, Springer Lecture Notes in Physics **38** (Springer, Berlin), 1.

Misra, B., 1978, Proc. Natl. Acad. Sci. USA, **75**, 1627.

Misra, B., Prigogine, I. and Courbage, M. 1979, Proc. Natl. Acad. Sci. USA **76**, 3607.

Misra, B., 1987, J. Stat. Phys., **48**, 1295.

von Neumann, J., 1955, Mathematical Foundations of Quantum Mechanics (Princeton University Press).

Nicolis, G. and Prigogine, I., 1977, Self-Organisation in Non-Equilibrium Systems (Wiley, New York).

Noyes, R.M. and Field, R.J., 1974, Ann. Rev. Phys. Chem. **25**, 95.

Penrose, O. and Coveney, P. 1994, Proc. R. Soc. London A **447**, 631.

Penrose, O., 1979, Rep. Prog. Phys. **42** 1937.

Penrose, R., 1979. In: General Relativity: an Einstein Centenary Survey, S.W. Hawking and W. Israel (eds), (Cambridge University Press), p. 582.

Penrose, R., 1989, The Emperor's New Mind (Oxford University Press).

Petrosky, T., Prigogine, I. and Tasaki, S., 1991 Physica **173A** 175.

Prigogine, I., 1947, Etude thermodynamique des phénomènes irréversibles, (Desoer, Liège).

Prigogine, I., George, C. and Henin, F., 1969, Physica **45** 418.

Prigogine, I., George, C., Henin, F. and Rosenfeld, L., 1973, Chem. Scr **4**, 5.

Ruelle, D., 1986, Phys. Rev. Lett. **56** 405.

Ruelle, D., 1989, Elements of Differentiable Dynamics and Bifurcation Theory (Academic Press, New York).

Scott, S., 1991, Chemical Chaos (Oxford Univeristy Press).

Speziali, P. (ed.), 1972, Albert Einstein and Michele Besso, Correspondence 1903–1955 (Hermann, Paris).

Turing, A., 1952, Phil. Trans. R. Soc. Lond., **B237**, 37.

Wheeler, J, and Zurek, W., 1983, Quantum Theory and Measurement (Princeton University Press).

Wehrl, A., 1978, Rev. Mod. Phys. **50**, 221.

Winfree, A., 1987, When Time Breaks Down (Princeton University Press), p. 163.

chapter 6

PAST EVENTS NEVER COME BACK

XAVIER DE HEMPTINNE
Catholic University
Leuven

Time in Contemporary Intellectual Thought
Patrick Baert (Editor)

CONTENTS

6.1. Introduction

Time and space are fundamental concepts, which continue to resist all acceptable definitions. No discourse, by philosophers or scientists throughout the ages, has been able to force either notion into the strict intellectual construction required by human thinking. Both concepts lack the absolute frame of reference which is experienced in daily life and on which Newton built his classical mechanics of motion using time and space as the variables. Scientific interpretation imposes restrictions, but the picture inherited from mechanics, where the variables are interconnected by suitable differential equations, remains an unsatisfactory theoretical construction.

Newtonian mechanics has been revisited by Einstein with his theory of relativity, which gives the universe a curved non-Euclidean geometry. While this elegantly disposes of the absence of an absolute frame of reference for space, time remains a problem. Observers moving with respect to one another appear to live each with his or her own time scale. Minkowski unravelled the paradox by assigning properties that are mathematically connected and similar to that of space to the concept *time*.

While cosmologists continue to dispute mathematical models for the universe, it is clearly felt in daily life that time and space behave very differently indeed. Space concerns distances between objects while time is the field in which duration of events is measured. In contrast to our spatial environment, time has a direction. As it is said: *Time flies like an arrow*. Past and future are different and can never be made to coincide. Nature is by essence irreversible. In that context, Heracleitos, the ancient philosopher of Ephese, claimed Πάντα 'ρεῖ all things flow; all things pass. Space does not have this property.

The present contribution concerns the problem of the irreversible evolution of phenomena at daily life level: its meaning, its mechanism and its origin. Although restricted in its philosophical ambition, the subject is extremely instructive as soon as we try to relate theoretical predictions to experimental facts. Although there has been continuous activity in this field since Boltzmann's attempts to rationalise dynamics using Newtonian mechanics, there has been an increase of interest in recent decades, spurred by the development of the mathematical theory of chaos.

In current literature, the word 'chaos' has different meanings, usually not well distinguished from each other. This work also aims to clarify which is related to loss of memory of past events.

In the local domain considered here, Newtonian mechanics is valid as has been proven over the years. Its possible quantum mechanical extension will be neglected in a first step. Newton's general equation for dynamics, relating the acceleration a of any object with mass m to a force F is:

$$ma = F. \tag{1}$$

According to Newtonian mechanics, the detailed evolution of space co-ordinates in the course of time (trajectory) depends on the initial conditions assumed to the motion, i.e. the spatial and velocity co-ordinates at the presumed initial instant ($t=0$).

Without external intervention, forces are explicitly time invariant. Acceleration itself, as a second derivative of position with respect to time, is invariant under time reversal symmetry. It means that the artificial change of variable t into $-t$ changes neither its value nor its sign so that this mathematical operation keeps the conclusions unchanged. Hence, according to the basic equation, Newtonian mechanics is perfectly time reversible.

In general, reversibility holds for any isolated system where the forces between the constituents have a zero sum. The laws of motion are indeed symmetrical with respect to inversion of the parameter *time*. Hence, no matter how intricate may be the particular trajectories of the system as time evolves, they preserve the memory of the initial conditions. This conclusion contradicts our general experience concerning the macroscopic property of nature: systems removed from their equilibrium state tend more or less quickly to lose the memory of their past history and spontaneously and irreversibly to reach their equilibrium state.

In taking isolated conditions as the basic hypothesis, Boltzmann was confronted with this incompatibility between the irreversible character of macroscopic dynamics and the time reversibility of Newtonian mechanics. To escape this contradiction, he assumed that the system would, by some unknown mechanism, reach and maintain what he called *molecular chaos* between its components. This chaos is a condition where no correlation whatever exists between individual particle motions. The mechanism for removing correlation of molecular motion was left unspecified but it was conjectured that the great number of collisions or interactions between the system's components and the complexity of the mechanics involved would be sufficient to account for it. In this, Boltzmann was violently attacked by his colleagues Zermelo and Loschmidt. Collisions, no matter how complex and numerous, are mechanical events responding perfectly to Newton's laws of mechanics which dictate their strict symmetry properties, in particular that with respect to time reversal.

Despite the early time controversies, Boltzmann's proposal remains today the basis for most fundamental investigations concerning the irreversible character of nature and the quest for theoretical predictions of its phenomenological consequences (transport coefficients). Clearly, the debate is not closed (Prigogine et al., 1988; Lebowitz, 1993).

The vast contemporary literature replaces Boltzmann's initial molecular chaos assumption by more detailed arguments derived from mathematical developments on ergodic theory, which address the concept *deterministic chaos* (at microscopic level) (Sinai, 1976; Eckmann et al., 1985; Ruelle, 1989).

The adjective *deterministic* points to evolutions where each state has a unique consequence. As such, it is opposed to the words *stochastic* or *random*. Newtonian mechanics for isolated systems is strictly deterministic at microscopic level. This property suggested to Laplace the dictum "given precise knowledge of the initial conditions, it should be possible to predict the future of the universe".

Chaos is defined in the literature as the property of motion where long-term behaviour is unpredictable. It must be emphasised that, given perfect knowledge of the initial conditions, a deterministic dynamical system is perfectly predictable. In putting forward deterministic chaos, contemporary literature ascribes unpredictability to very sensitive and unstable dynamics coupled to uncertainty concerning initial conditions. Would therefore unpredictability and its consequence, irreversibility, be a logical inference of our personal lack of knowledge? This contradiction, amplified by the extreme positivistic attitude of some modern schools of mathematics, where formalism is preferred to experimental logic, generates a feeling of uneasiness in the scientific community, often hidden, but sometimes expressed formally (Dorfman et al., 1995).

Ambiguous semantics is the gateway to misunderstandings. Most controversies arise from unsettled disagreements in fundamental definitions. When focusing on time dependence, words like *irreversibility, isolation, equilibrium,* need further clarification.

The colloquial meaning of irreversibility implies the absence of spontaneous recurrence of particular conditions that would have been valid at some past instant. It will become obvious that this definition is far too weak. Our experience of nature suggests a stronger definition, where the word refers to complete loss of correlation or memory in going from past to future.

Daily experience teaches that perturbed macroscopic systems (consisting of many particles) tend to relax more or less rapidly until they reach an equilibrium state. The present chapter focuses on the time dependence of this evolution. For this to be discussed, an accurate definition of equilibrium is also required. It will soon become obvious that quantifying the equilibrium state of any macroscopic system is a vain exercise if the properties of the surroundings are ignored. The necessary intervention of the environment in relaxation dynamics and in the ultimate equilibrium conditions contradicts the generally accepted isolation paradigm implied by deterministic Newtonian mechanics at the microscopic (atomic or molecular) scale.

In the next section, strong experimental evidence for stochastic intervention of the surroundings will be highlighted on the basis of Joule's experiment. Section 6.3 lists a number of objections raised against the non-isolation paradigm and refutes them. In section 6.4 we shall demonstrate that the dynamics involving stochastic exchange with the surroundings are not purely Newtonian. Thermodynamics is the appropriate tool for describing the interplay between systems removed from equilibrium and their environment. This will be generalised in section 6.5 to the particular conditions valid out of equilibrium. Section 6.6 will examine the role of quantum mechanics.

6.2. Relaxation scenario

Discussion of irreversible macroscopic dynamics is traditionally illustrated by the observation of the spontaneous expansion of a gas initially compressed into a fraction of what will become its final volume. The system is assumed to be at equilibrium before the experiment. At the end of the process, when the ultimate equilibrium state is reached, the gas is distributed homogeneously throughout the complete volume. It is easy to convince oneself that, starting from the latter expanded state, the system does not compress again spontaneously to its initial conditions. This is considered to be the sign of irreversibility. The scenario is a simplified representation of the Joule-Thomson classic experiment (1852) which, however, was not designed for quantitative investigation of the time dependence of the process. By considering only initial and final states, the experiment gives no more than a hint of the existence of a direction to the variable *time*.

The purpose of Joule and Thomson was to measure heat produced and exchanged with an external reservoir during spontaneous expansion. For an ideal gas of non-interacting particles, if no mechanical work is allowed to be performed while the system reaches its final equilibrium state, the total exchange of heat with the surroundings will be zero. This fundamental phenomenological result led to the hasty conclusion that irreversible expansion from the initial state to final equilibrium does not involve the surroundings. With the definition of isolation as the condition of a system where neither heat, nor energy under any form (work), nor matter (closed system) is exchanged with the environment, and generalising the conclusion, it has been claimed that irreversible expansion and dispersion of the gas throughout the volume towards final equilibrium is a genuine property of isolated macroscopic systems. From then on, the assignment of a correct mechanism to the process and the justification of its time dependence are considered to be the sole remaining open questions.

In their experiment, Joule and Thomson coupled their system to a reservoir representing the surroundings. The assumption of isolation is therefore incorrect.

The presence of this supplementary device allows the exchange of random fluctuations, which may concern energy or momentum. Zero total energy transfer is obtained only by averaging on a time basis much longer than the system's high interaction frequency with the reservoir. If, for historical reasons, it is felt that the word 'isolation' should still be used in the context described above, it must be qualified by the adjective 'weak'. 'Strict isolation' should be reserved for objects that are left completely alone.

Let us repeat the experiment under conditions that separate the overall process into its parts. To that end, we examine the effect of rupturing an air-inflated balloon inside either an acoustic reverberation hall or an anechoic chamber. The same picture is obtained by performing the experiment in a room stripped of all curtains, rags or soft tissues on the walls or in the same room, but with the walls covered with soft material. The experiment is no different from Joule's, except that possible intermediate steps of the global dynamics are made observable as the modified fate of acoustic transients.

In the two cases, excess air contained in the balloon disperses irreversibly throughout the rooms. Initial and final conditions are identical in the two cases, as are the nature and physical properties of the gas. The intermediate dynamics appears however to be very different indeed. In the acoustic reverberation hall an acoustic perturbation is created and, the better the walls' reflecting quality, the longer it lasts, limited only by well-known sound absorption. By contrast, in the anechoic room, the perturbation vanishes promptly.

The indisputable experimental fact that the global relaxation dynamics of an expanding gas towards its final equilibrium conditions can be modified by changing the kind of object (curtains versus hard walls) which represents the unavoidable reservoir with which the system interacts, indicates that the process consists of at least two distinct elementary steps. At least one of these depends strongly on the nature of the walls. The weaker the walls' isolating character (soft material), the faster is the global relaxation. The prominent role of the environment is thereby emphasised.

The experiment asserts further that, next to dispersion of the initial perturbation (air compressed in the balloon) throughout the system, full relaxation implies intervention of the walls where correlated acoustic motions must be dissipated. Complete isolation is impossible. In nearly isolated conditions, the second step controls the dynamics but, if the system is strongly coupled with its surroundings, internal redistribution of density and thermal perturbations becomes rate determining.

The two steps of the global process are very different in their dynamics. Depending on the system that is considered and on the quality of the interaction with the surroundings, they may be almost concomitant. For simplicity, we shall discuss them here as if they were separated in time.

As soon as the membrane separating the two parts of the initial system has been ruptured, a stream of gas is ejected from the higher-pressure compartment, creating a collective and correlated motion of the particles. The system behaves as if it were isolated. If expansion had been performed against a moving piston in adiabatic conditions (no heat exchanged with the environment), work would have been done and therefore energy would have been delivered to the outside world. If the gas were to expand in a vacuum, the same work would have been performed by the system on itself. A jet would have been created. In the simplified experiment proposed here, where expansion occurs in a low pressure environment, a shock is produced. In both cases, the system provides energy adiabatically for this collective motion. Since the total energy is conserved during this first step (isolated Hamiltonian system, i.e. Newtonian mechanics), that stored now in the collective mode has been taken from the initial thermal supply.

On reaching the wall opposite the rupture, if this is hard, rigid and strictly immobile so that collisions are perfectly elastic, the jet or the shock are reflected and the initial collective motion progressively becomes an acoustic perturbation with the same energy. The spectrum and phases of this motion (coherence) constitute the memory of the initial conditions and of the shape of the reverberating walls.

Depending on the shape of the room where possible acoustic resonance might occur, initial expansion and dispersion is irreversible only in Poincaré's sense. This means that the global trajectory does not allow concentration of the particles back into their initial volume for a reasonable period of time. The memory of the initial conditions is however still present in the collective motion, no matter how intricate (chaotic) the individual trajectories may be. The process is apparently irreversible but in fact is not so. Let us call this *weak irreversibility.*

Relaxation of the coherent or collective component of the motion starts now. Energy accumulated first in the jet and later in the acoustic perturbation is to be reinjected into the thermal energy bath. The mechanism involves collisions of the system's particles with boundary atoms. Thermal (random) motion of the wall atoms is by no means correlated with that of the system's particles. The trajectories following surface collisions are therefore completely uncoupled with the incoming ones. The events cause stochastic jumps between trajectories. The loss of time correlation near the surface is transmitted to the interior of the system as soon as a particle leaving the surface collides with particles in the bulk.

With ideal gases (hard spheres allowed), loss of coherence thermalises the initial collective motion and returns its energy to the thermal bath. When the collective motion is completely neutralised, the thermal energy of the system (its temperature) has regained the value which it had before expansion, in full agreement with Joule's observation. In the same time, information about the initial conditions is completely lost. This is the sign of *strong irreversibility.*

Permanent rapid stochastic intervention of the environment, leading to random transitions between trajectories, blurs the exact conditions of the system in terms of the actual positions and velocities of its constituent particles on a longer time scale. In contradiction with recent literature, the resulting situation is by no means to be interpreted as 'deterministic chaos', which indeed preserves strictly individual trajectories in the course of time. Instead, the mechanism generates exactly what Boltzmann looked for as 'molecular chaos'. Due to the wealth of different trajectories, all randomly accessible by action of the surroundings, motion can only be discussed in terms of probability distributions of the possible trajectories.

Reference to probability distributions in the context of relaxing objects is not new (Prigogine et al., 1988). However, the literature refers to uncertain knowledge of initial conditions (at $t=0$) and not to stochastic mechanical jumps as mentioned above.

6.3. Objections

The dominant role of the surroundings in the time evolution of macroscopic systems is not readily accepted by the scientific community. Laplace's comment saying "given precise knowledge of the initial conditions, it should be possible to predict the future of the universe" remains profoundly rooted in the heart of many theorists. They are not keen on abandoning determinism and, at the same time, losing tight internal control on dynamics. The objections most often cited are listed below together with counter-arguments.

6.3.1. Transport coefficients

Transport coefficients are major parameters governing the dynamic properties of fluids (gases and liquids). They describe how thermodynamic forces give rise to corresponding flows. Most important are viscosity and thermal conductivity. With multi-component systems out of equilibrium, the diffusion coefficient describes how fast one component flows with respect to the other. Sometimes, flows of different properties are coupled. A typical example is the flow of matter driven by a temperature gradient.

An often expressed objection to our interpretation of Joule's experiment concerns the implication to the transport coefficients. The laws of hydrodynamics predict that sound is dampened by viscosity and thermal conductivity, both properties independent of the nature and shape of the fluid's container. Viscosity and other transport effects occurring in the bulk of a fluid are experimentally verified. It is said that their intervention in the phenomenological laws of hydrodynamics (Navier-Stokes equations) does not require the boundaries to be specified.

The objection treats the transport coefficients as bulk phenomenological parameters, thereby showing confusion between the phenomena and their mechanism. Let us develop this matter by considering the property 'viscosity'. This is the transport coefficient for shear momentum through the fluid (ratio between the rate of transport of shear momentum across the system and gradient of shear velocity, as forced on the system by some unspecified means); as such, it is indeed a bulk property. At the same time, the word more generally expresses viscous flow itself; this implies that what is to be transported is supplied at some places and removed elsewhere. Without the presence of suitable sources and sinks, represented by appropriate boundaries, walls, or analogous interfaces, the very concept is meaningless. In fact, the objection mentioned above is unfair; if it is correct that the differential equation of hydrodynamics do not mention explicitly the presence of boundaries, their solution impies boundary conditions to be specified.

How is viscosity measured? This may be done, following Couette, by studying the fluid bounded by two parallel plates moving in opposite directions and driven by measurable forces or, following Poiseuille, by measuring the flow driven by a pressure gradient in a capillary. Determination of the value of the coefficient of viscosity is unthinkable without the presence of the boundaries (pair of plates or capillary) on which mechanical force is exerted and measured. Uncoupling viscous flow from the boundaries or walls that are necessary for it to manifest itself is therefore incorrect.

It is sometimes objected that transport coefficients, like viscosity and heat conduction, depend on the collision frequency in the bulk, suggesting that they are by no means related to surface effects; with liquids, the discussion is more complex. Let us therefore consider gaseous systems at moderate pressure, where deviations from the ideal state are negligible. In such systems, the numerical values of the coefficients do depend on the average collision cross section. We must stress that the collision cross-section appears in the denominator of the relevant expressions. Considering that the average collision frequency increases with increasing collision cross section, it is clear that increasing collision frequency reduces transport efficiency, in contradiction with the objection attributing viscocity to collisional effects.

6.3.2. Mixing

Mechanical description of macroscopic systems of particles requires a geometrical construction where positions and velocities (better: momenta) of all the particles may be represented and in which the relevant trajectories unfold in the course of time. This construction is called the phase space (not to be confused with configuration space, limited to the position co-ordinates).

Much scientific material has been accumulated in the last two decades on so-called mixing properties of a number of model systems. Starting from simple initial configurations in phase space ($t=0$), mixing is the property according to which the dynamics would spread the particles progressively as uniformly as possible over the accessible domain. The Sinai billiards and the Lorentz gas are very popular research subjects in this context (Sinai, 1976; Spohn, 1980; Cornfeld, et al., 1982). They concern computer simulations of parallel beams of non-interacting particles assumed to be moving with fixed velocity in a space containing convex obstacles with which collisions are taken to be elastic (deterministic). Besides the conservation of energy on impact, the dynamics implies conservation of the velocity component parallel to the surface at the collision site (specular reflection). The positions (periodic or random) and the shape of the obstacles determine the reorientation of the particles' velocity at each impact. The convex character of the obstacles results in a complex dispersion of the initial beam in all directions of the configuration space thus destroying the initial collimation.

Many authors stress the importance of mixing on the evolution of representative points in phase space for the restoration and maintenance of chaos or the establishment of ergodic distributions in macroscopic systems.

It should be stressed, however, that the mathematical definition of mixing involves a phase space, a measure on it, and a group of transformations implying the complete set of dynamical variables. When applied to isolated Hamiltonian systems, published demonstrations never embrace the complete set of canonical variables (positions and momenta) as required by the mathematical definitions. One of the variables of the motion, the magnitude of the velocity, is indeed invariant under the prescribed dynamics. Hence, when applied to dynamics of systems of particles, mixing eludes systematically one of the degrees of freedom in phase space, in contradiction with the fundamental theorems involved. The conclusions are therefore unacceptable.

The mechanical effect of such filters on the motion of particles is pictured exactly by the effect of shining a laser (coherent) beam on diffusing objects. Diffusion spreads the light in all directions but the coherence of the motion is by no means affected (as it would be if it were changed into 'thermal' light). Strict conservation of temporal coherence can be demonstrated by causing this diffused light to interfere with the incident beam or with another laser beam as in the production of holograms.

Some authors (Balescu, 1975; Prigogine et al., 1988) insist on the mathematical mixing properties of the so-called baker transformation. It is said that, by repeatedly folding a system on itself, as a baker does with dough, initial inhomogeneities are progressively neutralised. In physics, adopting this conclusion is equivalent to cutting off arbitrarily the higher frequency domain of a spectrum because the wavelength would have become too short for the resolving power of the instrument which the observer happens to be using. This is objectively unacceptable since it submits physical reality to subjective implications.

6.3.3. Renormalisation

Some theoretical trials for justifying the irreversible character of isolated macroscopic dynamics suggest a renormalisation of the system's parameters. The arguments are developed systematically for $N \rightarrow \infty$ and $V \rightarrow \infty$ (Balescu, 1975; Goldstein et al., 1975). It is claimed explicitly that *this is the only precise way of removing unessential complications due to boundary effects, etc.* (Lanford, 1975). Infinite systems are regarded as approximations to large finite systems. Such limiting conditions are often labelled 'the thermodynamic limit', (not to be confused with the same terminology used in hydrodynamics where it characterises a system of which the physical dimensions exceed the mean free path by orders of magnitude).

With Hamiltonian (Newtonian) dynamics, forces are backed by reactions equal in magnitude and opposite in direction. During collisions, the reaction to the force acting on one particle and carried along by its collision partner may be thought of as if shared by the $N-1$ remaining particles. In systems interacting with the surroundings, the reaction is taken care of by dissipation to the boundaries. If we deliberately suppress the role of the boundaries and leave the remaining particles to handle the reaction, which would then be eliminated because its individual incidence on single particles is diluted by their great number, we incorrectly reject an infinite sum of infinitesimal contributions which add up to the value of the reaction force.

6.4. The dynamical equation

It is now clear that the mechanism by which past events are forgotten as time goes on is related to chaotic evolution. The first question in the debate concerned the kind of chaos implied (deterministic or stochastic). Phenomenological arguments given above, based on variation of the rate of loss of the memory of earlier events, illustrate that deterministic chaos (hypothetically fully isolated systems) cannot lead to the observed irreversible behaviour of nature. However, nothing contradicts the establishment of molecular chaos by stochastic interaction with the fluctuating environment. In this section, we evaluate quantitatively the relaxation dynamics.

From Newton and his followers (Lagrange, Hamilton, etc.) we have learned that the dynamics of particle motions involves their positions and momenta (velocities). This collection of canonical co-ordinates defines the phase space, the points of which represent the complete variety of conditions the system may assume. Starting from some initial set of values of the canonical co-ordinates (at

$t=0$), the equations of the motion describe how these co-ordinates change in the course of time.

With isolated systems, deterministic equations of motion describe the trajectory to be followed by a representative point in phase space. This covers only a portion of phase space in which many trajectories coexist. An elementary theorem of deterministic mechanics states that different trajectories in phase space never cross and that jumps between trajectories are forbidden. In contrast, if systems are allowed to undergo stochastic interactions with the surroundings, these events interrupt existing trajectories and start new ones, with possibly very different conditions. We have stochastic jumps.

Stochastic jumps occur whenever a particle collides with a boundary. In realistic conditions (macroscopic systems) corresponding interactions are so frequent that usual physical measurements on the system, where some time averaging is necessary in order to eliminate fluctuations, easily cover many accessible trajectories. Apart from exceptional cases, single trajectories are indeed extremely short-lived features and are therefore usually irrelevant. In contrast, some of the trajectories available in phase space are more probable than others; the same conclusion holds with individual positions along phase space trajectories (set of co-ordinates). We therefore need to study the time-dependence of probability densities or distribution functions of accessible positions in phase space.

A simple example, that of a particle translating back and forth between walls, should clarify the problem. At time $t=0$, it is supposed that the velocity of the particle is v (momentum $p=mv$; kinetic energy $E=mv^{-}/-\geqslant>$ While moving at constant velocity on its initial trajectory, the particle hits the wall. From there, it is reflected but certainly not in the same way as a light beam from a mirror. Depending on the particular motion of the wall atom involved in the collision, the kinetic energy may be modified; it may be increased or decreased. By collision with the wall, the initially sharply peaked probability distribution of energy becomes diffuse, depending on the wall temperature; in fact, the wall and system temperatures equalise.

Not only is the kinetic energy involved; the average direction of the new reflected trajectory also depends on the motion of the wall atoms at impact. While the particle may leave the collision site in any direction, it is expected that, on average, the new trajectory adopts preferentially (highest probability density) a direction corresponding to the average motion of the wall itself.

This simple example highlights the two constituent parts of the equation of the motion, generally valid for all macroscopic systems interacting with their surroundings. At first, in the time separating collisions with the walls, the dynamics strictly follows the laws of deterministic mechanics, with conservation of energy and momentum. The second part concerns impact with the boundary (creating so-called boundary conditions). The former trajectory disappears as in a sink and is replaced by a new one, as if a source were located at the same place. Hence, by writing f for the probability density in phase space, the equation of the motion reads

$$\frac{df}{dt} = [f, H] + J. \tag{2}$$

Symbol $[f, H]$ is the Poisson bracket describing the conservative contribution to the motion

$$[f_N, H] \equiv \sum_r \left(\frac{\partial H}{\partial p_r} \frac{\partial f_N}{\partial q_r} - \frac{\partial H}{\partial q_r} \frac{\partial f_N}{\partial p_r} \right), \tag{3}$$

while J is a source/sink term which explicitly expresses the action of the environment. The equation should be considered to be averaged over the stochastic interaction frequency.

Though difficult to use, mainly because of its generality, this equation allows a very simple discussion of time-dependent phenomena. In stationary conditions f is time independent ($df/dt=0$). If, on average $J=0$ (and $[f, H]$ too), the system is at equilibrium and behaves as if it were not interacting with the environment. If the latter conditions are not true, we are in stationary conditions. If $df/dt \neq 0$, we have a transient state. The balance of the two contributions gives the rate of relaxation of the transient; as neither component is negligible, it is difficult to obtain accurate results with this equation. However, in all cases the main conclusion to be drawn from the existence of the sink/source contribution is the prevailing local equilibrium between the system and its surroundings in the boundary regions for all exchangeable properties. This is the analogue (and justification) of the boundary conditions of conventional hydrodynamics.

In hypothetically isolated conditions, J vanishes and the mechanics is purely deterministic and conservative. There is no relaxation.

6.5. Thermodynamics

In the previous section, the distribution function f was not specified. Accurate conclusions concerning the time dependence of relaxing systems are readily obtained by first studying the shape of f in general.

6.5.1. Entropy

Referring to earlier work by Carnot and Joule, Clausius (1854) invented the concept of entropy as a special thermodynamic function of state that would characterise macroscopic systems. The change of entropy under modification of a system's conditions was found to depend on the kind of process involved. If this is reversible, we have $\Delta S = Q/T$, where Q is the heat absorbed by the system from the surroundings during the process and T its temperature. By contrast, if the process is irreversible or spontaneous, Clausius observed that $\Delta S > Q/T$. Entropy, as a function of state therefore occupies a key position in discussion of irreversibility in macroscopic dynamics and evolution in the course of time.

According to Clausius, a process is said to be reversible (quasi-stationary) when it is conducted either in such a way that collective or coherent motion is never allowed to develop or when it has been allowed to relax completely (very slow modification of the system's properties). For historical reasons, collective motions are not welcome in traditional equilibrium thermodynamics. The current proposal to generalise thermodynamics to non-equilibrium systems should help reconsideration of this limitation.

In 1877, Boltzmann derived an expression that relates experimental entropy to statistical properties. This reads

$$S = k_B \ln W(M), \tag{4}$$

where W is the measure of the portion of phase space occupied by the particular condition of the system. Another expression for the same symbol is *the number of different phase space trajectories complying with the given set of mechanical extensive constraints*. This depends on the nature and the values of the constraints imposed on the system's dynamics. Hence, if we consider a gaseous system in an immobile container of volume V, consisting of N_r particles of type r, the total kinetic energy E being exclusively thermal (no additional collective or coherent motion), the set of constraints is the usual collection (V, N_r, E) of microcanonical variables.

6.5.1.1. Equilibrium

Using for the set of extensive constraints the compact notation $\{X_j\}$, the entropy differentiates as follows

$$dS = -\sum_j \xi_j dX_j. \tag{5}$$

The set of partial differentials $\{\xi_j\}$ is, by definition, the set of intensities or intensive variables conjugate to the microcanonical variables $\{X_j\}$. The distinction between intensive and extensive variables is important.

Rewriting the formal expression and giving to the variables their usual names, Gibbs' traditional equation is retrieved. (μ = chemical potential).

$$dS = \frac{1}{T} dE + \frac{p}{T} dV - \sum_r \frac{\mu_r}{T} dN_r. \tag{6}$$

Equilibrium conditions (no collective motion) having been assumed at the very start, the last expression is an equality.

6.5.1.2. *Non-equilibrium*

In non-equilibrium conditions, the complete description of a system's properties requires specification of the additional constraints represented by collective or coherent motion in which part of the total energy is stored. In the example of a spontaneously expanding gas, this was first the jet, followed by the acoustic perturbation. Their presence therefore implies additional extensive properties which must be considered in the expression for the entropy. By differentiation of the entropy with respect to the relevant variables, new conjugate intensities are obtained, typical for the non-equilibrium perturbations.

For simplicity, we consider the particular case of a jet defined by its total momentum **P**. Gibbs' expression for dS must then be amended by adding a new contribution $(-\xi dX)$ which, in this particular case, is $-(\mathbf{v}/T)\cdot d\mathbf{P}$, where **v** is the average (collective) velocity of the jet (de Hemptinne, 1992). The momentum in the jet being $\mathbf{P}=Nm\mathbf{v}$, the additional contribution may also be written $-(1/T)\,dE_{co}$, where E_{co} is the energy in the collective motion. Hence, in the particular non-equilibrium condition assumed we have:

$$dS=\frac{1}{T}\,dE+\frac{p}{T}\,dV-\sum\frac{\mu}{T}\,dN-\frac{1}{T}\,dE_{co}. \tag{7}$$

E is now the total energy, the sum of the thermal and the collective contributions. This conclusion is generally valid for all kinds of collective motion.

If the last term had been omitted, if our only information were that the system is definitely in a state of non-equilibrium without our knowing what additional perturbation was relevant, the equal sign would have to be replaced by $\geqslant$ as Clausius prescribed in his definition of the entropy. This confirms Carnot's statement (Carnot 1824) according to which energy in a collective mode may decrease ($dE_{co}\leqslant 0$) and thermalise, while the reverse is impossible.

The expression shows that suppression of the collective mode characterising the state of non-equilibrium maximises the entropy. The mechanism involved to reach that maximum is that discussed above, namely decorrelation of the system's internal collective motion by interaction with the surroundings.

6.5.2. *Distribution function*

It has been stressed above that a correct definition of the distribution function f requires a thermodynamic debate. This involves maximising the entropy under the given conditions (existence of the collective transient).

Let us consider an arbitrary system of particles at a given instant in unspecified conditions out of equilibrium. The motion of its particles is characterised by many kinds of correlation. In the examples above, stress has been laid only on collective translation. Vortices, internal rotations and vibrations, and a wealth of other motions may contribute to the non-equilibrium conditions. We call the particular state of the system a *fluctuation*.

The common tendency of all collective motions is to thermalise. The fundamental mechanism is the same: interaction with a correlation-destroying neighbourhood; the efficiency of the interaction will depend on the kind of motion involved. Some perturbations relax promptly, others last longer. Depending on the time-resolution the observer chooses to consider, which will be in all cases longer than the average interaction frequency with the surroundings, the memory of the fastest components of the initial fluctuation disappears and only the very few longer-lived correlations remain. These are the additional extensive constraints to be considered in maximising the entropy.

Maximisation of the entropy as given by Boltzmann's formula, taking account of all the extensive constraints, including those imposed by collective motion, leads readily to Gibbs' expression for the entropy, based on the distribution function f.

$$S=k_B\int (f-f\ln f)\,d\Gamma. \tag{8}$$

The variables in the maximisation are the set of intensities conjugate to the initially introduced extensive constraints ($d\Gamma$ is the elementary measure in phase space). It is no longer the energy that is important, but the temperature. Energy fluctuates and is therefore known only as an average quantity, while the temperature is defined unambiguously by the surrounding reservoir. The same rule holds for all other constraints strongly coupled to the surroundings.

The function f must now be incorporated in the dynamic equation describing evolution with time of the intensities characterising the state of non-equilibrium the observer has chosen to investigate by selecting an appropriate time resolution. The procedure is straightforward in fluid dynamics where it leads to results in perfect agreement with experiment, both in stationary and in transient conditions (de Hemptinne, 1992).

6.6. Quantum mechanics

In microphysics (molecular and sub-molecular level), it is known that Newtonian mechanics does not work and must be replaced by quantum mechanics. The most striking property of quantum mechanics is, for many, Heisenberg's uncertainty principle which states that position (q) and momentum (p) are defined with an uncertainty connected by the relation

$$\delta q \, \delta p = h. \tag{9}$$

For any individual degree of freedom, the space occupied by a single quantum state in phase space equals indeed Planck's constant h (Pathria, 1972). Some authors erroneously replace the equal sign by $\geqslant$, thereby increasing the impression that quantum mechanics is dominated by uncertainty. This is sometimes taken as the origin of the kind of chaotic uncertainty necessary to justify irreversibility of spontaneous processes.

The formal starting point of quantum mechanics is Schrödinger's equation. Only its time-dependent version is relevant here. Let $\mathcal{H}$ be the Hamiltonian operator. Its structure contains a special contribution for kinetic energy added to the potential field interacting with the particles. Integration leads to a set of functions $\Psi(q, t)$ of position and time (the wave function). The time-dependent Schrödinger equation reads

$$\mathcal{H}\Psi(q, t) = -\frac{\hbar}{i} \frac{\partial \Psi(q, t)}{\partial t}. \tag{10}$$

The square of the wave function is usually interpreted as a probability density in configuration space. Mention of probabilities gives a supplementary hint in the direction of fundamental uncertainty, be it only in configuration space.

It can be demonstrated that, for isolated systems, Schrödinger's equation is symmetrical with respect to inversion of the variable t. This indicates that quantum mechanics alone by no means justifies the irreversible property of nature and that Heisenberg's uncertainty principle refers to another reality.

Schrödinger's equation yields stationary states. Transitions between them may be allowed in certain conditions but this implies normally emission or absorption of electromagnetic radiation. If the field, together with a radiating particle, is enclosed in a cavity and supposed to be isolated from the outside world, the solution of Schrödinger's equation is a permanent Rabi oscillation back and forth between the previously mentioned stationary states. There is no relaxation (Sargent III et al., 1974; de Hemptinne, 1985).

Decay by emission of radiation, which represents relaxation of excited states, implies that the system would be accessible for exchange of radiation with the outside world. The time-dependent Schrödinger equation confirms that the outgoing field is phase-matched to the motion of the radiating particles (coherence). At the same time, in addition to the ubiquitous background radiation, the field accessible for re-absorption is the total incoherent thermal radiation issued from the collection of external sources which constitute the surroundings. As in the classical case, one result of exchange of radiation with the outside world is complete loss of memory of the initial phase information. At equilibrium, the temperature of the set of radiators equals that of the surroundings (de Hemptinne, 1991).

In fact, quantum mechanics bridges the motion of particles and the properties of radiation fields. It appears that incoherence of thermal sources of radiation is the analogue of molecular chaos of particle motion.

6.7. Conclusions

Irreversibility, time's arrow and its origin remain the source of much scientific discussion; this started before Boltzmann. The dispute has been nourished by inaccurate definitions and hasty conclusions concerning a number of concepts such as the word irreversibility itself, isolation, relaxation and equilibrium. The controversy has been amplified by ambiguous interpretations of simple experimental facts (de Hemptinne, 1997; Kumičák et al., 1998).

Arguments based on observation and theoretical discussions strictly compatible with confirmed laws of mechanics (at microscopic level) have been used in this work to emphasise that the equilibrium state cannot be defined without taking account of the conditions valid in the environment. Relaxation dynamics represents exchange with the surroundings, that is export of coherent (collective) information compensated by stochastic import of thermalised (incoherent) information.

Simultaneous discussion of the properties of the system and its surroundings makes the use of thermodynamic arguments necessary. Intensities, which are directly related to the probabilistic concept of entropy, are defined for every exchangeable property.

It is frequently objected that interaction of the walls and the enclosed system itself follows Hamiltonian dynamics. It would therefore suffice to define the macroscopic system as the addition of the enclosed system and its walls in order to build a (strictly) isolated system where the laws of Hamiltonian dynamics would be strictly valid. The discussion concerning irreversibility would then occur at this higher level, where the global system would be isolated. This is not the case, because the walls themselves interact with a broader environment, moving the problem one step further.

There is an arbitrary choice in the definition of boundaries to systems. It depends on how far mechanics is allowed to take care of correlated reactions to forces. In the domain where Hamiltonian dynamics is valid, forces acting on the components of the system sum identically to zero. Some forces, however, clearly have an external source. Their reaction cannot be included in the dynamic equation unless they are labelled 'stochastic' because they are completely uncorrelated. They are the cause of irreversibility.

We are tempted to extrapolate questions and conclusions to the universe itself, assuming certain alleged cosmological properties, often without proof. This must be strongly resisted. Science has no information on the extreme properties of the universe and conclusions based simply on the extrapolation of arguments valid at our observational level to domains beyond our reach are meaningless and invalid. This is why the final source of irreversibility will never be unveiled.

References

Balescu, R., 1975, Equilibrium and Non-equilibrium Statistical Mechanics. John Wiley, New York (N.Y.).

Carnot, S., 1824, Réflexions sur la puissance motrice du feu. Reprinted by J.-P. Maury, Presses Universitaires de France, Paris 1986.

Cornfeld, I.P., Fomin, S.V. and Sinai, Ya.G. 1982, Ergodic theory. Springer, New York (N.Y.).

de Hemptinne, X., 1985, Thermodynamics of laser systems, Phys. Rep. **122**, 1–56.

de Hemptinne, X., 1991, Matière et lumière, Rev. des Questions Scientifiques, **162**(3), 289–299.

de Hemptinne, X., 1992, Non-equilibrium Statistical Thermodynamics applied to Fluid Dynamics and Laser Physics. World Scientific, Singapore.

de Hemptinne, X., 1997, The Source of Irreversibility in Macroscopic Dynamics, Ann. Fond L. de Broglie, **22**, 61–85.

Dorfman, J.R. and Gaspard, P., 1995. Chaotic Scattering Theory of Transport and Reaction-rate Coefficients, Phys. Rev. E, **51**, 28–35.

Eckmann, J.-P. and Ruelle, D., 1985, Ergodic Theory of Chaos and Strange Attractors, Rev. Mod. Phys., **57**, 617–656.

Goldstein, S., Lebowitz, J.L. and Aizenman, M. 1975, Ergodic Properties of Infinite Systems. In: J. Moser, (ed.), Dynamical Systems, Theory and Applications, Lecture notes in Physics, **38**. Springer-Verlag, Berlin.

Kumičák, J. and de Hemptinne, X., 1998, The Dynamics of Thermodynamics, Physica D, **112**, 258–274.

Lanford, O.E., 1975, Time evolution of Large Classical Systems. In: J. Moser, (ed.), Dynamical Systems, Theory and Applications, Lecture notes in Physics, **38**. Springer-Verlag, Berlin.

Lebowitz, J.L., 1993, Macroscopic Laws, Microscopic Dynamics, Time's Arrow and Boltzmann's Entropy, Physica A, **194**, 1–27 and Boltzmann's Entropy and Time's Arrow, Physics Today, **46**(9), 32–38.

Prigogine, I. and Stengers, I., 1988, Entre le Temps et l'Éternité. Fayard, Paris.

Pathria, R. K., 1972, Statistical Mechanics. Pergamon Press, Oxford.

Ruelle, D., 1989, Chaotic Evolution and Strange Attractors: The Statistical Analysis of Time Series for Deterministic Nonlinear Systems. Cambridge University Press, Cambridge, England.

Sargent, M. III, Scully, M.O. and Lamb, W.E., 1974, Laser Physics. Addison-Wesley, Reading, (Mass.).

Sinai, Ya.G., 1976, Introduction to Ergodic Theory. Princeton University Press, Princeton (N.J.).

Spohn, H., 1980, Kinetic Equations from Hamiltonian Dynamics: Markovian limits, Rev. Mod. Phys., **53**, 569–615.

chapter 7

TIME'S POISONED ARROW: RECONSTRUCTING EVOLUTIONARY HISTORY

ADRIAN FRIDAY

University of Cambridge

Time in Contemporary Intellectual Thought
Patrick Baert (Editor)

CONTENTS

7.1. Introduction

Some perspectives:

> "This is later. That was then."
> Lester Young, quoted by Nat Hentoff in Porter (1991, p. 73).

> "The fascinating thing about telling stories is that they start with the end."
> Appleby, Hunt and Jacob (1994, p. 263).

> ". . . that you may command certain men not to teach false doctrines any longer nor to devote themselves to myths and endless genealogies. These promote controversies . . ."
> 1 Timothy 1: 3–4.

Confidence in our ability to reconstruct even our own recent history has taken a savage battering of late: just what, indeed, in these poststructuralist days, is 'history'? The study of the evolutionary past has not been immune to such crises of confidence. The vastness of the geological time-scale, and the sheer diversity of organisms, extinct and extant, ensure that there certainly is grandeur in this view of life, but can we ever expect, in principle, to be able to reconstruct evolutionary history with conviction? This question is not trivial, because all biologists are evolutionary biologists in the sense that the organisms on which they work are the products of evolutionary processes. Biology undeniably has a historical dimension, and explanations that ignore this fact risk being found incomplete. (Harvey and Pagel, 1991 and Eggleton and Vane-Wright, 1994 amply demonstrate the extent to which a consideration of evolutionary history is generally important in biology.)

Evolutionary biology has suffered from a debilitating incapacity to determine whether or not the events of evolution display any law-like behaviour, and, in particular, to decide how rapidly those events occurred, given the vast spans of time that were available for them to do so. A particularly exhilarating cascade of new ideas characterized the scientific (and personal) battles about evolution in the late nineteenth century, and there have, of course, been very many subsequent, fundamental advances: in genetics, especially at the molecular level; in the dating of past events; in the recovery of fossil evidence of extinct organisms. However, despite, perhaps even because of, new knowledge in these disparate areas, the reconstruction of evolutionary trees remains an area of intense controversy. In science this usually means that someone, somewhere has a great deal to lose. No wonder, then, that the controversies of evolutionary biology are so fierce.

With the emergence, in the second half of the last century, of the elements of an essentially modern approach to the study of embryology, on one hand, and the fossil record, on the other, it became an overwhelming temptation for some (including that gaseous vertebrate, Ernst Haeckel) to see in the sequence of events revealed in embryological development clues to, if not a repetition of, the sequence of events in evolutionary history. Physics in this century has struggled rather more publicly with the difficulties of reconciling the very small and the very large, and with knitting together in a seamless, theoretical whole the unobservables of quantum mechanics and the unimaginables of cosmology. In comparison, a solution to the biologists' problem with time seems as though it ought to have been easier to achieve. Indeed, it probably seemed that this was so to biologists, themselves; but it has not been, at least so far. No sooner had Charles Darwin published his sketched account of how biological adaptations might have been brought about, gradually, over geological time, than Haeckel not only argued that a profitable search might be made in the embryological history of extant organisms for clues to their evolutionary history, but he also, with a quite unseemly haste and with an enthusiasm bordering on the obsessive, published elaborate trees of life showing the origin and divergence of animal lineages over time. Attractive though Haeckel's ideas about the parallels of embryological and evolutionary history might originally have been (and at least some of his contemporaries, to their credit, resisted his ideas) much effort has been expended over the last few decades in re-defining the relationship between events observable over the short time-span of an individual's development and the events that must have taken place over the vastness of geological time to make that individual organism into what it is today. The former does not simply recapitulate the latter. (For recent discussion, see Minelli, 1998, and the references there.) This is a pity, since the capability to read off the hard-won triumphs of evolution, in strict sequence, in the stages of development would, undoubtedly, be valuable.

A moment's reflection should raise a suspicion, at the least, that life might very likely not be that easy. There is no reason to suppose, for example, that any stage of an organism's development is somehow immune from the moulding of natural selection; no reason, therefore, to suppose that the characteristics of adult ancestors would be, as it were, frozen (certainly not frozen unmodified) as early stages in embryological development. It is ironic that the word 'evolution' was earlier used to describe the unrolling of features during development and only later came to be applied, in its modern sense, to the processes leading to the acquisition of characteristics over geological time.

The parallel with physics can be taken further. Up to, or rather down to, a point, small-scale and short-lived events in physics can be visualised (one hesitates to say 'observed') experimentally. Sub-atomic particles can be caused to betray their behaviour by manipulating them, if necessary over and over again, in accord with ideas suggested by a body of theory. On the other hand, the events of cosmology are, in principle, more problematic, many having taken place in the long-distant past. One cannot readily contemplate engineering their repetition. In biology, the details of an organism's embryology can not only be directly observed, but can also be manipulated. Again, on the other hand, the events of evolution are history: they occurred (surely, at least this is beyond reasonable doubt) but they are just that, history, and as such are unique. For this reason, much was made, at a time when his ideas were particularly appealing in some ares of biology, of Karl Popper's views on historicism. (Indeed, at one point, Popper, himself, entered the debates with a consideration of the status of natural selection. His contribution provoked a number of august rejoinders and he subsequently changed his mind.)

Given that the course of evolutionary history cannot simply be read off in an organism's development and that the past cannot now be manipulated, the question remains: how can it best be established? The observation that there was a unique, historical sequence of events looks initially as though it might provide a crumb of comfort, in that, perhaps there is a way in which the data that we can gather about evolutionary history can be made to tell us whether or not we are near to or far from the correct solution. It would seem to go without saying that the fossil record must, in some way, have a crucial role to play in establishing the nature and sequence of evolutionary events. Again, however, there are difficulties. For one thing, the fossil record is clearly incomplete because a back-of-an-envelope calculation suggests that only a fraction of the organisms that have ever lived are preserved as fossils. In these circumstances, it would not be surprising if there were gaps in the record making the assembly of unbroken lineages of succession quite difficult to achieve by the mere juxtaposition of forms sufficiently similar to one another that they could be justifiably regarded as ancestor and descendant. This difficulty certainly seems as though it ought to be a matter of concern for a neo-Darwinian 'gradualist' and, indeed, the whole issue of rate of change of the characteristics of organisms in the evolutionary past has been the subject of intense debate over the last two decades.

Stephen Jay Gould and Niles Eldredge recently celebrated the twenty-first anniversary of the publication of their controversial paper on patterns of change in the fossil record (Gould and Eldredge 1993). In the original account they argued against the primacy of 'phyletic gradualism', that is against the idea that the prevailing manner of evolutionary change is by accumulated small steps, occurring more-or-less continually for long periods of time. The theory Eldredge and Gould proposed as a complement to phyletic gradualism, that of 'punctuated equilibrium', allowed that the apparent sudden appearance of new species in the fossil record and the apparent subsequent persistence of species for long periods with little or no apparent modification were real rather than simply the consequence of imperfections in the record. Their model suggested that 'stasis', that is those long periods of geological time when apparently little or nothing happens, paradoxically gives valuable information about evolutionary processes. Stasis seemed especially remarkable when maintained throughout periods of profound climatic change, suggesting that its maintenance could be active. As Gould and Eldredge (1993, p. 223) put it, "How odd, though, to define the most common of all palaeontological phenomena as beyond interest or notice! . . . And, even worse, as palaeontologists didn't discuss stasis, most evolutionary biologists assumed continual change as a norm, and didn't even know that stability dominates the fossil record." On the other hand, speciation, the generation of new species, might appear almost instantaneous on a geological time-scale, and early opposition to the theory of punctuated equilibrium was partly a result of a misunderstanding of this issue. It was not being proposed that speciation really was sudden or saltational, but simply that the resolution of the fossil record was such as to prevent the observation of change still gradual in terms of the generation time of the organisms concerned. In other words the issue was one of scale.

An additional major element in the theory of punctuated equilibrium, as it is currently constituted, concerns the nature of macroevolution, that is the patterning of large-scale events in evolutionary history. The Darwinian mechanism for such events involves an extrapolation of microevolutionary events, of the generation of adaptations by the action of natural selection on individual organisms within lineages of species. In contrast, from the point of view of the theory of punctuated equilibrium, if species persist, with relatively little transformation, for long periods of geological time, then it is possible to view the events of macroevolution as occurring predominantly by differential survival of lineages. Some lineages might speciate more frequently than others, others might simply survive for longer. The emphasis is here rather on species sorting. Gould and Eldredge (1993, p. 224) emphasise that they are suggesting ". . . not a claim for the falseness or irrelevancy of microevolutionary mechanisms, especially natural selection, but a recognition that Darwinian extrapolation cannot fully explain large-scale changes in the history of life."

Stephen Jay Gould, most notably in his book *Wonderful Life* (Gould, 1989), has also contributed greatly to the discussion of another venerable evolutionary issue that has periodically resurfaced, each time with a vengeance: the issue of 'progress' in evolution. It has always been tempting for gradualists to describe putative evolutionary series in terms of progression, thus suggesting that, the further one gets from an organism at the beginning of a lineage, both in form and, of course, in time, the more 'evolved' the relevant stage is. For an advocate of punctuated equilibrium the comparison would be between lineages; some earlier, some later. In both cases such considerations raise profound questions about, for example, whether 'evolution' (whatever, if anything sensible, the word means when used in this way) has intrinsic directionality. There is also the related issue of complexity: do evolutionary lineages (again, either internally, or in comparison with one another) become progressively (that word seems inevitable) more complex over time? Just what might reasonably be meant by 'progress' in a biological context? This question is phrased in such a way as to suggest to the non-biological reader the degree to which most evolutionary biologists currently mistrust the concept of 'progress', especially. It was not always thus. It is not just that biologists now aspire to be conspicuously liberal in thought, and that they (mostly) wish to be seen to reject the taint of imperialism and Social Darwinism. It is more that we have become concerned to reject scrupulously any interpretation that suggests a 'meaning' to the changes bought so dearly with the profligate sacrifice of so many organisms over such a long period of time. The bleakness of such a view may be uncongenial, but its acceptance gets us rather smartly out of a number of difficult problems of interpretation of evolutionary history. There is a cost to pay, of course, in the need to give full recognition to the possibility of our own essential insignificance. If the particular outcomes of evolution, those organisms existing at any point in past time, and, especially, now, in the present, are no 'better' than those that went before, what does that say about humanity? This is either a fine mess or the beginning of true humility, depending on how you look at it.

7.2. Reconstructing evolutionary history

To return, however, to the issue of how best to reconstruct evolutionary history. It seems to us now quite natural to approach the relationships of organisms through the metaphor of an evolutionary *tree*. It seemed so to Charles Darwin, whose rough drawing of an imaginary tree of relationships, in one of his notebooks, has become an icon of evolutionary biology. The tree in question was sketched in 1837 and indicates how well-advanced was Darwin's thinking at this relatively early stage in his scientific life, although it now seems likely that he was not alone at the time in recognizing the usefulness of a tree diagram. The historian of science, the late Dov Ospovat, convincingly argued that Darwin was later strongly influenced by an essay on classification written by the French biologist Henri Milne-Edwards. In his essay, Milne-Edwards presented a diagram of vertebrate classification that provides a graphic illustration of the nesting of some categories within others. This diagram has further complexities. Those who wish to see it, and to find further discussion, may most conveniently refer to Panchen (1992, pp. 20–23 and Fig. 2.6). Panchen's book also provides a convenient source for Darwin's early sketch of a tree diagram (Panchen's Fig. 3.2) the tree diagram from the *On the Origin of Species* (Darwin 1859) discussed below (Panchen's Fig. 3.3) and Haeckel's earlier and later trees (Panchen's Figs. 3.4 and 3.5). Darwin speculated on why it was that such an arrangement as Milne-Edwards' might seem somehow a natural one, and conceived what he, himself, rather portentously and uncharacteristically, named his 'Principle of Divergence'. According to this Principle, organisms would change over time in such a way that they would become increasingly more unlike one another: in other words, would diverge in their characteristics. Such a principle probably strikes most who read the *Origin* as eminently plausible. So important was the idea to Darwin that he chose to exemplify the Principle in a diagram. It is the only figure provided in the great essay. That it was the only one is, in itself, rather surprising, because Darwin was a very visual man, for whom biological objects clearly had an aesthetic appeal as well as an intellectual message. It is true that Darwin himself viewed his greatest work as a mere summary, and a somewhat hasty one at that, written to be maximally accessible to specialist and non-specialist alike (and in this, of course, he succeeded spectacularly). It was written without constant recourse to documentation and footnotes, as would have befitted a formal, scholarly work with a case to argue, especially a case of such magnitude and originality. It is still surprising, however, that only the one figure appeared; but this certainly gained in impact by being so conspicuously isolated. The famous diagram shows lineages of organisms diverging over time. Some of the lineages become extinct and a few negotiate all the imagined hazards to arrive at the present. The form of the time-scale Darwin used is of interest: it has a rather modern feel to it because it shows time divided into discrete sections (mimicking geological strata) in a manner recalling the division of time in some more recent, probabilistic models of evolutionary processes.

Although there is reason to believe that Darwin, not unnaturally, had his usual doubts that his ideas would meet with a generally sympathetic response, and, of course, publication was eventually urged upon him as a consequence of Alfred Russel Wallace's parallel development of a theory about the mechanism of evolutionary change, nevertheless the odd instance of self-congratulation creeps onto the pages of the *Origin*. It is important to realize that Darwin's educated readership did not need every element of his theories rammed home to them: doubtless some, who knows how many, were aching to be convinced. One might almost be forgiven for assuming, from some more popular accounts of events after the publication of the *Origin*, that Thomas Henry Huxley was the only person to welcome whole-heartedly the ideas expressed there. This was quite clearly not the case, although a good number of biologists (and others) expressed their disquiet at some elements of the argument (and, indeed, Huxley had some reservations about detail). Among the more wounding of the informed critics was the botanist (and, incidentally, phrenologist) H.C. Watson. Watson had advised Darwin on the matter of endemic floras and, indeed, had visited him at Down House to do so. When the *Origin* was published, Watson criticised the emphasis Darwin put upon evolutionary divergence. His criticisms were based on his own experience of comparing plants from different biogeographical regions. This had caused him to be impressed by the extent to which the same characteristics had apparently evolved independently in lineages separated in time and space. In other words, Watson was impressed by the extent of evolutionary convergence. In later editions of the *Origin*, Darwin replied to a succession of his critics. His reply to Watson is made with such apparent assurance, almost insouciance, that it causes one to consider seriously the possibility that Watson had caused Darwin to become really quite worried by the issue. Darwin's answer was essentially that convergence was, when all was said and done, rather infrequent and when it did occur it was clearly recognizable for what it was! Why should this issue, which seems at first sight, perhaps, to be a historical backwater of no particular current relevance, merit consideration? In order to illustrate its continuing fundamental importance in evolutionary biology, it is necessary to return to the question: how best can we reconstruct evolutionary history?

In considering this question it is instructive, yet again, to start with an observation about Charles Darwin. It is striking that he never once published a 'real' tree representing his views on the course of the evolutionary history of a particular group of organisms; not even of the barnacles, a group with which he had had to become so thoroughly familiar in order to establish his credentials as a serious, scholarly scientist, despite remaining an amateur in a new world of emerging professionals. Darwin's reticence in the field of the reconstruction of evolutionary history appears even more striking when his Principle of Divergence, of which he was so unaffectedly proud, is considered. The Principle leads rather directly to an approach to such reconstruction; so directly that it seems unlikely that Darwin, himself, was completely unaware of it. However, if he really could not see quite how to go about constructing evolutionary trees, that would, of course, naturally have prevented him from trying: he was a man who clearly liked to be sure of his ground, scientific, moral, financial and social. Two other possible reasons come to mind. One is that he quietly admitted the force of the criticisms of Watson and others, and came to see that the task might be more fraught with hazard than he initially, optimistically, supposed. The other possible reason is that Darwin felt that the task was quite feasible, indeed, merely a matter of tediously cranking the handle of machinery already effectively established for those with eyes to see, but that others should carry out the work. All these reasons might be held to gain some support from the view expressed by Darwin (in a letter to T.H. Huxley, who *did* have a go at making trees) that "The time will come, I believe, though I shall not live to see it, when we shall have fairly true genealogical trees of each great kingdom of Nature." (This quotation, incidentally, has been taken as an inspirational text by Morris Goodman, editor of the new journal *Molecular Phylogenetics and Evolution*, launched in 1992.) The 'fairly true' is interesting in the light of Watson's criticisms, and certainly others, notably the comparative anatomist St G. Mivart, were pleased to point out the frequent, apparent occurrence of convergence, for reasons of their own (Desmond, 1982, p. 183). If Darwin did, indeed, intend to leave the field to others, his comment to Huxley suggests that this was at least in part because the task struck him as just plain difficult. As Desmond (1982, pp. 156–157) makes clear, Darwin was at best only very timidly encouraged by Haeckel's attempts, first published in 1866, a mere seven years after the publication of the *Origin*, itself (and before Mivart's most vociferous criticisms). Haeckel's outrageously optimistic trees, which Darwin referred to as 'tables of descent', have been reproduced quite often recently, for example, on the cover of the book *Molecules and Morphology in Evolution: Conflict or Compromise* (Patterson, 1987), but this is not because they are held to be 'timeless' versions of their art. Rather, they seem to us now to be all too typical of an authoritarian approach to the subject, and authoritarian, what is more, without entitlement, because they were not constructed using any clearly communicable methodology that could be used by other workers on other groups.

Just how the Principle of Divergence leads so directly to the possibility of reconstructing evolutionary trees is easy to appreciate and is related to one of the most important, and controversial, advances in biological systematics since the work of Linnaeus. In order to explain these issues it is necessary to introduce some of the properties of trees. If organisms become mutually more different from one another over time, and none of the changes are overwritten by subsequent ones, then if the differences between pairs of organisms at the tips of the tree are scored and the resulting distances are inspected, these distances will be found to obey simple algebraic rules that enable recovery of the branching structure of a tree diagram representing the relationships of the organisms concerned. In fact, under these conditions, the recovery of the correct, unique tree is fairly trivial. An example may help here. Consider four imaginary species, A,B,C and D, at the tips of an evolutionary tree, as in Fig. 7.1. The four species are located, in this example, in the present, represented by the plane in which all terminal branches of the tree have their tips. Since the tree is an evolutionary tree, a time-scale is shown alongside.

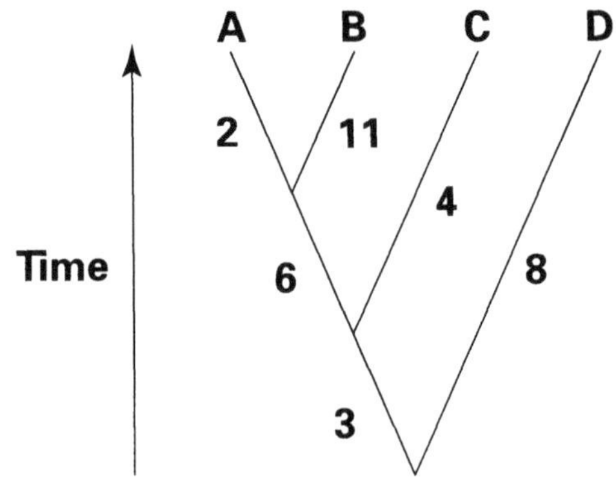

Fig. 7.1.

Following the prescription for analysis just described, the numbers of evolutionary events that took place on each segment of the tree (shown by the numbers against the branches) are summed on the path between each pair of species. For example, the distance between species A and species B is 2+11=13; and the distance between A and C is 2+6+4=12. This arithmetic, carried out for each species-pair gives the following matrix of distances (Fig. 7.2).

B	13		
C	12	21	
D	19	28	15
	A	B	C

Fig. 7.2.

Note, incidentally, that the tree distance between A and B is greater than that between A and C. This might initially seem paradoxical, because a glance at Fig. 7.1 suggests that, in terms of recency of common ancestry, A and B can be described as more closely related to one another than either is to species C. This observation raises another interesting point: is any of the species A, B and C more closely related, in this sense, to species D than are the others? The answer is no, because the lineages leading to all three species, A, B and C, diverged from the lineage leading to species D at the same point (in time), which happens, in this simple, four-species case, to be the root of the tree. Again, this might appear paradoxical, because species C is clearly less distant, in terms of summed tree distance, measured in numbers of events, from species D than are species A and B. These observations underline the fact that closeness of 'relationship' is here being defined in terms of recency of common ancestry, rather than in terms of degree of difference. The elements of paradox are provided by the fact that numbers of events are, in this example, not related to time (at least, not in any simple law-like way that can be discerned). This is appreciated most simply by realising that the time since species A and species B diverged from one another is necessarily the same for both (this, of course, is a property of sidereal time, itself, and has nothing especially to do with evolutionary biology) and yet the numbers of changes on the relevant branches differ.

The construction of this particular example, then, reflects an unruly behaviour of change over time. This behaviour means that species that are more closely related to one another might look less like one another (that is, be more different or distant from one another based on a metric of events that represent changes in their observable characteristics) than species that are less closely related. Nevertheless, as observed above, a fairly simple analytical operation leads to the capacity to retrieve the correct (in this case, of course, known) relationships of the four species; but with one missing element, however: the position of the root cannot be determined simply from the properties of the distances. Further information is necessary of a sort that will be considered later, below. Given the undefined relationship of change to time, which was built into the example, this should not appear unreasonable. (Actually, there is a, rather weak, relationship between change and time, in that amount of change always increases with time and never decreases: species always, therefore, get more dissimilar to one another with elapsed time, even though, as demonstrated, the amount of dissimilarity between pairs of species is not rigidly proportional to elapsed time: divergence, again). If anyone has doubts about the uniqueness of the (unrooted) tree reconstructed from the distances, he or she should try constructing what, in this contrived example, are know to be incorrect unrooted trees. The correct unrooted tree is shown as tree (a) in Fig. 7.3, below, whereas trees (b) and (c) are incorrect, as is tree (d). Tree (d) simply expresses the fact that the four species are all

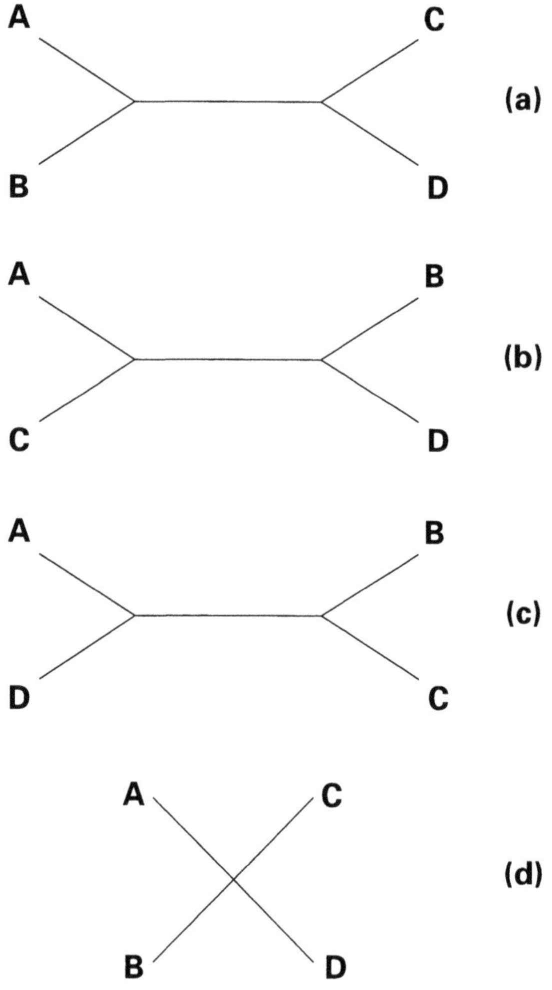

Fig. 7.3.

different from one another, but, of course, says nothing about their relative degrees of closeness of relationship in terms of recency of common ancestry.

Tree (d) also exemplifies a branching pattern that is not dichotomous; that is, with two lineages 'going out' and one 'coming in' (on a rooted tree). Non-dichotomous branching will be examined again later.

Most people accept tree diagrams of evolutionary relationships, drawn against a time-scale, quite happily, even when they are not aware of, or even remotely interested in, the technical details of the properties of such diagrams, or in the principles of reconstruction of evolutionary history. There is an aptness about this form of representation that is somehow satisfying. Doubtless this aptness has much to do with the use of similar forms of diagram to depict lines of descent in human families. We appreciate the nature of a family tree and readily extend the principles embodied there to the depiction of relationships on a rather grander time-scale. To the extent that both tree diagrams depict, geometrically, the flow of genetic information, they are, indeed, similar. However, it is the aim of a family tree to apportion genetic responsibility, so to speak, in an explicit way, in that every individual must have two (and only two) known parents, no matter how embarrassing the identity of those parents might be. The same principles are true, of course, of genealogical trees for (say) cattle and horses (although here, admittedly, the degree of potential embarrassment is rather less). The fact that it is the flow of genetic information that is depicted remains true even where there is introgression: brother-sister matings that produce offspring can still be effectively represented on such diagrams. There are, however, further parallels between the two varieties of tree diagram. For example, on a family tree, some branches may stop short of the present. An individual who does not mate, or whose matings result in no offspring, has no line proceeding from them. On an evolutionary tree, a lineage may become extinct, and hence be depicted as a branch that stops abruptly somewhere before the plane of the present. (Amusingly, it is usual to draw family trees with the 'ur' individuals at the top and their descendants nearer the bottom of the page, whereas evolutionary trees are generally drawn with their root at the bottom and their branches extending up the page: 'the ascent of man'? Those who use molecular data to construct trees often draw them sideways, probably largely because there is a premium on efficient use of journal space in this fast-moving field! There is clearly an interesting sociology here.) The tree in Fig. 7.4 illustrates a case in which two lineages, leading to species A and C, have become extinct. In addition, the three labels, E, F and G, indicate the presence of ancestral species at the three branch-points.

It is instructive, in pursuing the relationship between family trees and evolutionary trees, to consider what a small segment of a conventional evolutionary tree might look like if inspected at high magnification, as it were. A complete evolutionary tree would include all relevant individuals that ever existed. Some, naturally, would have become extinct without progeny; all would become

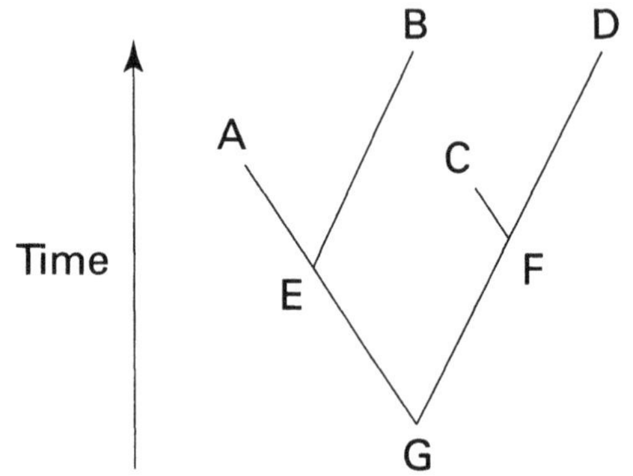

Fig. 7.4.

extinct as individuals. John Maynard Keynes (had he had the opportunity) might have added 'in the long run', but 'long' is, of course, not very long at all on an evolutionary time-scale. However, no one expects that an evolutionary tree, as it is generally understood, will faithfully include all individuals of all species that have been involved along the various evolutionary lineages: the tree diagrams would then be hopelessly complicated, even if we had more than the smallest fraction of such information available to us (which, of course, in general we do not). The complexities of individual ancestor-descendant relationships are normally submerged, then, in an evolutionary tree, which depicts only long-term flow of genetic information: the continuity that matters. This is, of course, immediately apparent when one realises that most evolutionary trees have taxonomic units at least as large as species at their tips, and those units are frequently genera, families, orders, classes, or even phyla (the word is Haeckel's). The evolutionary tree thus emerges as a map whose connecting branches, which have directionality in time, depict the way in which genetic information is passed on, often tenuously, from generation to generation, in lineages that sometimes split and diverge, and that sometimes become extinct. There certainly is 'grandeur in this view of life': somehow what is going on accords with behaviour we see in the wild, in our domestic animals, and in ourselves. The evolutionary tree is a framework for the reconstruction of events that we suspect, on the basis of experience and experiment, must have been happening on a vast time-scale: events that are now unobservable directly, and which can only be made to make sense by embedding them within the theoretical framework of the tree.

It is again helpful to return to the procedural matters of how best to reconstruct the course of evolutionary history, using the two components already introduced to frame a synthetic model: the tree structure as a map of directional flow of genetic information, and the Principle of Divergence as a statement about the behaviour of change in the characteristics of organisms. Surprisingly, it was not until the 1950s and '60s that a rigorous methodology for reconstruction was worked out in a formal manner, by the German entomologist, Willi Hennig. The publication, first in German and later (in amended form) in translation into English in 1966 (Hennig, 1966) of his work on what he called 'phylogenetic systematics', initiated several decades of lively, largely constructive, and often vituperative debate about the reconstruction of evolutionary history. (Those who require a full, practical treatment of modern 'cladistics', as this approach has come to be called, will find it in Forey and others, 1992.) It is implicit in his title *Phylogenetic Systematics* that Hennig was proposing a system of classification of organisms based squarely upon their evolutionary relationships (that is, in terms of their recency of common ancestry with one another). Hennig's methodology, in a modified and elaborated form, has represented the prevailing dogma for many, probably most, comparative morphologists over the last two decades, at least as far as the determination of evolutionary relationships is concerned. There remain systematic and evolutionary biologists who continue to argue that there is no compelling reason why classifications and trees of evolutionary relationship should reflect one another exactly. These biologists would, therefore, reject phylogenetic classifications in favour of special purpose, synthetic classifications that take into account (in ways that are not always easy to communicate) other aspects of organisms than just recency of common ancestry. Phylogenetic systematists, however, continue to maintain the primacy of a diagram of evolutionary relationships, and have further argued that that diagram, and a classification based explicitly upon it, represents the optimum basis for storage of comparative information about organisms in an information theoretic sense.

Hennig's achievement was that he pointed out that the course of evolution could be charted by the reconstruction of the innovations that occurred intermittently over time. Arguably his most important contribution was the demonstration that similarities between organisms could, in the context of a tree model, be of different kinds. Consider, for example, the character *number of legs*, and the states of that character in three familiar animals, a cow, a whale, and a monkey. The cow and the monkey both have four legs, whereas the whale has apparently lost its rear legs and retained only the front ones, which are used as flippers. For the purpose of this example, we shall conveniently ignore the occurrence of 'vestiges' of the previously-present hind legs, and, as a less evolutionarily biased observation, suggest that our chosen whale can, legitimately, be said simply to have two legs.

For this character, therefore (and we are considering only this single character to start with) the cow and the monkey are indisputably more similar to one another than either is to the whale. Accordingly, one might attempt to depict the evolutionary relationships between these three animals, again in the sense of recency of common ancestry, as in Fig. 7.5.

All is not well, however, with this argument: a problem is revealed by attempting to map onto the tree diagram the state which the character *number of legs* has in the various parts of the tree. From the evolutionary value judgement

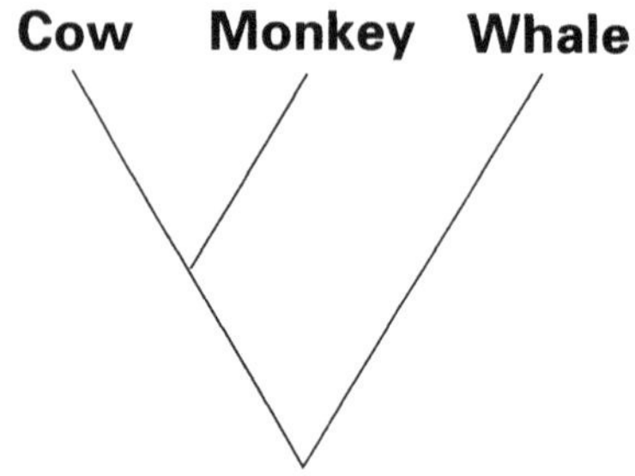

Fig. 7.5.

already put forward, the assignment of states at the nodes, or branch-points, of the tree might look like that in Fig. 7.6.

In assigning these states, we might make use of some previously proscribed knowledge about the vestiges of the rear legs: the state '4' at the root of the tree reflects the use of the word 'lost' to describe what happened to the two rear legs during the evolution of whales from (surely) four-legged ancestry. The main point to be made about Fig. 7.6 is that the cow and the monkey are similar, certainly, but only in the sense that they have both retained the state that was already presumed to be present at the root of the tree. In other words, these two animals are not united by the possession of an innovation: in fact, nothing has happened on the left hand side of the tree leading from the root to the cow and the monkey. Nothing in terms of innovations, that is: it is sometimes easy in this special context to lose sight of the fact that what has happened is the continual passing on of information leading to the development of four legs, with a fidelity that is quite astonishing: this we describe as 'nothing happening'! However, on the tree, the whole action in terms of innovations takes place on the branch leading from the root to the whale. On this branch we can mark, using some appropriate convention, a change from four legs to two. The conclusion must be that the acknowledged, perceived similarity between the cow and the monkey is invalid as supporting evidence for the tree in Figs 7.5 and 7.6. This result, initially surprising to those new to the

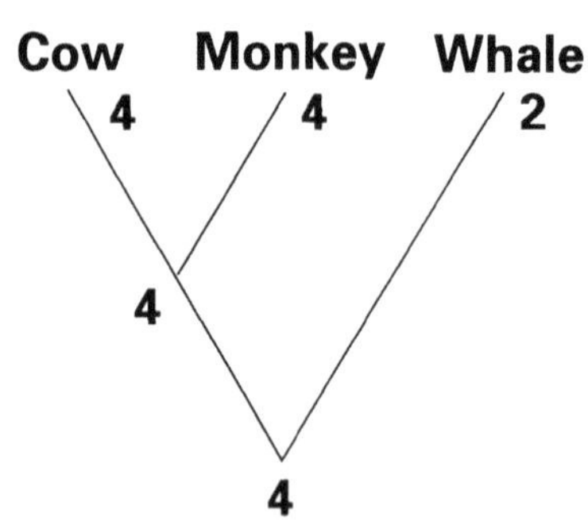

Fig. 7.6.

concept that similarity, in this context, is not as straightforward as it might appear, is central to Hennig's methodology. It is worth emphasising, parenthetically, that there is actually quite compelling evidence to suggest that cows and whales are more closely related than either is to monkeys, and that, therefore, the relationships shown in Figs 7.5 and 7.6 really are wrong.

If the character *number of legs* does not allow us to support the relationships of Figs 7.5 and 7.6, does it tell us anything about the other possible hypotheses of relationship between the three animals, including the hypothesis we currently believe, on extrinsic evidence, to be the correct one? The answer is no, because the change from four legs to two legs must occur once on each of the possible tree patterns for these animals, and each time it has to be placed on the branch leading to the whale, no matter what that branch is connected to. This is demonstrated in Fig. 7.7.

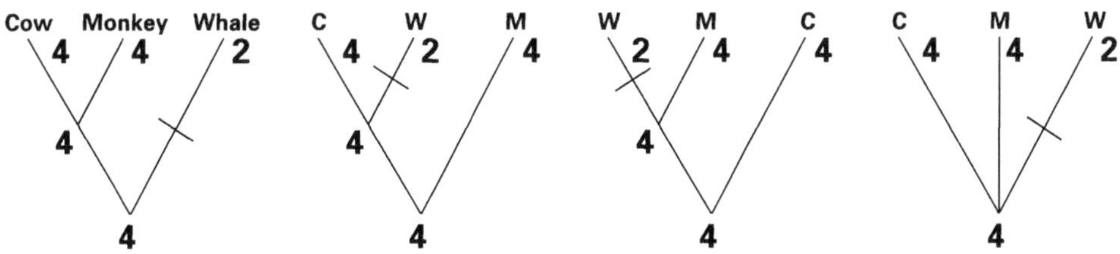

Fig. 7.7.

This figure emphasises, again, the fact that despite the similarity between cow and monkey in the number of legs they have, this similarity has no information about the relationships of the three animals. (On reflection, it might be thought of some interest that we feel able to recognize the very character, *number of legs*, in all three animals, suggesting that at least they may all belong to a larger grouping with a similar way of approaching life's vicissitudes: however, this information has not helped us to establish their relationships *to one another.*)

How, though, can we really be sure that the 'vestiges' of the rear legs are not rudiments on their evolutionary way to *becoming* 'full' legs? Just suppose that we have good reason, again on some extrinsic evidence, that the direction of change is, rather, from two legs to four legs. In other words, we reverse the 'polarity' decision we previously made. In these new circumstances, change from state '2' to state '4' can be marked on a set of possible trees like those of Fig. 7.7. The results are shown in Fig. 7.8.

Note that three of the possible patterns dictate two, independent instances of a change from two legs to four legs. Somehow the first pattern, the one in which cow and monkey are shown as more closely related than either is to whale, now seems to be the preferred pattern because it involves only a single change. In other words, a pattern previously rejected now unites cow and monkey by the possession of an all-important innovation. Of course, in this example we may feel

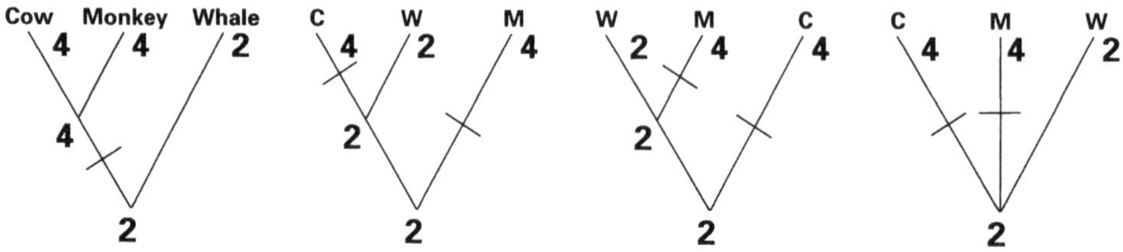

Fig. 7.8.

uneasy about the credibility of our new polarity decision, but the example is chosen as an illustration of how the same similarity between cow and monkey can be interpreted, first to make no supporting statement about their immediate closeness of relationship, and then, with a reversal of the 'polarity' decision, to provide support for just that grouping. Clearly, the ways in which polarities are determined in real analyses are of fundamental importance, and this topic will be mentioned again, below.

It is crucial to the argument to ask why it is that the trees costing two changes in Fig. 7.8 somehow strike us as less satisfactory solutions. Each of these trees implies that the change from state two legs to state four legs has occurred independently on different branches, thus resulting in a resemblance between cow and monkey that is misleading if the pattern of either of those trees were true. Such a resolution of the data appears unreasonable, given that they can be mapped onto the preferred tree without any such complicating interpretations. The preferred tree has a certain economy of representation of the data. Is this not a good thing, considering that the winning tree of Fig. 7.8 might have been constructed as in Fig. 7.9?

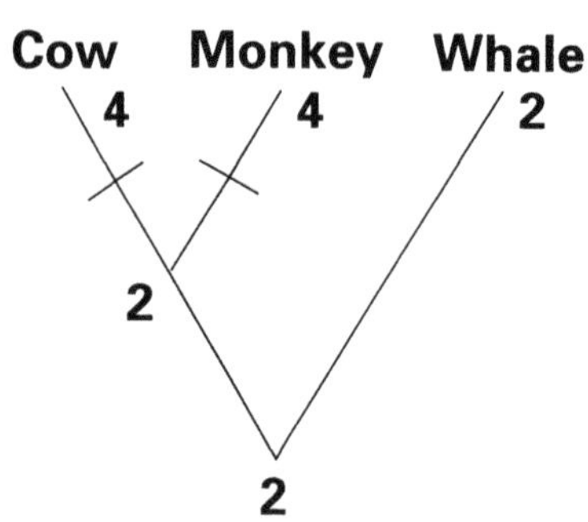

Fig. 7.9.

Here, two changes have been wilfully reconstructed where we know that we could get away with reconstructing only one (as in Fig. 7.8). Yet, what evidence is there that this is not a more accurate depiction of past, unknown reality? On that dangerous line of argument, Fig. 7.10 might also be worthy of serious consideration.

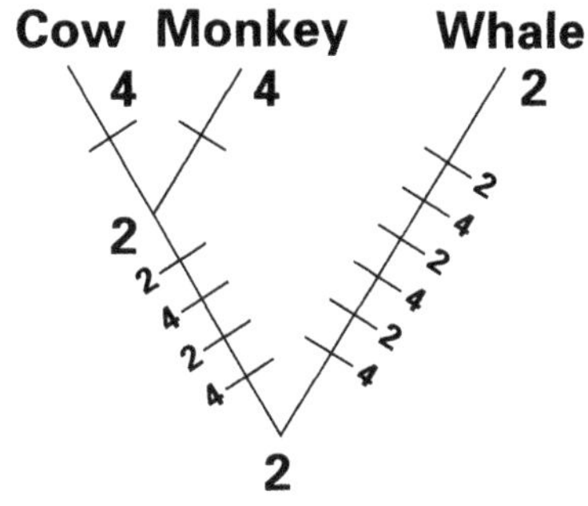

Fig. 7.10.

However, Fig. 7.10 now appears to have entered the realm of the absurd: there is surely no justification for 'imagining' all these events for which there is no evidence from the observed data. (But are not the data, themselves, observed under conditions in which we inevitably have preconceptions about what constitute data at all; and have we not already exercised a degree of imagination in reconstructing even the minimum number of unobserved, and unobservable, events to 'explain' those data?) A feeling for economy of explanation might suggest that, in general, an acceptance of the least number of reconstructed changes necessary to account for the observed data leads to a more satisfying and 'reasonable' solution. Before exploring this issue further, an additional complication deserves consideration. So far, only a single character has been used in the examples. What new difficulties are introduced when several or many characters are involved, as is the case in real analyses of this sort?

Again, consider a character with two states, this time the states will simply, and more neutrally, be represented as 0 and 1 (the character is a 'binary' character). Assume that the polarity of change has been established as $0 \rightarrow 1$, and that three species, A, B and C, have, respectively, states 1, 1 and 0 for this character, character I. By the logic used above, the tree that minimizes the number of changes that it is necessary to invoke is shown in Fig. 7.11.

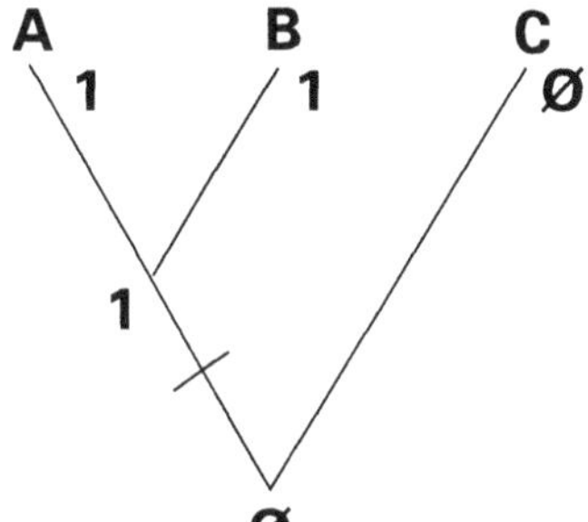

Fig. 7.11.

As in the previous example (of Fig. 7.8) other possible tree patterns require a minimum of two changes.

Now consider a second character, character II, again with states, 0 and 1, and the same polarity of change, 0 → 1. This time, however, the three species, A, B and C, have states 1, 0 and 1, respectively. By the customary logic, the favoured tree for character II is shown in Fig. 7.12. It is immediately obvious that there is a conflict between the characters I and II.

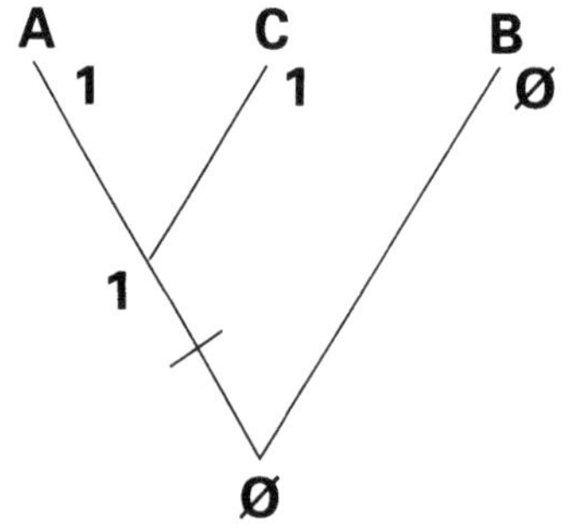

Fig. 7.12.

This conflict cannot be resolved, given the data available, in such a manner that both characters lend support to the same tree pattern. The addition of a third character might result in an agreement with one of the two existing characters, and that agreement might hint at the possibility of a 'majority' decision. On the other hand, of course, it is plausible that the third character might favour a third tree pattern. The favouring of new patterns cannot go on for ever, because there is only a finite number of tree patterns for a given number of species; although, this number can be very large indeed for quite a small number of species. For example, for just ten species there are 34,459,425 possible, rooted patterns to consider, even when only dichotomous branching is allowed. The incredulous, but traditionally inclined, might try drawing them on the back of a (large) envelope.

What does character conflict mean? When encountered, it tells us immediately that not all change obeys the Principle of Divergence (assuming, of course, that no errors have been made in scoring the data and determining polarities). This is important, though perhaps disappointing, information to have. On reflection, however, conflict is not a surprising finding when it occurs, as is almost invariably the case, with real data. Bearing in mind H.C. Watson's criticisms of Darwin's Principle, it might well be expected that different organisms could independently acquire the same state for a particular character, over the course of time and, possibly, under the influence of similar selection. There is, after all, no obvious mechanism to provide an absolute prohibition on such occurrences. Darwin's own speculations on the mechanisms underlying divergence suggested merely that such behaviour might be expected to be the rule, with deviations from it as exceptions. Certainly, the procedure of reconstructing a tree reveals a minimum level of these exceptions, but does that necessarily mean that it is reasonable to assume that this minimum level reflects past reality? A great deal of argument has been provoked by this and related issues. Most argument has been about the interpretation of the concept of 'parsimony', that is, the preference for a tree pattern that involves the smallest number of reconstructed changes. As demonstrated above, parsimony, in this sense, is apparently relevant at two levels that can be distinguished in the examples given: first, in the mapping of changes onto a particular tree pattern; and second, in the choice of a tree pattern, from a range of competing patterns, on the basis of the fewest number of reconstructed changes, overall. (An extensive discussion of some of the issues surrounding parsimony, and a route to further literature, is provided by Sober, 1988: the arguments have a good deal of life in them yet.) It is true that if an underlying model of evolutionary change that is based on the property of divergence is favoured, then use of the sort of methodology described will be justified when all change is, indeed, divergent. Under these conditions, the solution to the tree problem is, as previously demonstrated, fairly easy to find. However, when there is some character conflict (when, in other words, change was, demonstrably, not all divergent) it is more difficult to know how best to proceed. The mere occurrence of character conflict is certainly not enough to justify the reconstruction of events for which we have no need (that is, evidence) on consideration of the data. However, it has been argued that it is sufficient to raise doubts about the wisdom of the tactics based on minimization. In the presence of evidence for at least some non-divergent events, we might question whether or not we are wise, or entitled, to assume that reconstructing the minimum will get us closest to the (unknown) past reality. Granted, one should never give up a beautifully simple model just because the data indicate that a few, minor exceptions must have occurred. One might choose to think of the evolution of characters as a basically straightforward process with error. Quite frequently, however, the level of the exceptions, the error, is rather too high for comfort, and our confidence in the applicability of the model is correspondingly less: often the model appears implausible in these circumstances.

7.3. The disappearance of time?

Throughout the preceding discussion, the assumption has been made that we are indeed attempting to reconstruct the events of the evolutionary past: that we can, as it were, cheat time and inspect history, if not directly, at least at one remove, and so closely that it makes no odds. What, however, if we do not wish to assume, a priori, that evolution (in the sense of descent with modification, no matter what the mechanisms) necessarily occurred? This suggestion might, at first sight, seem an outrageous one: did Charles Darwin live in vain? On more settled reflection, such an attitude might come to seem more appealing. The 'fact' of evolution has often been (and still is) questioned by some. If it were possible to make minimal, or even no, assumptions about evolution and yet consistently find properties, as a result of comparing organisms, that point convincingly to the past operation of particular sorts of evolutionary processes, then such an outcome would simultaneously provide powerful support for the fact of evolution and be evidence for the particular nature of the processes by which it was brought about.

At this point we can profitably return to some of the logic that led Darwin to his Principle of Divergence, with its attendant driving mechanisms involving adaptational innovation via natural selection. The diagram of Henri Milne-Edwards, which worked so powerfully on Darwin, showed the groups of a classification nested inside one another. The Linnaean hierarchy had already expressed the similar concept of nested sets. Such approaches to classification (and there were many other similar ones) seemed to make sense, but why should they appear so satisfactory, so 'natural'. The clue comes as soon as one recognises that nested sets and tree diagrams can be formally interconverted. Consider, for example, the two diagrams in Fig. 7.13. These diagrams are quite clearly related to one another, and can be interpreted as providing similar information about 'relationship'.

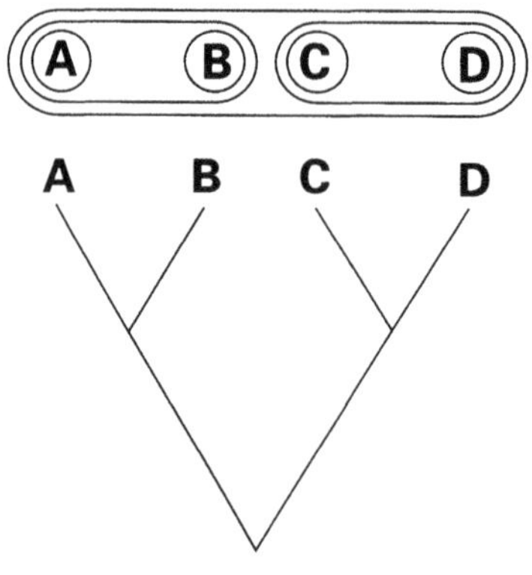

Fig. 7.13.

It is very instructive, here, to consider how one might include extinct organisms in the sort of analysis that has been described. Initially, on Fig. 7.4, as in Darwin's original, sole figure in the *Origin*, lineages were drawn against a time-scale. In introducing Hennig's ideas, there was an element of convenient artifice in the choice of only extant species for the examples of Figs 7.5 to 7.10. If the tree diagrams produced as a result of Hennig's methodology are, effectively, representations of set membership ('relationships') supported by character states ordered by polarity decisions, then extinct taxa would be treated no differently to extant ones. In other words, all species would appear at the tips of branches, at the top of the trees (the top as drawn here). In that case, it appears that the top of the tree cannot be a plane representing the present, in time, and the branch-points are not, therefore, representations of the divergence of lineages, in time, but alternative ways of ensuring that set boundaries are shown. They do not, therefore, represent the flow of genetic information. The conclusion is inescapable: the tree diagram resulting from the application of Hennig's methodology is not an evolutionary tree. This apparently startling, and rather unsettling, result is less so if one realises that this, primary, form of tree diagram, usually termed a 'cladogram', can be converted into an evolutionary tree, but at the cost of extra assumptions. There are, for example, no ancestor-descendant relationships on a cladogram. That is not the same thing as saying that such relationships do not (did not) exist, in reality: it is, however, the case that the concept of an ancestor-descendant relationship is quite unnecessary, indeed undesirable, when the awful purity of the cladogram is acknowledged.

7.4. Pattern and process

The disappearance of time from an analytical methodology pioneered under the banner of *phylogenetic* systematics proved a difficulty for many. Accusations of essentialism were levelled at those who subscribed to the purist formulation, and the whole argument, and many other connected issues, is chronicled and discussed at length by Mark Ridley (1986) and by David Hull, in his historical, sociological and philosophical study of the clash of sectarian ideologies (Hull, 1988). The movement to defend the primacy and sanctity of the cladogram came to be known as 'pattern cladistics' (sometimes, 'transformed cladistics') for obvious reasons: proponents suggested that the exploration of *patterns* of relationship (in the sense of set membership) was logically antecedent to a consideration of what sort of processes might have operated to give rise to those patterns. It was held to be no good to make assumptions about processes of evolutionary change, use those assumptions to reconstruct the pattern of relationship, and then proudly step back and announce that the patterns discovered provided support for our beliefs about the nature of the processes. Put like this, the case for pattern cladistics looks compelling. It is especially amusing to recall that Darwin's own logic might appear at one point to have proceeded from the pattern (Milne-Edwards' diagram of relationships) to the process (his own Principle of Divergence, and, eventually, his sole figure in the *Origin*). Such a view of Darwin's train of thought seems, however, naive: Darwin's ideas about process seem to have formed early in his scientific life and, doubtless, pattern and process were just as inextricably intertwined in his thinking as they still are in ours. Viewed from such a perspective, the attempts by pattern cladists to purge the exploration of pattern of relationship of all taint of mechanism appear new and heroic. Their point of view appears either essential to the rebuilding of a true phylogenetic systematics on the firmest of foundations, or, to many critics, methodologically vacuous and an unnecessary (and even dangerous) denial of all, or at least much, that we think we already know about evolution. Indeed, one of the battle-cries of pattern cladistics is "What do we know about evolution?" (that is, *really* know beyond reasonable dispute). This question raises the largest of issues: the phrase 'beyond reasonable dispute' begs questions about the nature of scientific method, and could not, therefore, be more fundamental, but it does not appear to be a question that is peculiar to evolutionary biology.

One important area does need some clarification as far as the practice of pattern cladistics is concerned. Although most of the operations described in the methodology for the construction of a most parsimonious cladogram can be re-framed, or, at least, re-named, in such a way that any reference, explicit or implicit, to evolution is avoided, the determination of polarity of state change might appear to be more problematic. Surely, the determination of polarity implies change over evolutionary time? This is not necessarily the case, however, if it is accepted that polarity can find expression in a static directionality. This might, for example, be in the form of more general to less general states of a character, or a state possessed by a group believed to be outside the study group giving way to another state within the study group. Physical scientists and mathematicians are happy to work with such time-free directionalities, why not biologists?

If we can acknowledge that the construction of a cladogram can be carried out without recourse to any of the baggage of evolutionary process, and even that this might be a desirable way to proceed, there are still two main objections to the approach of pattern cladistics. The first is that no comparative analysis of character state data can logically be carried out in the total absence of some sort of framework (model). The search for pattern, that is, for nested sets, is, after all, a search for a very particular sort of structure; and is not the particular sort of structure that we expect generated by our beliefs about the behaviour of character states over evolutionary time (see, for example, Roth, p. 304 in Hall, 1994)? A possible rejoinder, on behalf of pattern cladistics, to this criticism is that, in principle, all sorts of other structures could be tried to see whether or not comparative data fit consistently more comfortably with them rather than with nested sets. This suggestion might seem otiose and, in any case, alternatives to nested sets have already been tried, in immediately pre-Darwinian years for example, and they have been found wanting (by most people's criteria). The answer to the criticism is, therefore, that such a pattern does seem to emerge quite consistently as the most appropriate (in the sense that it has the best 'fit', with the maximum number of characters exhibiting congruence with one another). Pattern cladists might use this finding to claim that classification by nested sets is empirically justified, at least so far and for the particular groups of organisms studied, or that it is likely to be so justified at some point in the future. The second main objection to the ideas of pattern cladistics follows directly upon this possibility.

If it is the case that classification by nested sets be considered to have been empirically justified, either already or at some time in the future, it would seem legitimate to use the finding and to ask the question, How many potential candidates for *processes* of evolutionary change are thereby ruled out and how many remain plausible? Of course, it would be prudent, if we are in search of generalities about processes, to examine many such cladograms for different groups of species. The level of generality of the conclusion would then be related to the range of organisms for which we had cladograms available. Because cladograms represent nested sets, we would surely learn that a tree model is a promising component of any sensible model of evolutionary process (at least as far as eukaryotic organisms are concerned). It would be comforting to have such a vindication, but this could not, perhaps, be held to be an altogether surprising result. We might also expect to learn that, if convincing most parsimonious cladograms are, on the whole, possible to obtain from many or most sets of data, character conflict can be reduced to relatively low levels and hence there is no need to suppose, when considering a possible process of character state change, that this has been other than mostly divergent. But surely, because of the way a cladogram is constructed, involving as it does the minimization of character conflict, such a conclusion would hardly be a matter for great surprise.

It has been argued that the invocation of parsimony, in the sense of preference for most parsimonious cladograms, is uncontaminated by evolutionary theory, because it merely reflects a commitment to the explanation of data by the simplest set of hypotheses. On this argument, one might count the congruence of most characters as one hypothesis, and then each instance of character conflict would need a separate hypothesis (often rather perjoratively described as 'ad hoc') to explain it. It is difficult to see how elements of process can be completely avoided here, because the sorts of explanations that might be called upon are likely to make reference to the idea of adaptation. Admittedly, again, adaptation (the fit of an organism to its circumstances) can be viewed as a static phenomenon, something to be observed without reference to the evolutionary processes that might have brought it about, but this defence is verging on the contrived. At the very least, the methodology of cladistics, pattern or otherwise, appears biased in favour of particular models of process, when the moment comes to make the conceptual connection.

Perhaps there is only one way to maintain the purity of the pattern, and that is to refuse to consider that the methodology is anything more than a form of classification, an optimal storage system for information but with nothing to say about process whatsoever. It is doubtful that such a resolution would appeal to more than a small minority of systematic biologists. Indeed, throughout the period when time was threatening to disappear from the study of systematics, people were still studying the fossil record, still searching for ancestors, still hazarding reconstructions of the evolutionary past. It might have seemed to those outside evolutionary biology, however, that some of its practitioners had suffered a severe loss of confidence during this period. In reality this was not the case. Those who defended the high ground of pattern cladistics presented a challenge, as Hennig had originally done, to an existing woolliness of thinking about just what it was possible to learn with confidence about evolutionary history. Argument continues, but it is certainly the case that controversies debated inside evolutionary biology are not the same ones that evolutionary biologists find themselves periodically embroiled in with those outside the discipline. Those outside who challenged the very idea of evolution appear to have watched almost with disbelief as evolutionary biologists flew at one another's throats apparently with rather more energy than they expended on countering the arguments of their external critics. But that, surely, is the sign of a discipline in the very rudest of health.

The majority of practitioners do appear still to see cladistic methodology as a means to an end; that is, a means to statements about recency of common ancestry and, ultimately, about ancestor-descendant relationships: in other words, to evolutionary trees. The most important question now again becomes: does the methodology deliver defensible evolutionary trees? Does it do *this* job well? We are back to considering how we would know, given that the events in question are in the past. Under renewed scrutiny, this question begins to reveal itself as a question of a sort again certainly not restricted to evolutionary biology or the historical sciences: rather, it emerges as a question about how good a model might be in the light of some data to which it is held to be relevant. The ability to reject a model and, in principle, to replace it with a better one is, of course, an important component of scientific methodology in general. This is an area of current interest and development as far as evolutionary trees are concerned, and some encouraging progress is now being made. Essentially, one makes an intelligent guess at the sorts of behaviour of characters that might have taken place over evolutionary time, and having framed these in terms of models of process, one then asks whether or not there are properties of the data that force the rejection of these models. One of the healthiest features of this type of approach is that it forces the explicit revealing of the assumptions that are being made. Ideas can be traded between what is (tentatively) regarded as known and what is to be estimated, and convictions about the degree of realism inherent in the various models must be only lightly held. Naturally, there is also the customary trade-off between increasing realism of a model and its simplicity. At some point, a model may become so complex that it ceases to have any power as a simplifying, explanatory and predictive agent, and becomes merely a re-description of the data.

Once again, these issues are familiar scientific territory and by no means unique to the reconstruction of the evolutionary past.

7.5. Coda

At about the same time that Hennig's ideas were beginning to become generally known to systematists, a parallel methodology for reconstructing the past was being introduced. This approach had its basis in the statistical techniques of population genetics and made use of models of process of a probabilistic nature, in which a dependence of change on elapsed time was explicit. In this statistical methodology, it was natural to speak of *estimation* of trees and, indeed, the favoured approach to estimation made use of the Method of Maximum Likelihood introduced by the geneticist, R.A. Fisher. The mathematics behind the probabilistic models is not trivial and, after pioneering work by A.W.F. Edwards and L.L. Cavalli-Sforza, the procedures (in their present, widely-available form) owe their existence to progress made by E.A. Thompson and, especially, J. Felsenstein. In a series of original and illuminating papers, Felsenstein introduced techniques for the phylogenetic analysis of different sorts of comparative data: for morphological characters; for continuous measurement characters, including gene frequencies; and for molecular sequences. Simultaneously, he pursued a running battle, always enlightening and entertaining and occasionally abrasive, with those advocating cladistic approaches of various sorts. Felsenstein's critique of cladistic methodology was dominated by the attempt to formulate these approaches within a framework of statistical estimation. This meant that elucidation of the model or models underlying cladistic analysis was a necessary stage in the discussion.

In pursuance of these aims, Felsenstein demonstrated that, under certain conditions, the tree of highest likelihood, for some data, had a branching pattern identical to that of the most parsimonious tree, for the same data. This could be achieved when likelihood analysis was conducted using a simple probabilistic model of time-dependent change, with the limiting condition of very low rates of change. As those rates of change come close to zero, time effectively drops out, which should not be surprising given the comments made above about the absence of time-dependence in cladistic methodology. If such a model were to be used as a basis for the generation of simulated change on an imaginary tree, it would have the property of enforcing near totally divergent behaviour: if any change at all, on a given segment of the tree, is of very low probability, then the probability that the same change will occur independently on another segment of the tree will be vanishingly small. (The fact remains, however, that, as noted previously, real data are not found to behave in this way.) J.S. Farris, who is one of the most articulate and cogent advocates of cladistic methodology and who has taken issue with a number of Felsenstein's arguments, has, indeed, himself used a similar limiting approach under a probabilistic model of state change in order to provide a justification for the use of parsimony.

Perhaps the road forwards leads inevitably to an engagement with models of evolutionary processes: their generation, testing, rejection, replacement and refinement. Although both morphological and molecular data are likely to figure in these analyses, the latter will probably continue to provide the main impetus. This is partly because of the sheer quantity of molecular data: a statistical treatment seems encouraging in these circumstances. It may be that quite simple, descriptive models of a probabilistic kind will prove to be acceptable when evaluated with real molecular data. In other words, despite the probably very complex nature of the processes of molecular evolution, some simple, law-like, large-scale properties may make molecular data, at least, amenable to historical analysis. These avenues are now being very actively pursued, and time will tell. Of course, at present molecular sequences can be determined for extant organisms but, with difficulty, from only very few fossil remains. In the case of comparative morphological data, if it should transpire that here no strong generalities usable in the reconstruction of evolutionary history emerge, despite intensive examination, then it may be that, for these data, it is correct to *aim* at precise resolution (there is no justification for sloppy thinking or methodology) but necessary, with regret, to accept a fuzzy outcome, much as at present. Morphological comparisons can, however, be made across the whole span of the fossil record, and it is unrealistic to imagine that people will stop hazarding reconstructions of the course of evolutionary history using such data: the interest is just too great. For extant groups, trees based on both molecular and morphological data can be compared. Such comparisons are becoming frequent (Patterson, 1987) but it is too soon to draw conclusions. Nevertheless, in many cases, agreement is encouraging. Of course, in these cases, *both* could be wrong, but surely life could not be so cruel that we are prevented from cheating time in this way altogether?

References

Appleby, J., Hunt, L. and Jacob, M., 1994. Telling the Truth about History (Norton and Co, New York).

Darwin, C., 1859. On the Origin of Species by Means of Natural Selection, or the Preservation of favoured Races in the Struggle for Life (Murray, London).

Desmond, A., 1982. Archetypes and Ancestors (University of Chicago Press, Chicago).

Eggleton, P. and Vane-Wright, R. (eds) 1994. Phylogenetics and Ecology (Academic Press, London).

Forey, P.L., Humphries, C.J., Kitching, I.J., Scotland, R.W., Siebert, D.J. and Williams, D.M., 1992. Cladistics. A Practical Course in Systematics (Clarendon Press, Oxford).

Gould, S.J., 1989. Wonderful Life (Hutchinson Radius, London).

Gould, S.J. and Eldredge, N., 1993. Punctuated Equilibrium comes of Age. Nature **366**, 223–227.

Hall, B.K. (ed.) 1994. Homology. The Hierarchical Basis of Comparative Biology (Academic Press, London).

Harvey, P.H. and Pagel, M.D., 1991. The Comparative Method in Evolutionary Biology (Oxford University Press, Oxford).
Hennig, W., 1966. Phylogenetic Systematics (Illinois University Press, Urbana).
Hull, D.L., 1988. Science as a Process (University of Chicago Press, Chicago).
Minellli, A., 1998, Molecules, Developmental Modules, and Phenotypes: A Combinatorial Approach to Homology. Molecular Phylogenetics and Evolution **9**, 340–347.
Panchen, A.L., 1992. Classification, Evolution,and the Nature of Biology (Cambridge University Press, Cambridge).
Patterson, C. (ed) 1987. Molecules and Morphology in Evolution: Conflict or Compromise? (Cambridge University Press, Cambridge).
Porter, L. (ed.) 1991. A Lester Young Reader (Smithsonian Institution Press, Washington).
Ridley, M., 1986. Evolution and Classification: The Reformation of Cladism (Longman, London).
Sober, E., 1988. Reconstructing the Past. Parsimony, Evolution and Inference (MIT Press, Cambridge, Massachusetts).

chapter 8

TIME AND EVOLUTION

IAN TATTERSALL
American Museum of Natural History
New York

Time in Contemporary Intellectual Thought
Patrick Baert (Editor)

CONTENTS

8.1. Introduction

The arena of organic evolution is time; of that there can be no argument. No time, no evolution. But is it permissible for us to conclude from this that time by itself virtually implies evolution in the natural world – as our received notions of the evolutionary process suggest we might? Here we must surely answer in the negative. Our emerging understanding of the mechanisms of evolution points rather towards the conclusion that time and change are synonymous only in a limited way; that evolution is not a process that sweeps smoothly and steadily through time. Evolution is instead a multi-layered affair in which time plays an obvious role but of which discontinuous events, and periods in each species' history that approximate non-change, are also essential components. In this review I shall examine how our notions of evolution have themselves evolved, and present my current viewpoint on the matter.

Charles Darwin's notion that life has an evolutionary history was born in great part of his realization of the immensity of geological time. His early interests were as much geological as biological; and he was particularly influenced by the work of Charles Lyell, who did as much as anyone to mold the study of geology into a recognizably modern form. Lyell was a 'uniformitarian', believing that the Earth's surface had always been shaped by the very same agencies that we can see at work today. For him, the modern landscape was recognizably the result of such processes as the sculpting by erosion of uplifted plateaux into dissected mountain ranges, and the building-up by deposition of great river deltas. The uniformitarian view still provides the underpinning of modern geological studies – although we now know much more, of course, about the vast range of different agencies that are actually involved – and it implies an immenseness of time that was totally foreign to the world view of the middle nineteenth century in which Lyell and Darwin lived and worked. Mountain ranges are levelled more or less grain by grain, just as deltas are built up; and geologists, contemplating for example the huge thicknesses of sedimentary rocks exposed at the Earth's surface, could not for long avoid the conclusion that vast spans of time had been involved in their deposition.

Of course, a true appreciation of the Earth's astonishing age – some 4.5 billion years – had to await the introduction in our own times of chronometric dating techniques; but it was already apparent to Darwin and to his like-minded colleagues that Earth history was sufficiently long to make abundant time available for the diversification of the organic world by natural means. It may also be that it was the very gradualness of sedimentary processes that led ultimately to Darwin's concentrated focus on the slow steadiness of the mechanisms governing the evolution of life (although his early prospections in South America had also led him to appreciate the importance of short-term events in the formation of the recent landscape).

8.2. Precursors of Darwinian evolution

Evolutionary notions go back, of course, to a time well before Darwin's 'joint paper' with his remarkable colleague Alfred Russel Wallace was read to the Linnaean Society of London in 1858. The great Swedish naturalist Linnaeus had already 'systematised' the diversity of life a hundred years before, expressly recognizing that the living world consists of nested sets of organisms with increasingly detailed patterns of similarity among them. Once the overall motif of natural diversity had been made explicit in this manner, the way was open for others to explore how this pattern might have been achieved other than by the Will of a divine Creator – which Linnaeus himself had been content to accept (though hardly as an unequal partner in the whole process; he is famous for his statement that 'God created; but Linnaeus classified.') Early attempts to factor time into the interpretation of the newly discovered fossil record were, understandably enough, made within a framework established by the ancient Biblical chronology (notoriously refined by Archbishop Ussher in 1650 with the calculation that Earth and humankind had been created in 4004 B.C.); and even an agnostic such as the Frenchman Georges Cuvier was politic enough to suggest that the last of the 'catastrophes' (wholesale replacement of faunas) that he saw in the fossil record equated with the Biblical flood. But even Cuvier, while abundantly aware that the fossil record showed numerous discontinuities, was ready to assume that, individually, species were fixed. They might be replaced, or simply go extinct; but one did not change into another. They stayed as they had been created.

Among established scientists it was thus left to Cuvier's compatriot and older contemporary, Jean-Baptiste de Lamarck, (1809) to abandon Biblical considerations altogether. Lamarck had discovered that some of the fossil molluscs he studied could be arranged into gradually modifying series over geological time. He identified such series as ancestor-descendant lineages, in which older species gave rise to younger ones. Time and change were thus interrelated at the level of species as well as that of faunas. Lamarck's fatal mistake was to champion a mechanism for this interrelationship that with hindsight we can clearly see is wrong. This is that organs develop or atrophy through use or disuse during the lifetime of an individual, and that such modifications are passed along to each individual's offspring. History has cruelly pilloried Lamarck for this error (which did not actually originate with him; it was, in fact, widely assumed), and for the resulting visions such as that of giraffes straining over generations to stretch their necks. Yet he had perceived three elements that are crucial to our modern understanding of evolution: that the Earth is in a constant state of change; that species respond to such change through adaptation; and that new species can arise from old ones. These three factors sum out to the process that has ultimately resulted in the entire spectrum of ancient and modern species diversity. All three elements involve the passage of time in large quantities; and neither the public nor science was ready for the conclusion that flowed from them.

8.3. Darwin and Wallace

Incipiently evolutionary notions were not unknown in England during the first half of the nineteenth century, but undoubtedly the greatest achievement of this period in preparing the way for the Darwinian viewpoint was the expunging of Biblical chronology from scientific discussion (although public opinion followed more gradually). What science and the public were less ready for, when Darwin was galvanized in 1858 by the reception of a manuscript from Wallace that mirrored much of his own thinking, was the idea that species were not fixed. And the need to counteract the theologically-based notion of the fixity of species dominated Darwin's approach to presenting his central perception: that the diversification of life has a history that depends on change in species.

It is presumably for this reason, as well as for the one cited earlier, that Darwin largely avoided the issue highlighted by the title of his great book, published in 1859, *On the Origin of Species by Means of Natural Selection*. For in it, Darwin was not principally concerned with species origin as such, but with the transformation of species over time. Darwin had somehow to destroy the idea of the fixity of species; and while he could not reasonably deny species reality in space (although he certainly had his doubts in the matter of their practical recognition), with his appreciation of the immensity of the geological ages he saw that he could do so in the dimension of time. This attack on species fixity was made possible by his notion of natural selection, which can be taken to imply slow, gradual, generation-by-generation change over the eons.

In 1798 Thomas Malthus had published his famous *Essay on the Principle of Population*, in which he described how the potential growth of populations far exceeds the potential growth of resources available to support them: organisms can reproduce in geometric ratio, while food resources tend to remain constant. In the case of humans, Malthus argued, even the best technology could only assure an arithmetic increase in food supplies, while, if unchecked, population numbers would continue to boom. Malthus had proffered this idea in the context of a plea for population control in the face of increasing misery among England's lower classes; but both Darwin and Wallace seized on it as the key to understanding a basic sorting process in nature.

Malthus had noted that human populations did not in practice double every 25 years, as his calculations suggested they should; they were, instead, restrained by starvation, disease, wars and a host of the other social ills against which Malthus was protesting. It was Wallace's and Darwin's genius to realize that this observation applies with equal force to all organisms; for in every natural population many more individuals are born than ever survive to reproduce. Wallace reckoned that, other things being equal, a single pair of birds could potentially give rise to as many as ten million descendants in a mere fifteen years. The more conservative Darwin used the example of slow-reproducing elephants, concluding that in 750 years a pair of elephants could have 19 million descendants. Yet the world is crammed neither with birds nor with elephants; something must be controlling their numbers. And it was this something to which Darwin gave the snappy name of 'natural selection'.

Natural selection works like this: all individuals in any population differ from each other, if only slightly, in heritable characteristics. Those individuals who survive to reproduce most successfully will generally be those best adapted to their environments; those less well adapted will be pruned from the population by natural forces, and will thus fail to pass on their unfavourable characteristics to the next generation, or will do so in lesser numbers. As the generations roll by, favorable adaptations will in this way become commoner in the population, while their less advantageous counterparts will tend to fade out. Over long stretches of time this inevitable natural process of winnowing will transform the population, until eventually a new species emerges. Variation amongst individuals will always be present, however, so that should conditions alter, selection can turn around and drive population change in a new direction. Natural selection is thus a blind, mindless force that lacks any intrinsic direction; but it is nonetheless deterministic in the limited sense that it lies at the heart of adaptation, the fitting of organisms to their environment. And, along with heritable variation in populations, the key to the working of natural selection is time. The natural world is unimaginably diverse, with millions of species (nobody knows even approximately how many) representing variations on a host of adaptive and architectural themes. If natural selection depends on the slow accretion of minor changes, then vast amounts of time have been necessary to build up the variety familiar today. And by Darwin's and Wallace's day, the geologists had made that time available.

The central tenet of evolution is that all living organisms share a single common ancestry, and that the pattern of similarities among organisms reflects a pattern of descent. This obviously holds implications for humanity's place in nature, and it was for those implications (to which he had taken pains to avoid drawing attention) that Darwin found himself most widely and roundly denounced immediately following publication of the *Origin*. Nonetheless, even though it has often been claimed that it was his proposal of a mechanism by which evolution might take place that allowed Darwin to implant his radical ideas with ultimate success, in the longer run it was natural selection that had a rougher ride than his notion of common descent. Once the initial shock had worn off, most scientists – and even most members of the public – turned out to be quite willing to accept that life had evolved. The blind force of natural selection was, however, a lot for a devoutly Christian society to swallow. The vast vistas of time implied by natural selection might have to be accepted in view of the geologists' findings; but might not the evolutionary process be interpreted equally well as the simple unfolding of a Divine plan? Opposition to natural selection also extended well beyond those who still wished to incorporate theological considerations into the interpretation of the natural world. Many returned to various forms of Lamarckism as an alternative to natural selection; and confusion still reigned when the science of genetics got underway at the turn of the twentieth century.

8.4. The rise of genetics

All of the early argument over the mechanisms of evolution was caried out in the absence of an accurate notion of heredity. It was enough for Darwin and Wallace to know that physical features of organisms are heritable; it was not essential to their argument to specify exactly how genetic transmission occurs. Indeed, Darwin's own speculations on the nature of heredity were grievously in error. It is hardly surprising, then, that the rediscovery in 1900 of simple genetics (the principles of which had actually been ascertained, but not accessibly published, over thirty years earlier) did not lead to immediate clarification of the evolutionary process. Indeed, for a time things became more confused than ever.

Unravelling the basic principles of inheritance involved the realization that physical traits are determined by the discrete hereditary units that we now know as 'genes'. Genes have two basic properties. First, each gene is passed on intact from one generation to the next whether or not it is expressed in the structure of the individual possessing it. Genes are paired (one 'allele' of each pair being inherited from each parent). 'Recessive' alleles are not expressed in the presence of 'dominant' ones, although they have equal chances of being transmitted to the next generation. Second, genes are independently shuffled around in every generation (though those lying on the same chromosome will normally be inherited together). What's more, it became evident that genes could change, via a process of spontaneous 'mutation'. In elucidating all this, the geneticists claimed the key to understanding the mechanisms of the evolutionary process. For if evolutionary change boiled down to the sum total of gene changes in lineages over time, its explanation was to be sought in the ways in which those changes accumulated. So far, so good. But beyond this point, ideas diverged dramatically. Some, for example, sought the origin of new species in spontaneous 'leaps' achieved by sudden major changes in the genetic material: a process known as 'saltation'. Others, realizing that most mutations are in fact rather small-scale in their effects, looked upon 'mutation pressure' (the rate at which mutations occur) as the driving force of evolution. Alternative notions also abounded. By the time the 1920s arrived, the wealth of evolutionary processes on offer had little more in common than that few of them found much room for natural selection as the prime mover of evolution; and if most evolutionary phenomena could be traced back to spontaneous changes in genes, the role of time in evolution remained equivocal.

In the following years, however, the burgeoning field of 'population genetics' began to clarify all this. Geneticists of all stripes came to understand that most mutations did not have large effects, and to see that minor mutations of the common kind are the wellsprings of the variation on which natural selection acts. Evolution thus became viewed as a matter of generation-to-generation modifications in the 'gene pool' (as the sum of all alleles in each population became known), under the guiding hand of natural selection. Once again, time became a specifiable factor in the evolutionary process. Mathematically-inclined population geneticists such as Sir Ronald Fisher (see Fisher, 1930) created the basis for this understanding, which was later refined by others whose concern was not with genes as such, but rather with organisms and larger-scale phenomena in nature. Ultimately, over the second half of the 1930s and into the 1940s, geneticists, naturalists, paleontologists and developmental biologists pooled their expertises to arrive at what became known as the 'Evolutionary Synthesis'.

8.5. The evolutionary synthesis

The emerging tone of the Synthesis was set in 1937 by the naturalist-turned geneticist Theodosius Dobzhansky, in a book called *Genetics and the Origin of Species*. Ultimately, Dobzhansky followed Darwin in viewing all evolutionary phenomena as resulting from gene frequency shifts in lineages governed by natural selection. He did so, however, with some reluctance. For while as an experimental geneticist he was able only to produce incremental genetic changes in his laboratory, he was also uncomfortably aware that major discontinuities indeed do exist in nature. Dobzhansky identified three levels at which the evolutionary process acts. First is the origin of genetic novelties. Second comes the ordering of those novelties in "molding the genetic structures of populations into new shapes." And third, there is "the fixation of the diversity already attained on the preceding two levels" (Dobzhansky, 1937, p. 13). The first level of action is accounted for by mutation and by the recombination of genes that occurs in each generation (for most characteristics are in fact determined by several genes). The second reflects the action of natural selection. The problem was that the third is not accounted for by either, and Dobzhansky himself pointed out that "the origin and functioning of the isolating mechanisms [that keep species apart] constitute one of the most important problems of the genetics of populations." (Dobzhansky, 1937, p. 14). This problem remains debated to this day.

Dobzhansky's pioneering work was soon complemented by the ornithologist Ernst Mayr (1942), who presented a systematist's point of view. Mayr pointed out that Darwin had not really addressed the problem (the origin of one species from another) that was implicit in the title of his 1859 book, and he was particularly concerned to emphasize the discreteness of species in nature. Nonetheless, he, too, ultimately conceded that natural selection and shifting gene frequencies lay behind the entire process of evolutionary change. Finally, the paleontologist George Gaylord Simpson brought his discipline into line in 1944. He was broadly content to follow the Darwinian view of slow, steady change in the evolutionary histories of the fossil mammals he studied, but drew attention to the fact that the fossil record did not in fact show the expected array of gradual transitions between related forms. Yet, as Darwin had done almost a century before, Simpson in the end accepted that the failure to find the anticipated intermediates was due to the inherent incompleteness of the fossil record; very few individuals are ever preserved as fossils, and fewer yet are ever found by scientists. Of course, Simpson was also acutely conscious that missing fossil intermediates in the case of major adaptive varieties of mammals could not be entirely explained away by the gappiness of the record. He compensated for this by the observation that evolutionary rates need not be constant: the appearance of the major mammal groups must have been related to episodes of accelerated change. Under this view time was related to evolutionary change, but not in a consistent way.

The founding documents of the Synthesis thus reflected an awareness on the part of their authors that gradual generation-by-generation change under natural selection could not encompass the whole evolutionary story; but in the decade following the end of the Second World War attitudes gradually hardened across the board. Evolution was a matter of gradual changes in lineages resulting from the adaptive pressure of natural selection, and that was that. Even speciation – the origin of a new species from an old one – became simply a special case of this same process. Simpson found a particularly ingenious way of equating the twin processes of gradual change and speciation. Borrowing the population geneticist Sewall Wright's metaphor of the 'adaptive landscape', whereby 'fitter' individuals cluster on peaks while less fit individuals fall fatally into the valleys below, Simpson suggested that, viewed in the dimension of time, this static landscape became more like a 'choppy sea'. Natural selection was constantly at work to keep each population atop the peak shifting beneath it; and from time to time a peak would divide, carrying the two resulting populations away from each other, on distinct adaptive paths. Ultimately, the degree of divergence achieved would be sufficient to make them reproductively incompatible with each other: they would have become different species. Numerous iterations of this process would produce new genera, new families, new orders, and so on. Once more, time and evolutionary change had become at least approximately synonymous.

During the 1950s the Synthesis, in its simpler, more confident form, became received wisdom throughout evolutionary biology. One reason for its success is that the gradualist metaphor for speciation is beguiling in its elegant simplicity. It was, of course, particularly seductive to the geneticists, whom it confirmed as the guardians of evolution's essential mysteries. Gradual change in gene frequencies was all; and genes were the business of geneticists. Systematists (students of the world's current biodiversity) also found this view congenial; for although they had to concede that species had no identifiable boundaries in time, they could still regard them as discrete entities in space – which was all that mattered to them in the conduct of their daily business. What is remarkable, though, is the readiness of paleontologists to go along with all of this. For, under the Synthesis, time, the dimension that was uniquely theirs, robbed paleontologists of their basic unit of study: the species. From this viewpoint species are no more than arbitrarily-defined segments of gradually-transforming lineages; and this formulation led to the incredible spectacle of paleontologists congratulating themselves on the very deficiency of the data available to them. Suddenly, the gaps in the fossil record had a useful purpose: they provided points at which species boundaries could be recognized, relieving paleontologists of the onus of deciding – as would be theoretically necessary with a complete record – between which parent and offspring to define each species boundary.

In hindsight, it is possible to see that the account of the evolutionary process provided by the Synthesis is at the very least incomplete. But at the same time it is easy to forget what a salutory development it was, clearing out of evolutionary biology a huge accumulated body of myth and false belief; and in doing this it cleared the way for an appreciation of the multilayered nature of the mechanisms of evolution.

8.6. Punctuated equilibria

Given that the Synthesis had relegated paleontology to the essentially clerical function of clearing up the details of evolutionary history while others concerned themselves with the processes underlying it, it is hardly surprising that the first intimations that the Emperor of the Synthesis was at best scantily clad, came from paleontologists. For many years, paleontologists had been dimly aware that fossil species often appeared to be 'real' in a manner that the Synthesis did not predict. Rather than gradually transform themselves over the ages, new species tend to show up quite suddenly in the fossil record, and to linger, recognizably themselves, for often quite extended periods of time. At last they generally just vanish, to be replaced by other species which might be closely related to them but equally might not be. This is a signal in the fossil record that cannot in the least be ascribed to that record's incompleteness.

This pattern was first specifically pointed out by the paleontologist Niles Eldredge (1971), who found that the North American trilobites – ancient sea-bottom-dwelling invertebrates – he was studying did not at all conform to the expectations of the Synthesis. Indeed, he was strongly impressed by the lack of change that these fossil creatures showed. In the American Midwest, for example, only a single detectable change occurred in his trilobites over a period of eight (now six) million years. A similar pattern obtained in New York, except that the change appeared earlier – and that at one site, representing an instant in time, he seemed to have caught the change in progress. To Eldredge, the sequence of events was clear. In New York, after millions of years of stability, an anatomical innovation had occurred – and a new species had emerged from the old. Millions of years later yet, an environmental change had allowed the new species to invade the Midwest, where in a single event it swept away the old species, which had been its parent. Eldredge was reluctant to squeeze this historical drama into the theoretical fabric of the Synthesis – for example, by invoking extreme fluctuations in evolutionary rates – and said so. The following year, with his colleague Stephen Jay Gould, Eldredge expanded this observation into a reappraisal of evolutionary concepts in general (Eldredge and Gould, 1972). The resulting theory of 'Punctuated Equilibria' stated that evolutionary histories are generally not ones of gradual within-lineage change. Rather, the history of species tends to consist of varying periods – sometimes extremely long – of anatomical stability, interspersed with brief episodes of evolutionary innovation.

Eldredge and Gould associated these periods of change with speciation; and ironically, in doing so they invoked the 'allopatric' concept of speciation that had been most fully developed by the Synthecist Ernst Mayr. Mayr it was who had explored in most detail Dobzhansky's concern about how the mechanisms that isolate species as reproductive entities come about. At least as far as the vertebrates he studied were concerned, Mayr concluded that speciation – the division of one freely (or at least potentially) interbreeding population into two or more which are not fully interfertile – requires physical isolation. If a small isolate is separated by some physical barrier from the larger 'parent' population, the way is potentially open for genetic changes to occur in the 'daughter' population leading to disruption of interfertility. Mayr favored the smaller daughter population as the target of such genetic changes because small gene pools are inherently less stable than larger ones. Indeed, the sheer genetic inertia of large interbreeding populations argues strongly against the fixation of genetic innovations within them.

As importantly, in geological terms speciation of this kind is a short-term event. Exactly how long it may take is unknown, as is the full spectrum of genetic mechanisms that might be involved; but Eldredge and Gould estimated between about 5,000 and 50,000 years – a mere eyeblink in the history of most species. This is probably not an unreasonable approximation. Thus the role of time in the evolutionary process changed again. Instead of being something that takes place inexorably over time – if at fluctuating rates – evolutionary change suddenly became something that happens in rapid spurts, while the vast majority of each species' history is characterized by non-change. 'Evolution by jerks', some unkind critic termed this process. Under punctuated equilibria, time is still the measure of evolutionary change in the aggregate – when comparing whole faunas over long periods, for instance; but when individual species or species lineages are considered, an indeterminate amount of time can be expected to lapse between episodes of significant evolutionary innovation.

It should also be abundantly evident at this point that the notions of punctuated equilibria and phyletic gradualism (the doctrine of the Synthesis) *predict* that we should find entirely different patterns of change in the fossil record. Despite its origins in the contemplation of one single lineage of long-extinct invertebrate, punctuated equilibria is not simply an ad hoc explanation of one example in the fossil record, or even of many; it embodies an approach to making sense of the entire history of life. As a general formulation, phyletic gradualism predicts that change in evolution should be a smoothly flowing continuum. Rates of evolution may alter, but the process itself is continuous; even speciation is a gradual process of divergence of populations along different pathways, and species themselves are merely notional links in a long chain. Such chains may occasionally terminate in sudden extinction, which in geological terms is an inescapably abrupt event; but in most cases extinction will be simply notional, as a parent species slowly and inexorably evolves itself into another (or others). The evolutionary history of any group will thus approximate to a tree, with a stout trunk and relatively few main branches which continually elongate. In contrast the theory of punctuated equilibria, which emphasizes the historic individuality of species, predicts that the fossil record will reveal numerous sporadic speciations, each potentially followed by competition between daughter and parent that may lead to the extinction of one. The pattern, therefore, will approximate more closely to a densely branching bush, with many branches and numerous extinctions which equate with its terminal twigs.

This is more than an academic consideration, because the two models dictate entirely different approaches to the study of the fossil record. If we work under the gradualist model, our main problem becomes one of discovery. If we crawl over enough outcrops and collect enough fossils, eventually the outline of the evolution of any particular group will reveal itself to us, as missing intermediates in the chain come to light. Here, calibration in time becomes essential; and indeed, many fossil phylogenies generated under this model consist of simply joining up fossil 'species' on a stratigraphic chart that shows the sequence of the rocks in which the fossils are found. Morphology, the true repository of information on relationships, becomes almost a secondary consideration. In dramatic contrast, punctuated equilibria proposes that what we see in the fossil record is a mass of species, of varying longevities, the relationships among which must be analyzed through the careful scrutiny of morphology. Here time takes a back seat to morphology; for if anatomical form changes only sporadically, then time is no longer an infallible key to relationships. To take one example, if both a parent and a daughter species have long spans in time, it is far from impossible that our field researches will turn up specimens of the daughter species that are actually earlier in time than known examples of the parent. Only careful analysis of morphology will show us what the actual relationships are. Time is, of course, still a factor – and an important one – but it has to be viewed with due caution in the interpretation of relationships.

It should also be noted that under the view articulated by Eldredge and Gould, the significance of speciation in the evolutionary process is turned upside down. Instead of being simply a passive result of a process of gradual change, speciation becomes as much its cause. All widely distributed species are variable, so local populations of the same species tend to differ from each other in minor ways. When geography or ecological change intervenes to separate a daughter population from its parent, the two will already differ somewhat on average. These differences will then form the basis for future differentiation between them. Eldredge and Gould considered that anatomical innovations are most likely to become fixed during the period of genetic instability associated with speciation; that, in other words, adaptive innovations are directly associated with speciation itself. How far this may be the case we will look at in a moment; meanwhile it's enough to note that, following a period of strenuous opposition, most paleontologists and other evolutionary biologists have become prepared to accept that punctuated equilibria is a documentable and widespread phenomenon of evolutionary history, even if it may not be exclusively the mechanism that has given rise to the patterns observed in the fossil record. It is important to emphasize that, despite early allegations to the contrary, the Eldredge/Gould formulation does not deny adaptation, or even natural selection. What it has done, though, is to change the context within which we need to consider them. At the very least, we now have a more complete account of the evolutionary process than we had previously, and a firmer structure on which to build and refine our hypotheses about how life has diversified. Compared to the received wisdom that preceded it, this new account assigns to time a more diffuse and multidimensional role in the processes that have governed the diversification of life.

8.7. The evolutionary process: a personal view

Naturally, debate on the nature of the evolutionary process has extended far beyond the matter of how major patterns have been generated over the history of life. Such diverse factors as the individuality of species and the role of behavior in evolution have also been subjected to intense scrutiny. Even a cursory review of this more specific literature is beyond my scope here; but what does seem in order in this discussion of the role of time in evolution is a general recounting of my own current views upon how the major mechanisms of evolution relate to each other. From this, the reader should be able to draw his or her own conclusions about how time and evolution intertwine. References germane to this viewpoint may be found in Tattersall (1994).

In trying to understand the complexities of the evolutionary process, it is necessary to realize that a hierarchy exists in nature. This starts with the genes, and proceeds up through the individual to local populations, species, and even beyond. Each of these levels participates in the evolutionary process, but each does so in its own way and has its own particular significance.

Mutations of the genes, and recombinations among them, provide the underlying variation upon which natural selection capitalizes. It is hardly necessary here to go into the variety that genes themselves present, beyond remarking that the individual stretches of the DNA strands of which they are composed range from virtual inactivity to the determination and regulation of extremely complex processes of development. Suffice it to say that most genes by themselves produce relatively minor effects. Genetic innovations, whether by mutation or recombination, rarely produce major changes in the organism possessing them; and when they do, they are rarely advantagous. Still, mutation and recombination are processes that are active in every generation, and they are the wellsprings of the variation in populations on which natural selection works. Natural selection itself acts mainly by eliminating deficient individuals from the reproductive pool; but it also promotes the transmission to the next generation of the genes of those individuals who have the highest rates of economic success. Successful new variants do not arise *for* any purpose whatever; they are not the result of 'selection pressure', but are the consequence of the random appearance of innovations that happen to be advantageous in a particular habitat.

Natural selection thus targets individuals. Adaptation, however, should be viewed as a population phenomenon. Nobody knows exactly how new variants become 'fixed' (i.e. the norm) in populations; but what is known is that local populations of any widespread species do routinely develop their own peculiarities. Local populations are able to do this, presumably, because their gene pools are relatively small, and thus less subject to the genetic inertia that constrains large populations. To be unaffected by this inertia, local populations must be isolated in some way from the larger population with which they are interfertile. It is not known how complete this isolation needs to be (it is unlikely in most cases to be total, although given the propensity of environments to be unstable, it is possible that effective isolation is commoner than we think); but it is evident that some limitation to 'gene flow' throughout the overall species population has to exist for speciation to work. In any event, it appears that it is at the level of small local populations that natural selection acts to promote adaptation to a specific set of local conditions. This makes sense in environmental as well as genetic terms, because any widespread species is likely to encompass a number of differing local environments; and if divergent adaptation by natural selection is to take place, it has to be in the context of a specific environment. Of course, the relative instability of small gene pools also admits the possibility of purely random changes, and the evolutionary importance of these should not be underestimated.

But even if local populations do develop their own peculiarities, it is nonetheless the case that as long as the chain of reproductive continuity remains unbroken it is potentially possible that at any time their distinctive attributes may be reabsorbed into the larger species population, and thus effectively be lost. Change in the course of a river or some comparable event may, for example, reestablish contact between a parent population and a daughter isolate, leading to reabsorption of the latter and the disappearance of its distinctiveness. This is where the event of speciation enters the evolutionary picture. Speciation is very much the 'black box' of systematics. We know beyond a shadow of a doubt that it occurs, but it does not appear that it can be localized in an event of a specific kind. Indeed, it is quite possible that speciation may be recognized infallibly only by its result, reproductive isolation; and even then, we do not know how absolute that isolation needs to be for it to be effective. What we do know is that populations do become isolated in this way; and we also know that such isolation is not necessarily associated with changes of the kind that we associate with adaptation through natural selection. Isolating mechanisms may be behavioral, or lie in chromosome alterations, or in single-gene changes; elucidating exactly what the causal range of these mechanisms may be is still something for the future. Nonetheless, it is sufficient for the moment to know that speciation does occur, and that it is not necessarily the result of accumulating adaptive innovations. After all, for the very same reasons of genetic malleability that adaptive change may occur in small local populations, chance events may affect such populations as well, in other ways.

The history of any species is thus typically one of divergence of subpopulations in relatively homogeneous local habitats – to which the latter will tend to adapt by natural selection. It is these small local populations that are the true engines of innovation in evolution; but they will remain ephemeral as distinctive units until 'validated', as it were, by the event of speciation – which may or may not have anything to do with any adaptive changes they have accumulated. Once speciation is achieved, the new species and its anatomical innovations are no longer at risk of absorption – though other risks, notably extinction, certainly still attend. This is the true importance of speciation in evolution: not in the creation of new 'species' as such, but by giving historical identity to one (or more) component population(s) of the parent species. Once this historic identity is achieved, the evolutionary process can move to the next level. For, should environmental changes occur that remove the previous reason for their isolation or quasi-isolation, new species are able to go out into the world without fear of absorption, and thus to enter into competition with other related species – including their parents. They are also at least potentially able to start the cycle of spread and local differentiation all over again. In view of all this, I am forced to conclude that in time, species as a whole behave very much as they do in space. Although, as the notion of punctuated equilibria suggests, the histories of species are marked principally by non-change, what is actually happening over such histories is a constant process – often over millions of years – of local differentiation. Sometimes this process will result in the generation of new species – with the concurrent persistence of the parent – but, since differentiation and speciation are not the same thing, it may equally well not. This is not exactly 'stasis' as envisioned by the authors of punctuated equilibria, but in terms of the signal this process would be expected to leave in the fossil record (in which only hard parts are preserved) it may well approximate the same thing.

Once new species become established, and move beyond their original restricted boundaries, competition with other species with similar lifestyles inevitably emerges as a major factor in the evolutionary process, as does coping with new predators, new habitats and so forth. And competition among similarly-adapted species, often close relatives, promotes a winnowing effect *among* species that is analogous to the winnowing of individuals by natural selection. It is competition at this level, indeed, that generates the evolutionary 'trends' that have so often been pointed to as evidence for the gradualist viewpoint. Such trends result as species with particularly advantageous adaptations (which they in turn pass along to their offspring) succeed disproportionately in the competitive game. Just as true evolutionary innovation is a sporadic process, though, the competitive success of species is episodic in nature. It merely gives the illusion of continuity in an incomplete fossil record because distinctions in hard-part morphology between closely related species are typically small. What is more, close scrutiny of most claimed examples of trends in the evolutionary record (the famous increase in brain size in human evolution, for instance), shows that such trends are actually rarer than is commonly supposed.

An inevitable concomitant of competition is extinction. Extinction may, of course, occur for a variety of reasons which often have little to do with excellence of adaptation, or its lack. The various mass extinctions that have – sporadically – dotted the evolutionary history of life have, for example, had little to do with how well the many thousands, even millions, of species concerned fit into their environments. For all seem to have been associated with major environmental changes that made the adaptations of most species irrelevant. But mass extinctions are only the most dramatic examples of a process that is going on all the time. The steady 'background' ticking of extinction reflects competition between species, but it also mirrors the fact that environments are not stable. Environments fluctuate continually, and often at a rate with which natural selection alone could never hope to keep up. Certainly, environments are rarely if ever stable over the long periods of time envisaged by the pure Darwinian view of evolution by natural selection at the individual level.

In the wake of mass extinctions, which have sometimes reduced the species count on Earth by as much as an estimated 90 percent, species numbers have typically recovered rapidly, followed by long periods of stability. Thus, while the cast of characters has typically changed at the 'background' rate, during most of life's history the total numbers of actors on the evolutionary stage has probably remained fairly constant. Consistent total diversity has resulted from the balancing of speciation by extinction, which 'prunes' an evolutionary bush which would otherwise grow wildly out of control; and, once again, it is the consequent disappearance of species from the fossil record through interspecies competition and the multifarious effects of random environmental fluctuation that allows the illusion of gradual change to persist. This illusion has, I think regrettably, tended to distract the attention of many evolutionary biologists away from the origins of biotic diversity, to an almost exclusive concentration on the 'evolution', within monophyletic groups, of particular physical characteristics or functional complexes. Understanding how organisms evolved as functional entities is, of course, important; but individual characteristics do not evolve, at least in the classic sense of the term. The problem is essentially the same as that of trends; and it, too, emerges from too closely identifying time with change.

8.8. Conclusions

Time and evolution, then, are inextricably linked; but, once we have abandoned the notion of evolution as the unique result of natural selection acting inexorably on individuals to effect population change, it is evident that the two are not linked in any straightforward way. We have seen in the preceding discussion that evolution results from a variety of agencies that act at different levels. Change at our first level – the emergence and the fixation within local populations of new genetic variants – is poorly understood, but seems to be a consistent theme. It is doubtful, however, that it is a simple result of 'selection pressure' and that it thus relates in a predictable way to time (the higher the pressure, the faster the change).

Whatever the case, in the instance of any particular species local diversification is likely, in geological terms, to be a short-term process.

Speciation, which gives the results of such accumulated variation historic identity, is a much more haphazard process; one that depends on genetic change unrelated to adaptive modifications. Moreover, its sequel, geographical spread and species-level competition, is subject to environmental fluctuations that occur on a sporadic and unpredictable schedule. Time is involved here, again, but not in any consistent manner. The same has to be said of species longevity. Species may persist for long periods with relative lack of change, or their existences may be fleeting on the geological scale. They have objective historical existences in time, but the length of their tenure, and whether they will be fecund or not, is unpredictable.

Since the history of life sums out as the total history of all the species that have ever existed, this lack of constancy among species relative to time makes the internal pattern of time vs. change in the evolution of life quite dramatically inconsistent. Nonetheless, there does seem to be some regularity in larger patterns revealed by the fossil record, notably in the tendency for rapid recovery of species diversity after large-scale extinctions, followed by stability. There is also some evidence for a degree of periodicity of extinction events; Raup and Sepkoski (1984), for example, observed that mass extinctions in marine environments tended to take place on a 26 million-year cycle. In the absence of any obvious terrestrial driving mechanism for such a pattern, it was suggested that extra-terrestrial agencies such as meteorite impacts might have been the cause of these extinctions; and this is certainly an attractive notion in the sense that events of this kind are governed by probability. Few of them, however, are known to have been marked by evidence of the geological perturbations which accompany bolide impacts (the demise of the dinosaurs – and much else – at the end of the Cretaceous being a notable exception). Most mass extinctions seem to have been linked one way or another with large-scale changes in Earth's ecosystem, albeit often governed by changes in the way that the Sun's energy has been taken up and distributed over the surface of the Earth.

The drama of life's history, then, is one that has played in the theatre of time, but with an inconsistent script. The Earth's ecosystem is a delicate complex of many different elements that is subject to influences of many kinds. Over the past $3\frac{1}{2}$ billion years the interplay of these influences has produced numerous environmental fluctuations. Such fluctuations have most often been regional, or local; but occasionally they have been global. The biota has responded in varying ways, depending both on the nature of those changes and on the size, structure and species makeup of the ecosystems affected. Time is essential for evolutionary change; but the result of the system's complexity is that the two are not synonymous. Given time, our ecosystem will change again; but (our own malign influence apart) there is no predicting how.

References

Darwin, C., 1859, On the Origin of Species by Natural Selection, or the Preservation of Favoured Races in the Struggle for Life (John Murray, London).

Dobzhansky, T., 1937, Genetics and the Origin of Species. (Columbia University Press, New York).

Eldredge, N., 1971, The Allopatric Model and Phylogeny in Paleozoic invertebrates. Evolution **25**, 156–167.

Eldredge, N. and Gould, S. J., 1972, Punctuated Equilibria: An Alternative to Phyletic Gradualism. In: T.J.M. Schopf, ed., Models in Paleobiology (Freeman, Cooper & Co., San Francisco), pp. 82–115.

Fisher, R.A., 1930, The Genetical Theory of Natural Selection (Clarendon Press, Oxford).

Lamarck, J.-B., 1809, Philosophie Zoologique (Paris).

Linnaeus, C., 1758, Systema Naturae, 10th ed. (L. Salvius, Stockholm).

Lyell, C., 1830/32, Principles of Geology, Being an Attempt to Explain the Former Changes of the Earth's Surface by Reference to Causes Now in Operation. (John Murray, London).

Malthus, T., 1798, An Essay on the Principle of Population, as It Affects the Future Improvement of Society. (J. Johnson, London).

Mayr, E., 1942, Systematics and the Origin of Species (Columbia University Press, New York).

Raup, D.M. and Sepkoski, J.J., 1984, Periodicity of Extinctions in the Geologic Past. Proc. Nat. Acad. Sci. U.S.A. **81**, 801–805.

Simpson, G.G., 1944, Tempo and Mode in Evolution (Columbia University Press, New York).

Tattersall, I., 1994, How Does Evolution Work? Evolutionary Anthropology **3**, 2–3.

Wallace, A.R., 1858, On the Tendency of Varieties to Depart Indefinitely from the Original Type. Proc. Linn. Soc., Lond. **3**, 53–62.

chapter 9

REAL TIME AND RELATIVE INDETERMINACY IN ECONOMIC THEORY

MARIO J. RIZZO
New York University

Time in Contemporary Intellectual Thought
Patrick Baert (Editor)

CONTENTS

9.1. Introduction

The twentieth century opened under the 'spell' of *Bergsonisme*. It now closes with a renewed and more sophisticated appreciation of the genius of Henri Bergson.[1] In the intervening years, however, Bertrand Russell's atemporalism influenced more than a few philosophers of science and the character of discussions about science. Whether it influenced the actual work of the scientific community cannot be univocally answered. Certainly, as we shall see below, the natural sciences recognized early in this century the importance of time and indeterminacy at the microphysical level. The social sciences, however, found in the work of philosophers like Russell a reinforcement of their nineteenth-century mechanical prejudices.

Bergson is pre-eminently the philosopher of time and of evolution. The two ideas are closely related. "The more we study the nature of time", Bergson tells us, "the more we shall comprehend that duration means invention, the creation of new forms, the continual elaboration of the absolutely new" (1911, p. 11). Bergson's philosophy was born partly as a reaction to the mechanical nature of Herbert Spencer's evolutionism. In the latter's approach, the evolution of genuinely new forms is an impossibility. Spencer's 'Principle of the Persistence of Force' means that the evolutionary process consists simply of the rearrangement of existing forms. True creativity is excluded. Furthermore, the mechanism by which these rearrangements came about can be modelled in a deductive fashion. The present is fully contained in the past. History is therefore a mere unfolding of the already-given.[2]

Russell's early philosophy of science was consistent with the Spencerian mechanism in many essential respects. But while Spencer did not fully appreciate that he had 'annulled' time (an odd accomplishment for an evolutionist), Russell wrote eloquently about the irrelevance of time in philosophy and in theoretical science. Consider the following from his essay *Mysticism and Logic*:

> [T]here is some sense – easier to feel than to state – in which time is an unimportant and superficial characteristic of reality . . . [A] certain emancipation from slavery to time is

[1] For some examples of the renewed appreciation, see Kolakowski (1985), Lacey (1989), Burwick and Douglass (1992), and Moore (1996).

[2] A summary of this view is contained in Taylor (1992, pp. 74–85).

> essential to philosophical thought . . . Both in thought and in feeling, even though time be real, to realize the unimportance of time is the gate of wisdom.
> . . . Whoever wishes to see the world truly, to rise in thought above the tyranny of practical desires, must learn to overcome the difference of attitude towards past and future, and to survey the whole stream of time in one comprehensive vision (1925, pp. 21–22).

At the beginning of the twentieth century, physicists believed that the fundamental laws of the universe were atemporal, deterministic and hence reversible. This view was shattered by Heinseberg's discovery in 1927 of indeterminacy at the microphysical level. Some philosophers have interpreted this as a mere 'uncertainty principle' resulting from interference of the observed by the observer or the process of observation. Thus there is an *appearance* of indeterminism arising from our ignorance of the true state of the elementary particles. If we had a complete accounting of all the relevant variables, the argument goes, we would find an essentially deterministic explanation of quantum phenomena. Nevertheless, there is persuasive evidence against this 'hidden variable' hypothesis.[3] In those cases where the experimental predictions differ, the results have usually been viewed as favoring the indeterminacy interpretation.

On the other hand, most contemporary economic theory has its origins in an application of mechanics to the study of society. This was explicitly recognized by one of the nineteenth-century pioneers of economics, W. Stanley Jevons:

> The Theory of Economy thus treated presents a close analogy to the science of Statical Mechanics, and the Laws of Exchange are found to resemble the Laws of Equilibrium of a lever as determined by the principle of virtual velocities. The nature of Wealth and Value is explained by a consideration of indefinitely small amounts of pleasure and pain, just as the Theory of Statics is made to rest upon the equality of indefinitely small amounts of energy (1965, p. vii).

The full implications of this approach were not widely appreciated until relatively recently. They include the self-same abolition of time, indeterminacy and novelty that characterized nineteenth-century physics (O'Driscoll and Rizzo, 1996, pp. 52–70).

Today there is unrest, if not a crisis, in economic theory emanating from the continued dominance of the nineteenth-century mechanical paradigm. Mechanical models that exclude change and novelty are problematic. This is because such models simply postulate what must be explained, viz., equilibrium. In the social world, as we shall see, equilibrium has a special meaning involving the dovetailing of individual plans. Since equilibrium in society has rather stringent knowledge requirements, it is far from obvious that such a state of affairs must actually exist. Therefore it is incumbent on the social scientist to show how an equilibrium could be generated. The static approach, on the other hand, must postulate either permanent disequilibrium or permanent equilibrium and cannot explain true movement from the former to the latter.

---

[3] For a brief survey of some of this evidence, see Suppes (1984, pp. 23–25).

Another line of thought – primarily a heterodox or minority viewpoint – connects the 'subjectivist' elements of economics with an internal conception of time. In philosophy the latter idea is closely associated with Bergson and Husserl (1964). More recently, in the natural sciences, it has been associated with Ilya Prigogine (e.g., 1986, pp. 243–45) who distinguishes between 'internal time' and 'mechanical time'. The latter corresponds to the aging of a system. From the human perspective, internal time is 'time consciousness' which is inextricably associated with change and the emergence of novelty. A concern with time consciousness flows naturally from those approaches to economics, such as that of the Austrian School, which emphasize the 'subjective' character of individual decision-making. Decisions are taken on the basis of knowledge, perceptions, valuations and expectations of the individual (Hayek, 1955). These are all contents of human consciousness. It is a short step from these thoughts to that of time. To see this, consider that the kind of decisions in which economists are interested are decisions to *act.* Action involves the purposeful transformation of a future state of affairs relative to what it would have been without intervention. Thus, action is future-oriented (Mises, 1966, p. 100). An actor must conceive of a future, better than the present, and then bring it about (or, at least, try to do so).

It is the *process* of acting, rather than the mental picture of completed acts, that discloses time as real duration or the continuous flow of novelty. This is time-as-lived rather than as thought. As we work through an action, we must conceive of a future different from the present and thus be conscious of a heterogeneous time-flow. We shall discuss this further below.

9.2. What is real time?

For our purposes 'real time' is time consciousness – consciousness of the *passage* from one state to another. This is the phenomenon Bergson called 'la durée réelle' or real duration and Husserl called 'internal time consciousness'. By the passage of one state of consciousness to another we must not understand discrete succession of sharply differentiated states. Instead we understand continuous phases that melt, as it were, into one another. Consider that the very idea of 'passage' requires a prolongation of the past into the present. Unless we remember the previous state there can be no sense of passage for otherwise each state is solitary or discrete. Memory of a previous state colors one's perception of the present so that to speak of a present isolated from the past contradicts the idea of time consciousness. Thus the present is not the instantaneous present of Newtonian physics. It is what William James inappositely called the 'specious present' – specious by reference to the purified present of the physics of his day.

We can contrast real time with the common external and mechanical notion of 'clock time'. The latter has nothing directly to do with time consciousness because it is, simply, a matching of physical motions. The clock's movements match the rotation of the earth on its axis. "We have therefore counted simultaneities; we have not concerned ourselves with the flux that goes from one [state] to another" (Bergson, 1911, p. 338). The flux, that is, the perception of succession, could increase or decrease in rapidity without affecting or being affected by clock time. Such changes in time consciousness are known to be induced by various drugs, including marijuana and hashish.

It is useful to consider the essential characteristics of real time more systematically at this point, and to contrast them with those of the static, Newtonian conception, of which clock time is one variant. Real time is: (1) dynamic, (2) irreversible, and (3) possesses a loose causal efficacy.

9.2.1. Dynamism

'Dynamics' often refers to the equations of motion in reversible Newtonian systems. The conception of time implicit in such dynamics is static. This is easily seen in differential equations because here we are dealing with instantaneous magnitudes such as velocity and acceleration. The 'succession' of instants in a Newtonian world is separated only by the differential of time (dt). Thus, essentially, we are dealing with the present. "You are therefore really speaking only of the present – a present, it is true, considered along with its *tendency*" (Bergson, 1911, p. 22).

The dynamics of real time, however, arises only in the context of becoming. Becoming is to be contrasted with juxtaposition in space. It is not simply being 'at' a certain condition at t_1, and at another condition at t_2. (This is the so-called 'at-at theory' of becoming pioneered by Russell and extended by Salmon (1984, p. 147–57).) The phases of a process of becoming are not discrete; they exhibit a quality of interpenetration. No phase has full meaning except by reference to the other phases. This can be understood in both a backward and forward-looking manner. In the former, the role of memory is crucial.

> We shall have to consider a moment in the unfolding of the universe, that is, a snapshot that exists independently of any consciousness, then we shall try conjointly to summon another moment brought as close as possible to the first, and thus have a minimum of time enter into the world without allowing the faintest glimmer of memory to go with it. We shall see that this is impossible. Without an elementary memory that connects the two moments, there will be only one or the other, consequently a single instant, no before and after, no succession, no time . . . To tell the truth, *it is impossible to distinguish between the duration, however short it may be, that separates two instants and a memory that connects them,* because duration is essentially a continuation of what no longer exists into what does exist (Bergson, 1966, p. 49, emphasis added).

From a forward-looking perspective, duration is inseparable from expectation. We can imagine the present enduring into a future only if that which is and that which putatively will be are bridged by expectation.

9.2.2. Irreversibility

If nothing of scientific importance hinges on a future state being after the present or the past being before the present (Russell, 1993, pp. 337–38), then time and its processes are reversible. Any state can be gone through more than once because its character is not affected by what has happened before or what is expected after. The history of such a system does not matter. Strictly speaking, it does not have a history because all of its characteristics exist or coexist at an instant in time. Any prolongation of this instant into many does not follow from the analytics of a reversible system. It is purely an ad hoc construction. J.C.C. Smart provides us with a compelling illustration of this:

> Suppose our four-dimensional geometry is interpreted as a geometry of space-time . . . Let 'cricket ball'$_4$ be the expression which in our four-dimensional representation refers to the cricket ball through its entire history. Then it makes no sense to talk of cricket ball$_4$ changing or staying the same . . . What we express in our ordinary language representation by saying that the spherical cricket ball *becomes* ellipsoidal we express in our four-dimensional representation by saying that the three-dimensional cross-section for $t = t_1$ is spherical and that the three-dimensional cross-section for $t = t_2$ is ellipsoidal. In both these last two uses of 'is' the 'is' must of course be timeless (1978, p. 164, emphasis added).

Since t_1 and t_2 exist simultaneously or are juxtaposed, it is possible to move from t_1 to t_2 and from t_2 to t_1. From the three-dimensional perspective, it is thus possible for a spherical cricket ball to become ellipsoidal and then for it to become spherical again. The cricket ball can go through the same state more than once.

Real time, however, is irreversible. This is because consciousness cannot go through the same state twice. Fundamentally, our consciousness of time is an experience of change. In a completely changeless universe there would be only one instant in which all characteristics are juxtaposed because the sense of before-and-after would be destroyed. Thus time and change go together. But there is a paradox here. *Consciousness endures by changing.* The sense of continuation must be based, as we have seen above, on memory. Yet memory is responsible for the sense of change – of novelty. The present is new by reference to the contents of memory. "What makes [the] present phase *novel*, that is, *richer* with respect to its immediately anterior phase? Precisely the fact that the antecedent phase is *still remembered*. . .." (Čapek, 1971, p. 128). More exactly, the person is new as he accumulates more and more experiences even if these experiences are unchanging in terms of objective measurements (e.g., constant visual or aural stimuli). He is thus at a new moment of his history. Perspectives will change if only because consciousness is older (Bergson, 1911, pp. 5–6). To use a mechanical analogue, the system is aging.

One may also understand the irreversibility of real time in another, closely related, way. Any present moment is affected, or more precisely, is defined in relation to the expectations held by the agent at its emergence. As the present recedes into the past, memory is now enriched by that particular fusion of expectation and primal experience. Now, strictly speaking, the same present moment cannot occur again because the background, the memory of the agent, is now richer or more comprehensive. Viewed prospectively, the remembered experience of the agent is inseparable from his prospective attitude or expectation. (In a sense, they are simply two aspects of the same phenomenon.) This means that the agent's expectational attitude is continually changing as he moves forward through time. As a consequence, each present experience must be different from its predecessor.

9.2.3. *Loose causal efficacy*[4]

In the static or mechanical time perspective, succession is illusory. It is simply an appearance based on human intellectual limitations. This was the position of Bertrand Russell who believed that "the apparent indeterminateness of the future . . . is merely the result of our ignorance" (1993, p. 238). Everything is, from the omniscient standpoint, coexistent. It is the task of philosophy and theoretical science, the argument claims, to embody, as far as practical, this ontological stance. From the abolition of succession it is a short distance to the abolition of time. While most contemporary economists are doubtless unaware of Russell's views, they embody his philosophy in the positive heuristic of their research program. Apparent changes are reinterpreted as cross-sections of a higher-order function. Thus, an agent's expectations may appear to change but the analyst who knows the structure of these 'changes' will see that they can be fully explained by changes in the data. In the terminology of Frank Hahn (1984, p. 55), the agent's theory is the same; it is just the messages sent by the economy that have changed. So change is thereby minimized to the status of (hopefully) only occasional exogenous shocks (Cowan and Rizzo, 1996, pp. 286–88).

If succession, on the other hand, is genuine and if it grows out of the past, then what is the epistemic relationship between past and present? Russell believed it was logical implication (1993, p. 216). As we have seen above, acceptance of this view effectly annuls succession and time. Implication is a timeless relation and it is a mere accident of our limited intelligence that prevents us from seeing everything at one grand instant. Since we reject the assimilation of causation to logical implication (Cowan and Rizzo, 1996, pp. 299–300), we are left with the following choice: either the present is a complete creation ex nihilo or it is loosely determined by the past. The latter is the position for which we, and Bergson before us, are arguing.

Creation ex nihilo must be outside of time. This is because it is inconsistent with any form of persistence of the past and hence of duration. To the extent that a past state persists in memory the novelty of present consciousness is limited. Memory persists *in* the new conscious state, and not simply as a contrasting

[4] For a discussion of the historical roots of this idea, see Čapek (1959, pp. 82–90).

image. If the latter were the case, there would be no continuity in our conscious life. The individual would experience re-creation at each instant. We would then have to agree with Russell that "[t]here is no logical impossibility in the hypothesis that the world sprang into being five minutes ago" (1995, p. 159). Or, we might add, one minute, one second, or, in the limit, that the world is only a *present* world. The claim that the world is entirely present is indistinguishable from the claim that it is outside of time and implies that the past is only a construction from present materials. All of these implications are untenable, for then we are left with the obvious, unanswerable question: Whence this illusion of change and of a real past (Bergson, 1911, p. 339)?

The indeterminacy of the future is a relative indeterminacy. Although the future is caused by the present, it is not necessitated by it. The view that the only alternative to strict causality is absolute indetermination is false. Even at the macrophysical level the evidence is against it. Consider:

> The predictability of the positions of any given macrophysical particle ... is only *approximate* ... The projected trajectory of a particle, which, in our macrophysical perspective, appears as a precise geometric curve with no transverse thickness, is, in reality, a *thin tube*, a *bundle of possible routes,* which, although very thin, still has transversed dimensions corresponding to the quantic indeterminations of the future positions ... In other words, it is only by virtue of our macroscopic myopia that the field of the diverse possibilities seems to shrink so that it appears finally as a precise infinitely thin line of 'the only possible route' (Čapek, 1959, pp. 88–89).

Of course, the looseness of causal processes is even clearer at the microphysical level where probabilistic forms of explanation are commonplace (Salmon, 1984, pp. 242–59). So the determination of the future is one of general patterns and not of precise details. As we have seen, the continuity of real time does not only permit loose or pattern causation but it *requires* it. The idea of a deterministic process is either an approximation or a contradiction. A process is extended in time and, as such, it embodies a form of 'memory'. Therefore, its novelty is both genuine *and* limited.

9.3. The 'extended present'

An example of the importance of the real time perspective for economic thought can be found in the work of the Austrian economist, Ludwig von Mises. Mises, it seems, was influenced by Bergson's philosophy, but this influence has yet to be documented fully. Mises understood that for purposes of explaining human action the present cannot be the mathematical instant. Such a present does not endure long enough to be utilized by the actor. "[F]rom the praxeological aspect there is between the past and the future a real extended present" (Mises, 1966, p. 100). The extended present encompasses what from the instantaneous perspective belongs to the recent past. "The present encloses as much of the time passed away as is still actual, i.e. of importance for acting" (p. 101). For the past to prolong itself into the present consciousness of the actor obviously requires memory. So the real extended present must embody memory of the past. But since all action is future-oriented, i.e. aims at improving the future state of affairs beyond what it would be without the action, the extended present must also encompass expectation (p. 100).

To see the importance of Mises's idea of the extended present for his treatment of an economic issue, consider his analysis of the role of product prices in the investment decisions of entrepreneurs (1966, pp. 336–37). He asks the question: Do entrepreneurs base their plans on the present prices of the products they intend to produce? Mises answers in the negative because these are really just-past prices. Since production takes time they are interested in the prices that *will* prevail when the products come to market. But these do not yet exist so the obvious question is the relationship between just-past present prices and future prices. Mises sees the "prices of the immediate past" as a "starting point of deliberations leading to forecasts of future prices" (p. 336). This, however, is 'only' a starting point. Future prices do not have a "direct causal relation" [Ibid.] to past prices. In other words, they are not tightly caused or necessitated by the past. There is a loose causal relationship that admits of favorable or unfavorable surprises in the eventual prices of the products. Nevertheless, there is no creation ex nihilo. The surprises or novelties facing the entrepreneur are limited for they "do not construct afresh every day a radically new structure of prices" (p. 337).

The Misesian real extended present is to be contrasted with the three atemporal approaches that pervade modern economics. The first is the instantaneous present of the Arrow-Hahn-Debreu model of general 'intertemporal' equilibrium. In this model all decisions are taken at a single instant for all future dates and contingencies. So, for example, the agent makes provision at that instant for umbrellas on December 25, 2010, if it is raining. While time in the calendar sense exists in this model, the passage of time is economically irrelevant. The future could be compressed or elongated at will without necessitating any essential changes in the model. In fact, analytically, there is no difference in Arrow-Debreu between a commodity at a future point in time and one at a distant geographical point. Since all decisions are made at the primordial instant, there is no re-evaluation of the probability of future contingencies as time goes on. The agents do not learn. Hence they do not endure.

The second approach sees the relationship between the past and present as one of logical containment. Here, while prices change over time, past prices logically imply (via some functional relationship) present prices for all time periods. So the present in rigidly *contained in* the past and thus no novelty is possible. Since the operation of 'logical implication' is timeless, such a model does not embody the passage of time in any essential way. A sufficiently intelligent observer in possession of the relevant formulae could see this universe at a single instant. The appearance of time is thus the result of human ignorance.

The third atemporal approach against which we can compare the Misesian extended present is simply the forward-looking version of the above containment model. In the rational expectations approach, future prices are implied by the complete array of present data. Therefore the future is not emergent but a simple displacement of already-existent parts (Bergson, 1911, p. 8). Once again, the 'duration' between the present and the future is inessential; it can be compressed to a single present instant without any essential change in the analysis. Time in these approaches is a purely ad hoc construction.

9.4. The plan: Newtonian or durational?

A plan is an integrated series of intended actions over time. As such it may seem to be the single most important place to start in an effort to understand the role of time in economics. While this will, in fact, turn out to be the case, the concept of a plan, as completed and as a mental picture of possible acts, embodies time in a static or Newtonian manner. Consider the plans of agents in an Arrow-Hahn-Debreu general equilibrium model. The plan consists of a mental picture of spatialized time in which the allocational (budgetary) decisions of individuals over time are identical to allocations over space. The only knowledge the individual has is that possessed at the primordial instant.

Although the plan-as-contemplated involves a static conception of time, the plan-as-created is a durational act. The *process* of planning is in real time because as we think we experience and we learn.[5] We could not go from a state of indecision to one of decision unless this were true. Ignoring the growth of knowledge would require that the plan be settled upon from the beginning as in the Arrow-Hahn-Debreu model. It is also true that since the plan will be used as a guide to action over time, there must be a process of implementation during which the agent experiences and learns. This again is a process in real time. So the plan as created (revised) and implemented is in real time, while the plan merely as an object of contemplation is static. This is an example of what Bergson meant by the statement, "We do not *think* real time. But we *live* it . . ." (1911, p. 46).

While the contemplated plan is essentially a static object of thought, it nevertheless contains a residue, as it were, of its durational origin. The individual knows that he will learn during the implementation of his plan. Therefore he will insist that it embody a certain degree of flexibility. A rigid commitment to a certain course of action may be optimal given the initial state of knowledge, but it may not be optimal when that knowledge changes. Flexibility in a plan is the outward manifestation of the fact that the plan will be lived as well as thought.

[5] An earlier presentation of this idea is in O'Driscoll and Rizzo (1996, pp. 62–64).

9.5. Coordination and time

The central importance of plans for economics appears in the interpersonal context. Since plans usually involve intentions to engage in market exchange, their effective implementation requires coordination among individuals. Thus, in a simple two-person situation, if *A* intends to sell a house for $100,000, it is important that there exist a *B* who intends to buy at that price. Otherwise, there will be discoordination of plans and the need for plan revision on the part of one or both parties. In a world of unpredictable change, plans are rarely, if ever, coordinated ab initio. Thus, a crucial phenomenon in economics is the process of plan creation, and revision engendered by the initial errors of discoordination.

Our task in this section is to examine the implications of real time for the analysis of the process of plan revision. We shall briefly discuss several approaches in ascending order of their ability to deal with real time.

(i) In his influential 1937 analysis of plans Hayek (1948) argued that if the plans of individuals are incompatible with each other, it is because some individuals are wrong in their predictions of the behavior of others. Sellers of houses may be overly optimistic about the prices that buyers are willing to pay, for example. Nevertheless, Hayek suggests, "it may be *inevitable* that in the course of his attempt [to implement his plan, the individual] will find the facts are different than he expected" (p. 52, emphasis added). Furthermore, "the relevant knowledge which he must possess that equilibrium [i.e. mutual compatibility of plans] may prevail is the knowledge which he is *bound to acquire* in view of the position in which he originally is, and the plans which he then makes" (p. 53, emphasis added). Hayek appears to be saying, first, that it is well-nigh inevitable that the individuals will learn that they were mistaken in the assumptions upon which their plans were based. He then goes on to say that the knowledge they need in order to have mutually compatible plans is knowledge they will certainly acquire as they discover their errors. Thus, sellers of houses will not only discover that they are asking too much but they will also discover the 'correct' asking-price – one that will render buyer and seller's plans compatible.

Hayek's deterministic 'process' is quite similar to, if less explicit than, that outlined in Erik Lindahl's 1939 essay. Here Lindahl specifies his deterministic research program:

> The first step in this analysis is *to explain a certain development as a result of certain given conditions* prevailing at the beginning of the period studied. These given conditions must be stated in such a way that they contain *in nuce* the whole subsequent development (Lindahl, 1939, p. 21).

The model must portray the process of plan revision as the necessary consequence of the conditions existing at the moment in time in which the discoordination obtained. These 'given conditions' are quite comprehensive. Thus,

> [I]f we know (1) the *plans* of the economic subjects concerned at the initial point of time, if we further know (2) how these individuals are likely to *change their plans* in the future under different assumptions, and if we have (3) enough knowledge of *external conditions* to make definite statements with regard to future changes in plans, and the results of the actions undertaken then it should be possible to provide a theoretical construction of the developments that will be the outcome of the initial position (Lindahl, 1939, pp. 37–38).

This model clearly does not imply a real-time approach for it explicitly endorses the 'container' view of causation which holds that the present can be deduced from the past. This was Laplace's view of the physical universe. In neither Lindahl's nor Laplace's framework can there by any genuine processes in time. As we have seen, in such approaches, the elapsed period of time is purely arbitrary. If *everything* that causes an effect is present at t_0 then so too will be the effect. If, on the other hand, not everything is present, then the state of the world at t_1 cannot be the deterministic consequence of the world at t_0.

(ii) If we go beyond simple mutual compatibility of plans to a fuller conception of economic coordination, Hayek understands that the deterministic quality of the process of plan revision must be attenuated. Consider the following example. Plans of buyers and sellers may be compatible as in the case of the house buyers and sellers discussed above. This does not imply, however, system-wide coordination of plans. For example, there may be potential buyers who are willing to pay more than the current buyers, or potential sellers willing to charge less than the current sellers. In either case there are unexploited opportunities for mutual gain and hence lack of complete coordination. Learning about these opportunities is not well-nigh inevitable because it is not knowledge that the individual is bound to acquire in the process of implementing his plans. One can successfully implement a plan without finding a better opportunity elsewhere. The individual "may learn of the new facts as it were *by accident,* that is, in a way which is not a necessary consequence of his attempt to execute his original plan . . ." (Hayek, 1948, p. 52, emphasis added).

Simply because a process of learning and of plan revisions is not deterministic does not imply that a near-coordinated outcome is impossible. In fact, Hayek believes that such an outcome is highly likely and that the world is characterized by a high degree of plan coordination. The process of plan revision is more likely to produce improvements in coordination the closer the system is to full coordination (equilibrium) in the first place (Hayek, 1976, p. 125; 1941, p. 23, n. 1). This is because, under conditions close to full coordination, errors will be fewer or smaller and hence (so Hayek apparently believes) easier to detect and correct.

If Hayek is correct that the economy is generally close to a state of full coordination at most times, then the underlying economic reality will yield more readily to the deductive method. The state of the world at t_0 will 'yield' a world at t_0 that is a tolerable approximation to the actual world. By definition of 'near coordination', the novelties added by time are relatively unimportant.

(iii) More recently, Hayek (1978) has conceived of competition in particular and the market in general as a 'discovery procedure'. In this view, the *essence* of economic processes is the discovery of new knowledge and the emergence of novel behavior and structures. If plan revision is associated with new discoveries, then the new plans of agents cannot be anticipated fully by all within the system. Learning processes in response to prior plan discoordination will themselves produce some (greater or lesser) discoordination as expectations are upset by the new behavior. Under these circumstances, the state of exact and full plan coordination cannot, *in principle*, be achieved.[6] This is because the very process that is responsible for whatever degree of coordination we enjoy is itself a discoordinating force (O'Driscoll and Rizzo, 1996, p. xxvii). In this conceptualization, economic processes cannot be modelled in a deductive manner for we must, to be faithful to reality, exhibit their essential novelty. This novelty implies that agents cannot forecast what they will learn, plan and do on the basis of current data. Consequently, plans of such real-time agents cannot be fully coordinated. This is the economic meaning of the general incompatibility of real time and equilibrium.

(iv) Although the later Hayek recognizes the indeterminacy of economic processes, his analysis is incomplete in a number of respects. There is no explanation of why or how a creative market order could still be stable or fairly orderly. (It is true that Hayek emphasizes the role of a predictable legal framework, but this does not directly address the orderliness of the market process itself.) The British economist G.L.S. Shackle gives us some indication of how we might proceed in arriving at such an explanation.[7]

Firstly, we must explicitly recognize that in a world of change or genuine "beginnings there is no complete inferential knowledge of the content of time-to-come" (Shackle, 1973, p. 55). Lindahl's ambition to model the world at t_0 so that it contains *in nuce* the world at t_1 is not faithful to reality and, even more damning, is incoherent in a world of real time. Of course, if this criticism is accurate, then a theory of economic processes based on the plan revisions of agents cannot be deterministic. There is no general theory of how agents respond to disappointments (Shackle, 1973, pp. 40–1).

Secondly, it is also important to realize that the novelty emerging from real-time economic processes cannot be completely unconstrained. For choice to be meaningful individuals must have some idea of "*what can follow what*" (Shackle, 1979, p. 20). Without some understanding of instrumental causal relations it is impossible to interfere in course of events to bring about a preferred result. This

[6] Further discussion of exact coordination is in O'Driscoll and Rizzo (1996, pp. 80–85).

[7] The following analysis of Shackle benefitted greatly from Currie and Steedman (1990, pp. 154–76).

understanding is based on how individuals "have seen the world to work" (Shackle, 1973, p. 62). Memory thus constrains expectations of future causal connections.

So Shackle is staking an intermediate claim between mechanism and absolute indetermination. Both the chooser and the economist "must view the world as a pattern of natural barriers rather than narrow and prescribed tracts. What *can* take place, he must suppose, is bounded but not prescribed" (Shackle, 1979, p. 20). This is an economic application of the intermediate philosophical position laid out by Bergson.

The implications of Shackle's view for interpersonal coordination are quite intriguing, although not adequately pursued by Shackle himself. The order of the market in real time must be flexible. It cannot inhibit change; in fact, it is an aid to the discovery of new knowledge. Therefore, this order involves only a certain degree of harmony or coordination of plans. These plans are, by necessity, imprecisely defined and map out paths that will, it is hoped, lead to profitable discoveries. Coordination is coordination in the pursuit of knowledge (Loasby, 1990).

(v) To make sense of the process of plan revision in response to disappointments we must break down the process into its component parts. In Hayek's formulation, plans rest on expectations of external events and the actions of other individuals. Usually such expectations are single-valued and thus, when they are correct, plans will dovetail exactly. Single-valued expectations are not only a simplification, however, they are an *oversimplification*. Individuals typically have an array of expectations each with an associated likelihood. (This is not to argue that they assume these arrays are exhaustive.) But if expectations are multivalued then the array might be 'correct', but plans may nonetheless fail to mesh exactly. The individual's plans are based on the best available view of the future – but that is never perfect. So when all systematic errors are eliminated an imperfect equilibrium will be attained.

Hahn (1984, pp. 55–6) uses a convenient terminology in which to discuss the components of plan revision. Hayek's expectations can be seen as generated by Hahn's 'theories' and Hayek's plans by Hahn's 'policies'.[8] A theory enables the individual to make forecasts on the basis of current data. A policy incorporates these forecasts in decisions about specific acts or a series of intended acts. Hahn's 'theories' and 'policies' are rigid, deterministic methods of generating specific expectations and plans. *Within* these frameworks there are changes in actual expectations and actual plans when the 'exogenous messages' received by essentially non-entrepreneurial agents change. This is not 'learning', Hahn admits. It is simply a mechanical adjustment to changed data.

[8] This connection was suggested by Loasby (1990, p. 46).

The question remains: If agents truly learn, as they will in real time, then can economists say anything about how plans are revised? Is there a nonmechanical or nondeterministic order that persists throughout the process of revision? We suggest that the answer is in the affirmative. Fundamentally, this is because without such an order individuals would be at a loss to forecast the ever-changing plans of others. If this were true, then market economies would exhibit little order or regularity, which is manifestly contrary to fact.

One of the most promising attempts to integrate real time and a process of plan revision is implicit in Loasby's (1990, esp. pp. 51–4) brilliant adaptation of Lakatos's (1970, pp. 91–196) methodology of scientific research programs to model the mindset of the economic agent.[9] The scientific research program (SRP) consists of a 'hard core' of metaphysical concepts or other basic concepts and propositions about the world. These are treated as irrefutable by the methodological decision of the scientist or agent. The hard core does not appear ready-made but is the product of a largely unexplored process of knowledge acquisition in time (Lakatos, 1970, p. 133, n. 4; Weintraub 1985, p. 112–14). Therefore, the hard core is the bequest of the past to current research efforts; it is a kind of collective memory, as it were.

The hard core is protected from refutation by a 'protective belt' of observational or other auxiliary theories and initial conditions. Therefore, a proposition in the hard core may be insulated from refutation by a 'fact' by questioning the observational theories that make the fact credible. This is part of the 'negative heuristic' of the SRP.

There is also a 'positive heuristic' or set of instructions on how to generate specific hypotheses within a SRP. In Loasby's adaptation, these hypotheses are the theories and policies discussed by Hahn. Forecast-theories are generated and later revised within the SRP. Thus, not only do expectations change but the method by which they are arrived at do also. Policies (and derivatively plans) will change partly as a result of changed expectations and expectational theories, but also partly as the result of a better coordination of means to ends.

The rules provided by the positive and negative heuristics guide the process of discovery. These rules do not generate a deterministic system for two reasons: (1) the list of heuristics is neither complete nor completable, and (2) the heuristics do not contain *in nuce* the discoveries yet to be made. Both the heuristics and the discoveries are emergent characteristics of the research program.

The research program is a structure that is compatible with the attributes of real time. In the first place, it provides a historical view of knowledge acquisition. The facts predicted by hypotheses (viz., theories and policies) of the SRP are novel in a limited way. They cannot be inconsistent with those facts or constructions in the hard core of the program. So the scientist's or agent's expectation of future

[9] See also the detailed treatment in Harper and Earl (1996, pp. 306–28).

discoveries is constrained by the past. Furthermore, the process of search for discoveries is itself constrained by the existing positive and negative rules for going forward. To the extent that the research program is successful it enables the agent to *anticipate* 'novel' facts, but only after the hypotheses are formulated in a temporally coherent process. Thus because the emergence of new hypotheses cannot be predicted, there will be an endless flow of fresh discoveries. As a consequence, the revision of plans cannot be deterministic as both the early Hayek and Lindahl suggested. Neither is it unbounded, however. The concept of a 'research program' provides a promising flexible structure that economists can use to describe plan revision and coordinating activity in real time.

9.6. Conclusion

During the twentieth century, mainstream economic theory largely ignored the possibility of an economics in real time. This occurred in large part because its practitioners sought to emulate the success of Newtonian physics. Nevertheless, as our selective overview has shown, there were strains of thought which provided the basis for a real-time reconstruction of economic theory. As we pass into the twenty-first century, it remains to be seen whether the real-time challenge will be heeded. There is a certain urgency to this because it is really the challenge of putting life back into economic theory. The way economists respond will determine whether we have an economics of automatons or an economics of real, enduring human beings.

References

Bergson, H., 1911, Creative Evolution (trans. Arthur Mitchell) (Henry Holt and Company, New York).

Bergson, H., 1966, Duration and Simultaneity (trans. Leon Jacobson) (Bobbs-Merrill, New York).

Burwick, F. and Douglass, P., eds., 1992, The Crisis in Modernism: Bergson and the Vitalist Controversy (Cambridge University Press, Cambridge).

Čapek, M., 1959, Toward a Widening of the Notion of Causality. In: Diogenes **28**: 63–90.

Čapek, M., 1971, Bergson and Modern Physics, Boston Studies in the Philosophy of Science, volume VII (D. Reidel, Dordrecht).

Cowan, R. and Rizzo, M.J., 1996, The Genetic-Causal Tradition and Modern Economic Theory, Kyklos. **49**, 273–317.

Currie, M. and Steedman, I., 1990, Wrestling with Time (University of Michigan Press, Ann Arbor, MI).

Hahn, F., 1984, On the Notion of Equilibrium in Economics, in: Equilibrium and Macroeconomics (MIT Press, Cambridge, Mass.), pp. 43–71.

Harper, D.A. and Earl, P.E., 1996, 'Growth of Knowledge' Perspectives on Business Behaviour. In: P.E. Earl (ed.), Management, Marketing and the Competitive Process (Edward Elgar, Cheltenham, U.K.).

Hayek, F.A., 1941, The Pure Theory of Capital (University of Chicago Press, Chicago).

Hayek, F.A., 1948(1937), Economics and Knowledge, In: Individualism and Economic Order (University of Chicago Press, Chicago), pp. 33–56.

Hayek, F.A., 1955, The Counter-Revolution of Science: Studies on the Abuse of Reason (The Free Press, Glencoe, Ill.).

Hayek, F.A., 1976, Law, Legislation and Liberty, vol. II: The Mirage of Social Justice (University of Chicago Press, Chicago).

Hayek, F.A., 1978, Competition as a Discovery Procedure. In: New Studies in Philosophy, Politics, Economics and the History of Ideas (University of Chicago Press, Chicago), pp. 179–90.

Husserl, E., 1964, The Phenomenology of Internal Time-Consciousness (trans. J.S. Churchill) (University of Indiana Press, Bloomington).

Jevons, W.S., 1965(1871), The Theory of Political Economy (Augustus M. Kelley, New York).

Kolakowski, L., 1985, Bergson (Oxford University Press, Oxford).

Lacey, A.R., 1989, Bergson (Routledge, London).

Lakatos, I., 1970, Falsification and the Methodology of Scientific Research Programmes. In: I. Lakatos and A. Musgrave (eds.), Criticism and the Growth of Knowledge (Cambridge University Press, Cambridge), pp. 91–196.

Lindahl, E., 1939, Studies in the Theory of Money and Capital (George Allen and Unwin, London).

Loasby, B.J., 1990, Equilibrium and Evolution (Manchester University Press, Manchester).

Mises, L. von, 1966, Human Action: A Treatise on Economics, 3rd edition (Henry Regnery, Chicago).

Moore, F.C.T., 1996, Bergson: Thinking Backwards (Cambridge University Press, Cambridge).

O'Driscoll, G.P., Jr. and Rizzo, M.J., 1996(1985), The Economics of Time and Ignorance (Routledge, New York and London).

Prigogine, I., 1986, Irreversibility and Space-Time Structure. In: D.R. Griffin (ed.), Physics and the Ultimate Significance of Time (State University of New York Press, Albany), pp. 232–50.

Russell, B., 1925, Mysticism and Logic (Longmans, Green and Co., London).

Russell, B., 1993(1914), Our Knowledge of the External World (Routledge, London and New York).

Russell, B., 1995(1921), The Analysis of Mind (Routledge, London and New York).

Salmon, W.C., 1984, Scientific Explanation and the Causal Structure of the World (Princeton University Press, Princeton).

Shackle, G.L.S., 1973, An Economic Querist (Cambridge University Press, Cambridge).

Shackle, G.L.S., 1979, Imagination and the Nature of Choice (Edinburgh University Press, Edinburgh).

Smart, J.J.C., 1978(1968), Spatialising Time. In: R.M. Gale (ed.), The Philosophy of Time: A Collection of Essays (Humanities Press, New Jersey).

Suppes, P., 1984, Probabilistic Metaphysics (Basil Blackwell, Oxford).

Taylor, M.W., 1992, Men Versus the State: Herbert Spencer and Late Victorian Individualism (Clarendon Press, Oxford).

Weintraub, E.R., 1985, General Equilibrium Analysis: Studies in Appraisal (Cambridge University Press, Cambridge).

chapter 10

TIME IN ECONOMIC THEORY

FRANK HAHN
University of Siena

Time in Contemporary Intellectual Thought
Patrick Baert (Editor)

CONTENTS

10.1. Introduction

In this chapter I propose to discuss how problems of time enter economic theory and how, at present, they are dealt with. It will be seen that much remains to be done both theoretically and empirically. However, the reader should not expect to find anything which is even remotely relevant to or connected with philosophical problems of time. Even more importantly, he should not expect a theory of economic history, although I shall have something to say on this topic. Economists have, with some exceptions, deliberately narrowed their focus so as to find questions which have some hope of being answered and answered in a manner that coheres with earlier answers. This of course means that the 'scientific' claims of the subject are quite limited, and their sphere of application even more so. The advantage is that there is little speculative chitchat. Economists, as Keynes recognised, are much more like dentists than social philosophers, although there are some who find this too restraining.

10.2. The basic framework

Everyone knows that the economic agent is taken to be *rational*: the agent knows what it wants and knows how to get it. The first part of this saying is summed up by the postulate that the agent has coherent preferences. Preferences are defined on a domain which may be very large: it may include the happiness of a grandmother as well as ice-cream. In general, most of economic theory restricts its attention to the sub-domain of goods and services. The argument is pragmatic: for much that interests us, we can take the excluded items as given or changing slowly. This of course means that grand theories of the future are beyond our reach. However it should be understood (although it often is not), that preference theory does *not* entail the view of humans as having no interest beyond goods.

Preferences are coherent if any two elements in the domain can be related by a relation: 'not preferred to', and if this relation is reflexive and transitive. One is speaking of a 'well integrated' personality and it is admitted that this is an idealisation. The second part of the saying is also straightforward: the agent knows how to get what it wants if, when it chooses, no other choice is both possible for it and preferred to the given choice. This very simple axiomatic construction turns out to have quite strong consequences and, in particular, leads to falsifiable predictions. However we shall see that it may not suffice for all problems involving time.

Consider first the 'commodity space', that is, the space of goods on which preferences are defined given all the things in the other part of the preference domain. It is clear that goods should be distinguished by their physical characteristics and it requires little argument to convince one that their location will also be an object of preferences. But we add two further characteristics: date of availability and 'state of nature' when available. The date of availability is clearly a relevant characteristic for preferences. Later we shall note that it is also relevant for production activity. The 'state of nature' is a technical construct (introduced by Savage, 1954) which describes the exogenous environment of the economy. At present I omit a formal definition and suggest that the reader think of it as events like 'rain at a date *t*' or 'sunshine at date *t*' or as 'healthy at date *t*' and 'sick at date *t*'. Clearly an umbrella in Cambridge on the first of January 1994 if it rains will have a different preference position from the same umbrella in the same place and at the same date if the sun is shining.

With this rich description of commodities an element in the commodity space is of course defined by its coordinates and so represents a collection of different physical goods at different dates and in different places under different states of the world. But this will not do for the theory unless the states of the world are known with certainty. Your actual willingness to exchange something for an umbrella available tomorrow in Cambridge if it rains depends on your belief as to the likelihood of rain. So your preferences must be thought of as preferences over gambles, that is, over the probability space of the commodity space.

I have just implied that the agent assigns probabilities to future states. This can be interpreted as a further implication of rationality where the latter now is defined by a richer set of axioms (De Finetti, 1937; Ramsey, 1926; Savage, 1954). These axioms convert beliefs into probabilities which obey the usual probability calculus. The axioms are not quite as immediately plausible as the ones we started with. In fact a joint axiomatisation of preferences and beliefs is offered. For instance, your degree of belief in *A* is related to the amount you would be willing to bet that *A* will obtain and that, in turn, is of course related to your preferences between what you bet and *A*. It is a very nice theory but this is not the place for a detailed account.

It will have been noticed that time has already led us away from the simplest choice theory. The fact that goods are available in the future rather than now is preference relevant. The further fact that the future is not known with certainty leads to an elaboration of preference theory and a theory of the representation of beliefs. One of the famous and much used refinements of the theory is due to Von

Neumann and Morgenstern (1947). By a further axiomatic elaboration they showed that preferences over gambles are representable by the 'expected utility' of the outcomes. To be precise, let x_s be a vector representing the payoff of a gamble in goods in state of nature 's'. Let λ_s be the probability assigned by the agent to s and let $u(x_s)$ be a function invariant under affine transformation (i.e. to choice of origin and scale). Then the expected utility, W, of the gamble is given by

$$W = \Sigma \lambda_s u(x_s).$$

The function $u(\ )$ is a representation of preferences over goods in state s. I ought to mention here that this theory has not done very well in experimental situations, and is now in the process of modification.

A further step is to bring time into the picture. (In the above expression I have taken the state of nature as referring to the same moment in time.) We do this by writing W_t as the expected utility at t, and then represent preferences over a stream of gambles by the discounted sum $\Sigma_0 \delta^t W_t$ where $\delta \leq 1$. Of course this requires intertemporal independence of preferences so that the bundle of good x_{st} (at s, date t) does not affect your preferences over bundles at $t+1$. So the axiomatic foundations are much less compelling than in the case where time can be ignored.

Now the rational agent who knows 'how to get it' is supposed to choose a place at an initial date which maximises $\Sigma_0 \delta^t W_t$ subject to the constraint that (a) at no date can more be spent or lent than is earned or borrowed and (b) that debts must be repaid. This asks a lot of the agent not only computationally but also of power to formulate beliefs and preferences coherently. But more is needed: the agent must know (or have beliefs) over how states of nature translate into prices since these are crucial parts of the constraints under which he maximises. The decision problem of a producer (firm) can be formulated in a similar way.

It is at this stage that economists have taken one of two decisive steps. The first is due to Arrow and Debreu, while the second is more recent and associated with Chicago. Arrow and Debreu assumed that at the date $t=0$ at a given location when the economy 'starts', all goods have markets. Recalling the characterisation of goods already given, this means that at $t=0$ there is, for instance, a market in Cambridge for the delivery of an umbrella in Cambridge at date $t+4$ if it rains. Nothing will be delivered if it is fine. That is, for every contingent future good there is a market at $t=0$.

The alternative (Chicago) step is not the postulate that all goods have markets at t but that agents know at $t=0$ all prices at every date t and every state of nature s. The uncertainty agents face is simply that which arises from the stochastic process of states of nature. This is the postulate of rational expectations.

Both of these moves make the agent's decision problem a great deal simpler and both have significant consequences for theories of market economies.

10.3. Arrow-Debreu and rational expectations

The consequence of the Arrow-Debreu (A.D.) postulate is to project the future into the present and vastly to reduce the costs of uncertainty.

An equilibrium for an A.D. economy is a set of non-negative prices for all the goods characterised such that when every rational agent has made the best choice amongst the goods available to the agent at these prices, demand equals supply for every good. A.D. famously proved that (under reasonable postulates) such an equilibrium exists. It was then also shown that under certain assumptions the equilibrium was *efficient* in the sense that, given the resources and technology of the economy, there was no reallocation of goods between agents (e.g. by an all-knowing benevolent dictator) which allowed some agents to reach a preferred outcome with no agent being forced to accept an inferior one. (Pareto-efficiency).

Ever since Adam Smith, economists have pointed to markets as the coordinating devices which allow decentralised decisions to have an orderly rather than a chaotic outcome. A.D. brought future and future-contingent goods into the ambience of this view – the future was coordinated just as the present is – by markets. Of course this also meant that present choices and allocations would be dependent on those over the future. For instance, if in some future date and state oil would be in short supply, then *cet. par.* the price for delivery of oil in that state and at that date ruling today would be high and that would cause agents to economise in the consumption of current oil, and the future supply of oil will be higher than it would otherwise have been. That is, the path of consumption of oil would be coordinated in a way that no agent would have an incentive to substitute oil consumption between dates or over states.

The A.D. economy is the pure picture of a decentralised market economy and, in it, time is of no particular importance. Exactly the same reasoning applies to future and present goods. Like idealised constructions in other subjects, it serves as a benchmark and as a starting point for further thought. But it does not describe any economy known to us, and there are good grounds for believing that no economy could correspond to the theory. At the very least it ignores important instances where market mediation is either impossible or inefficient. For instance, there are no property rights in clean air nor in the ambient temperature. But property rights are fundamental to markets and to exchange. There are problems with unborn generations whose preferences on current markets can only be indirectly represented by those now alive who may either be unwilling to do so or do so only imperfectly. Indeed the number of agents at any date and state can only be very indirectly brought under the influence of markets. But from the point of view of description, the objection is simply that the postulated market structure is much richer than the one that actually exists.

Arrow (1953) himself took the first step towards greater verisimilitude. He showed that many of the markets postulated were unnecessary for coordination over contingencies or over time. Here is a simplified version of the argument. Suppose that at $t=0$ there existed securities which agents can only buy and sell at $t=0$. Each security is defined by what it pays at each future date and state. (Suppose payment to be in gold.) If at each date the payoffs of the securities 'span the space of states of nature', that is, if one can generate by combining them in different ways any profile of gold payments over states *and* if agents have rational expectations (see above), then once again there will be efficient market coordination over time and over states. (Of course I am excluding cases like the ones I have discussed where markets cannot operate efficiently.) It is easily seen that the device of securities greatly economises in the number of markets required to operate at $t=0$. (If there are, say, two physically distinct goods all at the same location, two dates, today and tomorrow and two states of nature, then A.D. require $2\times2\times2=8$ markets. In the case of securities we require one for each state, that is two, and two current markets for goods, that is, four markets.)

This suggestion of Arrow leaves us with two questions. Are we willing to postulate rational expectations? and will there be enough securities for spanning? Rational expectations ask a great deal of agents – they must know market clearing prices at each date and state. The question arises: how do they come to know these prices? I shall later briefly discuss how answers to this question are attempted. But there is also a purely logical problem. It is known that many non-pathological specifications of an economy lead to there being many possible market clearing prices at each date and state. To know which of these will prevail requires agents to know what other agents believe and know. Indeed one must suppose that beliefs have been coordinated. Again one asks: how? We shall see that the attempt to answer these questions has rather profound effects on economic theory: it will not only be the future that casts a shadow over the present, but also the past.

In this part of the chapter I have given, I fear, a highly condensed account of how economists have attempted to circumvent the threat time poses to their canonical view of coordination through markets. They clearly have not been too successful in matching theory to fact, although some, especially American, economists take it on trust that the approximation is 'good enough'. I do not have this faith. But nonetheless I believe that something very valuable has been achieved. We now understand what would need to be the case if markets were to do the Adam Smith job in an intertemporal setting. Some of the basic issues have been made transparent and the stage has been properly set for the struggle for greater realism.

10.4. A simple intertemporal model

I now discuss a simple, but important, intertemporal model, called the 'Overlapping Generations Model'. It has quite a few things to teach us.

We are to suppose an economy with a constant population in which each agent lives for two 'periods'. Many elaborations are possible and 'two period lives' are not essential – I choose to stick to the simplest case. An individual born in period t is 'endowed' with a quantity of a single good. Write this endowment as e. The agent can consume that good or exchange it for money which he can store for the second period of his life and then consume. Let p_t, p_{t+1} . . . etc. be the money prices of the good in periods t, $t+1$ etc. and write m_t^t, x_t^t as the amount of money demanded and the good consumed by the agent born at t, and let x_{t+1}^t be the amount the agent plans to consume in the second period of life. Then the agent is constrained by his budgets

$$p_t x_t^t + m_t^t \leqslant p_t e$$

$$p_{t+1} x_{t+1}^t = m_t^t$$

He chooses his consumption for today and tomorrow in the manner already described where preferences are represented by the utility function $u(x_t^t, x_{t+1}^t)$.

While the agent born at t decides on consumption at t the agent born at $t-1$ will want to spend the money accumulated in $t-1$. So for equilibrium we require that the demand for the good at t should equal the amount of the good available, i.e. e. So

$$x_t^t + x_t^{t-1} - e = 0 \tag{1}$$

is a condition of equilibrium. But $x_t^{t-1} = m^{t-1}/p_t = m/p_t$ if the money stock in the economy is a given constant. Elementary theory of rational choice tells us that given e, x_t^t is a function of p_{t+1}/p_t – the relative price of future to present good. Assuming that prices are correctly predicted, we can now write (1) as

$$X\left(\frac{p_{t+1}}{p_t}, \frac{m}{p_t}\right) = 0$$

Or, taking m as fixed, more simply as

$$G(p_{t+1}, p_t) = 0 \tag{2}$$

In (2) we have a simple first order difference equation. Suppose p_0 is given. Then (2) determines p_1. But once we know p_1 we use (2) to give us p_2 and so on. Now notice a very interesting feature: it is the price expected for the next period that equilibrates demand and supply. When the analysis starts there is one old agent and one just born. The old agent's demand today depends on actions taken before the analysis started. In particular the p_0 is to some extent arbitrary.

Different values of p_0 will lead (2) to generate a different path of prices and since the economy is taken to extend into the infinite future we can always find a path which satisfies (2). (There are no terminal conditions.) We see then that even when we heroically stick to perfect foresight of prices the path the economy takes depends on these expectations. To put it the other way round: if we know that the correctly expected price is p_1 then (2) will tell us what p_0 must be in order for markets to clear at $t=0$. Indeed there is a continuum of equilibrium paths. (This of course would not be true for a finitely lived economy because if the final date is T, we cannot appeal to prices expected at $T+1$ to equilibrate markets at T.)

This kind of indeterminacy does not arise in the Arrow-Debreu model which is finite and it has been circumvented by rational expectations economists. This last I shall discuss shortly. But let me repeat the lesson of the simple model. The effect of the future, or rather the expected future, on the present leads, in open ended economies, to bootstrap equilibria and raises expectations to a central role in the economy. Keynes recognised this to be so (without the benefit of a complete formal model) and many economists have tried to escape this conclusion.

Before I leave the simple model which I repeat can be elaborated (and made less abstract), I need to notice two further implications:

(a) The equation (2) is non-linear. Non-linear difference equations can yield a large variety of paths, including chaotic ones, depending on the form of the function $G(\cdot)$. A number of cases of cycles and chaos have been investigated.

(b) The bootstrap element is nicely emphasised by what has come to be known as 'sunspot equilibria'. Let s_1 and s_2 be two states of the sun. Suppose that these states have no direct effect on either preferences or productivity. But assume that agents hold the belief that prices at s_1 will exceed those at s_2 ($p(s_1)>p(s_2)$). Sunspots are stochastic, and agents now decide in the face of uncertainty in a manner already described. Then it can be shown that in a number of cases there indeed exist equilibrium paths with $p(s_1)>p(s_2)$. Note that sunspots have 'real' effects only because they are believed to affect prices – without that belief, prices would be independent of sunspots.

10.5. Infinitely-lived agents

In the 1920s, Frank Ramsey wrote a celebrated paper on 'Optimum Savings'. (Ramsey, 1928). A simplified outline is this. The economy is run by a benevolent dictator who is also a Benthamite. He is concerned with 'happiness' of his people over the infinite future. For simplicity suppose there to be a stationary population, and let $u(c_t)$ be the utility of the representative person when consumption per head is c_t. Being Benthamite, the dictator is interested in

$$W=\int_0^\infty u(c_t)\,dt \tag{3}$$

which he wants to maximise subject to what is possible. Let K_t be the stock of capital per man at t and $\dot{K}_t$ its rate of increase. Assume that output per man (y_t) depends on capital per man according to the function $y_t=f(K_t)$. Then at any t, feasibility is given by

$$c_t+\dot{K}_t=f(K_t) \tag{4}$$

It is assumed that $K(0)>0$.

Ramsey, using the calculus of variations, then found the optimum path. This has to satisfy two conditions (I am assuming $u'(c)>0$, $u''(c)<0$): the Euler Lagrange conditions and transversality conditions.[1] I want to draw particular attention to the second of these conditions. They provide another anchor (K_0 is the other) for the differential equation yielded by the first and (4). Thus 'indeterminacy' which we met before is avoided.

Now Ramsey's solution has a 'price-dual'. That is, given the optimal Ramsey path of consumption and investment, we can calculate what path of prices corresponds to it in the sense that at these prices all agents in maximising their private benefits would indeed choose the path of consumption and investment prescribed by Ramsey.

Ramsey's paper has given rise to much useful and important economics, once it was realised that it could be adapted to deal with less global concerns. For instance, Ramsey's is the method in use to study the optimum depletion of an oil reserve, and many other specific intertemporal policy problems.

But in recent years it has also been used to escape the problems discussed in the previous section. This is achieved simply by assuming that the economy behaves 'as if' some 'representative infinitely-lived agent' acted as the benevolent Ramsey dictator. Once that has been swallowed, all sorts of fancy modification of the original Ramsey model is possible. One can include the disutility of work in the utility function $u(\cdot)$ and one can complicate (4) by allowing random productivity shocks (innovations, etc.) and introduce lagged responses. In that way with some ingenuity one can get cyclical Ramsey paths and one can then choose parameter values which make these paths not too different from observed ones. Note that this procedure essentially eliminates expectations of future prices from the picture, or rather ties them down to those generated by the Ramsey path. The model has no room for Keynesian bootstraps. Note also that all cyclical paths are optimal and so do not call for policy measures.

[1] They are: $-\frac{u''}{u'}\frac{\mathrm{d}c}{\mathrm{d}t}=f'(K)$ and $\lim u'(c_t)k_t=0$.

This account illustrates the length to which economists have been willing to go to escape the fickle hand of the future destroying the basic belief that market economies are simply a veil over the real economic possibilities of an economy which are given by preference, technological know-how and resources. That is, they are reluctant to leave room for expectations to yield paths which diverge from what in the 'fundamental' circumstances is optimal. This is not the place for polemics, and I leave the reader to judge whether the demonstration that we live in the best of possible worlds is convincing.

10.6. History dependence

I have already noted that in general there are multiple equilibrium paths. We have no a priori way of saying which, if any, will prevail. This, in recent years, has led to the recognition that one needs to look at equilibria as outcomes of some process. For instance the asymptotic distribution of some stochastic differential equations.

The processes, sometimes regarded as 'learning processes', sometimes as 'evolutionary', are not axiomatically based. The idea for instance is that expectations which are not systematically falsified will be reinforced while those that are will wither away. In the nature of things this is very hard theory to get to yield results, because what has to be learned by agents itself depends on the learning of others. Nonetheless some solid progress has been made, but much of it requires fairly advanced stochastic process theory. Evolutionary dynamics is somewhat more straight-forward once one has decided what constitutes 'fitness' in an economic context. For instance, one might try to model a process in which inefficient firms decay and efficient ones prosper. But 'efficient' here is not self-defining. Typically the process ('replicator dynamics') increases the fraction of the fit in response to 'success'. In game theory this kind of analysis has been used to pick out a particular equilibrium of a game and the analysis has been interpreted as showing the emergence of social conventions.

What is of interest here is that in general the convergence of a process to a particular outcome depends on initial conditions and, in that sense, on history. Ever since the middle of the nineteenth century, economists have worked under the influence of (classical) physics. The equilibrium of a frictionless pendulum, for instance, does not depend on its initial position. Thus economic theory became 'history free' although when pressed, most economists would have agreed that preferences or even technology at any time have their roots in the past. But they were taken as exogenous for purposes of analysis, and taken to change slowly relatively to the variables of interest. But the problem of multiple equilibria even when this route is followed, naturally and compellingly, leads to process analysis and so to a quite different reason why ‘history matters’. As this work proceeds it will become necessary for theorists to depart from their purely axiomatic foundations and to include the particular histories which are appropriate in the given context.

Some very interesting work has emerged in this area. One example concerns innovations. Suppose there are two new kinds of machines and that it is unclear (perhaps because of different expectations) which is more profitable to adopt. Adoption then is a chance event. Once a machine has been adopted more information as to its profitability is gained and the probability that this type will again be adopted (say by others) increases. There may also be increasing returns to adoption. The outcome of this adoption process thus depends on the chance ‘first’ adoption. There are now many other examples (for instance, why the keyboard layout of computers continues to be that of typewriters although this is arguably inefficient) (David, 1985; Arthur, 1989).

It is therefore not only the future but the past which needs to be explicitly brought into economic theorising. For an earlier generation of economists, say Marshall, this would not be a surprise. To the modern generation it is a not altogether welcome necessity, since it greatly reduces the reliance which can be placed on the axiomatic foundations and introduces an element (e.g. the exact form of a learning process) which in the present state of knowledge is pretty ad hoc.

There is one further remark that belongs here. If we consider a typical learning process, then agents can only learn from their observations which are here of course generated by their own actions. Thus nothing will be learned about outcomes along other paths which might have been taken (say because of different initial conditions). This in turn means that when the learning paths converge, the convergence outcome is one where agents do not know whether there would be more profitable actions available to them. If we call that outcome an equilibrium, as it is natural to do, then it need not be the equilibrium which I have described earlier in which agents have done best over the set of all actions open to them; here they have done best relatively to the beliefs which learning has induced. (There is now the possibility of ‘active’ learning, i.e. of experimentation, but a formal analysis of optimal experimentation leaves the conclusion intact.)

10.7. Growth

It is possible to describe steady state equilibria, and to prove their existence in suitable cases. If not stochastic, these are equilibria in which relative quantities of goods produced and used in production remain constant. These are pseudo-stationary state in the sense that if all outputs grow at the same exponential rate

g, the transformed variable obtained by multiplication by e^{-gt} is constant. The most celebrated study of such an economy is by Von Neumann, but he has had many successors. In a linear model Von Neuman showed that there was a maximal growth rate which would equal the rate of profit and that there would be associated prices with this path which were equilibrium prices.

It is not possible to give a proper account of what is a large subject, here. It must suffice to note that theories of steady states were motivated by the attempt to pinpoint the main determinants of an economy's growth. Clearly the amount of capital per man (see the Ramsey discussion) would play a role. But so would technical progress by which one understood increasing productivity of inputs. Solow's empirical studies suggested that it was the latter rather than the former which was more significant in determining growth. (However the empirical finding used a procedure not free from criticism.) By and large the rate of technical progress (and indeed also the rate of population change) was taken as exogenous – that is, beyond the grasp of economic theory. Recently this has changed and there are dozens of models in which this rate is endogenised by theories of R and D decisions, theories of education and training decisions and theories of the accumulation of knowledge. In almost all of these, either there is a return to the Ramsey approach (where a representative infinitely lived agent optimally decides on his accumulation of human capital (training) or on the optimum amount of R and D) or they are pretty ad hoc constructions with special functional forms designed to allow steady state growth.

One of the more interesting outcomes of this work has been the recognition that there may be multiple steady state equilibria and growth rates, so that it has become possible to recognise the plain fact that different economies grow at different rates. This of course, for the reasons discussed, brought history and process analysis back into the picture. As a matter of fact this multiplicity had been recognised much earlier by economists interested in economic demography. It was noted that the relation of population growth to real incomes was non-linear. To get high growth rates, capital had to be accumulated faster than population growth at some stage to get out of a low income low growth rate trap. Critical 'take-off' points were identified and economies classified by which side of this point they were.

All of these models provide some insight but they are not economic history. Not only do they neglect many elements historians would regard as important (e.g. institutions, religion, etc.), but they are, as a rule, highly aggregated and in any case preoccupied with steady states. The difficulty is that theories involving short periods of time may arguably neglect changes which over long periods may become central to understanding. Economists are no better equipped than others to provide more than suggestive speculation concerning these ambient social processes. Being used to a precise formulation of theories, they have thus simply neglected the 'non-economic'. This strikes me as satisfactory provided the neglect is recognised. Alas, that is not always the case.

10.8. Policy and welfare

I have argued that a-temporal and inter-temporal allocation of resources are to be distinguished mainly on the grounds that the latter are governed by (not necessarily correct) expectations and suffer almost certainly from a lack of market opportunities for coordination. To this must be added that some important intertemporal allocation problems involve the unborn. There are thus many opportunities for 'market-failure' which simply means that the market economy does not allocate intertemporarily, efficiently.

Now it so happens that economics has made valuable contributions to the discussion of such market failure and its possible cure. Indeed as far as I know they are alone in being able to formulate the problem with sufficient clarity and precision to allow it to be discussed.

I have already briefly noted how 'externalities' may lead to both a-temporal and intertemporal market failure. Take fishing as an example. The fish caught by any one boat increases the cost of fishing (after a point) for all other boats. This cost is not taken into account by the individual fisherman. He compares marginal *private* costs with marginal benefits while, for the economy, the externality he gives rise to should be added to private cost. This problem has straightforward cures. But there may be an additional problem if there is a danger of exhaustion or great reduction in the stock of fish. Then one encounters also an intertemporal externality. Note, even if each fisherman knows that the stock is being exhausted there is no incentive to reduce the catch (unless everyone does so). If an economist is given the appropriate biological data the pretty exact cure can be proposed. For instance, one introduces licenses for quantities of fish. These licenses should be tradeable (so that efficient boats can drive out inefficient ones). There are other possibilities – a tax on fish. The disadvantage here is an increase in the policy-maker's uncertainty (the exact reaction to taxes cannot be known) and a greater danger of exhaustion.

The above is a textbook example of how to deal with market failure in the case of renewable resources. A similar exercise is possible to deal with the problem of exhaustible resources if there is market failure in determining their rate of use. In all of these cases one first looks at the special case where authorities have full control and knowledge or equivalently of how allocation would proceed if there were no market failure. In general this yields 'shadow prices'. These are prices which fully reflect social opportunity cost. For instance, the price of a barrel of oil should fully reflect the reduction in maximal well being (in practice utility) of society of consuming it. In spite of difficulties and shortcomings of this approach which I shall now briefly discuss, the method is a very powerful one in orderly policy thinking.

(a) Evidently no policy maker knows the agents' utility functions although something can be deduced concerning them – for instance from the saving ratio and interest rates. Even so, there will often be clear indications of market failure. For instance one can show that intertemporal efficiency requires the real price of oil to rise at the rate of the interest rate. If it does not there may be considerations, e.g. market failure elsewhere, which could be used to argue against the conclusion of inefficiency. But the onus of proof is on those who argue for this escape route.
(b) The shadow price of oil at any moment depends on the stock of oil and on the probability of finding substitutes at any date. This requires uncertain technological information and its translation into beliefs. A policy maker's belief may or may not be better than that of market agents. But once again clarity has been brought into the discussion and shadow prices contingent on different beliefs can be calculated and discussed.
(c) The most troubling problem is what is called that of the 'second-best'. It is likely that there are lots of market failures in the economy and that they interact. For instance, a policy on oil will have consequences for transport which itself may have prices which deviate from their shadow values. To take account of all of these other 'distortions' when considering oil prices is extremely hard in theory and almost impossible in practice. But suppose that the 'pure' exercise leads one to conclude that oil prices should be higher. If that is opposed on the grounds that 'second-best' considerations have been neglected, one can see what it is the opponent has to argue and convince one of. This is all pure gain.

There are many other difficulties – for instance how to value future generations. But here too economists have succeeded in giving precision to the question and have, in my view, made serious discussion possible.

This whole area teaches one another valuable lesson. Exercises of the kind which I have described are often criticised as leaving out of account various matters taken to be relevant. This then is regarded as sufficient for dismissing them as 'abstract' and so not useful. Arguments of this sort cannot be taken seriously until one is shown how the neglected factors would, or at least could, affect the 'abstract' conclusions. Normally such arguments are not forthcoming. My own impression is that much that economists have done in the field of intertemporal market failure is pretty robust to doses of 'realism'.

References

Arrow, K.J., 1953, 'The Role of Securities in the Optimal Allocation of Risk-Bearing', in French: Econometrie: Proceedings of the Colloque sur les Fondements et Applications de la Theorie du Risque en Econometrie, Centre National de la Recherche Scientifique, Paris. English Translation: Review of Economic Studies (1964) **31**, 91–96.

Arthur, W.B., 1989, 'Competing Technologies, Increasing Returns, and Lock-in by Historical Events', Economic Journal, **90**(394), 116–131.

David, P., 1985, 'Clio and the Economics of QWERTY', American Economic Review Proceedings, **75**, 332–7.

De Finetti, B., 1937, 'La Prevision: ses lois Logiques, ses Sources Subjectives', Annales de l'Institut Henri Poincare, **7**, 1–68.

Ramsey, F.P., 1926, 'Truth and Probability'. In: Foundations of Mathematics and other Logical Essays, Kegan-Paul, 1931.

Ramsey, F.P., 1928, 'A Mathematical Theory of Saving', Economic Journal 38, December, 543–59.

Savage, L.J., 1954, The Foundations of Statistics, John Wiley, 1954.

Von Neumann, J. and Morgenstern, O, 1947, Theory of Games and Economic Behaviour, 1947, (2nd edition).

chapter 11

SOCIAL THEORY, COMPLEXITY AND TIME

PATRICK BAERT
University of Cambridge

Time in Contemporary Intellectual Thought
Patrick Baert (Editor)

CONTENTS

11.1. Introduction

This chapter discusses the issue of time in social theory in a manner accessible to non-sociologists. After distinguishing between various components of time (section 1), I elaborate on how they have occupied a central role in the history of social theory (section 2). Sociologists and social theorists may want to skip this historical section as they are likely to be acquainted with the issues raised. I subsequently outline my own view on how time may play a pivotal part in the restructuring of social theory (section 3).

11.1.1. Time in social theory versus sociology of time

Let me clarify what is meant by the terms in the title of this chapter. Elsewhere I have defined social theory as an abstract, general and systematic account of the workings of the social world (Baert, 1998). By describing social theory as 'abstract', I mean that it is distinct from the empirical study of society. Social theories might, of course, have some bearing on empirical research, but their main objective is to generate abstract propositions. By 'general' I meant that social theories purport to make statements applicable to a large number of social settings and to different societies. For example, the American social theorist Talcott Parsons intended to explain how consensus and predictability is achieved in any society – not just in the United States. By describing social theories as 'systematic' endeavours, a distinction is drawn between social theories and opinions or beliefs. Social theories are systematic in that they exhibit coherence and internal consistency. Even recent post-modernist attempts to undermine grand theory building are systematic in that sense.

There are two ways in which temporality enters the study of the social realm (see also Martins, 1974; Baert, 1992, pp. 4–22). First, there is the so-called sociology of time. It investigates the various temporal features of social life, ranging from class-related variations in perceptions of time, allocation of time in the household and at work, and socially induced rhythms of social life, to religious and secularised calendars (see for example Adam, 1990). Though often theoretically informed, the sociology of time is, like the sociology of the family or the sociology of work, primarily an empirical discipline. Second, temporal notions might play a role in theoretical debates and in theory formation. For example, structuralist inclined theorists like Claude Lévi-Strauss attributed epistemological priority to synchronic over diachronic analysis, and by doing so distanced themselves from the evolutionary framework of late nineteenth century social theory (Lévi-Strauss, 1993, pp. 1–27; 1994, pp. 312–362). To give another example, by advocating discontinuity as a methodological tool, Foucault's archaeological-genealogical history was directed against the 'traditional' search for continuity between past phenomena (Foucault, 1977b, pp. 140–164; 1989a, preface).

As social theory is taken to be distinct from empirical sociology, the sociology of time will not be addressed directly in what follows. I will instead focus on the extent to which temporal notions have influenced theoretical discussions and theory development. There are nevertheless two ways in which the empirical sociology of time overlaps with the topic covered in this paper. First, the empirical sociology of time arose partly out of theoretical concerns. In particular, Emile Durkheim's attempts to show the social nature of the Kantian elementary categories of thought (time, space and number) had a decisive impact on the empirical study of temporality (see Durkheim, 1915, 1963). Second, some scholars today argue that in order for social theory to become more sensitive to temporal issues, it needs to learn from the empirical research carried out in the area of the sociology of time. For instance, some claim that the 'temporalisation' of social theory can only be achieved by investigating how time is socially constituted (e.g. Elchardus, 1988). This depends on a questionable a priori Durkheimian position, and is not considered in what follows.

11.1.2. Five aspects of time

Before moving on to the history of social theory and time, it is worth outlining five ways in which temporality might enter theory formation, although this is not intended to be an exhaustive list.

There is, firstly, the well-known distinction between synchrony and diachrony. To study a social phenomenon synchronically is to 'bracket out' the lapse of time, and to take a snapshot of the social realm. In contrast, a diachronic study encompasses a certain period of time. For example, structuralist-inclined social researchers tended to attribute epistemological and methodological priority to synchronic over diachronic analysis. According to structuralist orthodoxy, the meaning of a social item or cultural artefact can be derived from its difference from other items or artefacts *currently* in use. Hence the lapse of time was considered to be irrelevant for uncovering meaning. (Saussure, 1959; Lévi-Strauss, 1993, 1994) In contrast, an example of diachronic analysis is Parsons' paradigm for evolutionary change. It draws upon analogies with biological evolutionism and cybernetics to reveal the mechanisms by which change occurs *over time* (Parsons, 1966, 1977). To give another example, Giddens' theory of structuration accounts for how social structure is continuously reproduced *across* time by competent individuals (Giddens, 1984; Cohen, 1989).

A second temporal issue concerns the precise status of the future. In particular, it is possible to treat the future as if closed or relatively open. A closed future is one which is already given, and which, if sufficient knowledge is obtained, might be predictable. For example, take a nineteenth century historicist view according to which the ineluctable laws of history overshadow human intervention, and according to which insight into these 'laws of motion' might allow for accurate foresight (Comte, 1969; Marx, 1990, pp. 221–247; see also Popper, 1986). In contrast, consider the future as relatively open; that is, not simply a *fait accompli*. The future is then to some extent 'established' or 'accomplished' in every present, and *a fortiori* unpredictable. Examples of social theorists for whom the future is relatively open are G.H. Mead and Niklas Luhmann. In spite of the disparate traditions to which they belong, both attempt to advance a general social theory which takes 'novelty' and 'emergence' as its methodological tools (Mead, 1959; Luhmann, 1976, 1979, 1995, pp. 278–356).

A third temporal dimension points out the relationship between the temporal flux and the invariable. Some notions in social theory refer to change or the temporal flux. Key concepts are social dynamics, social conflict, discontinuity, rupture, social transformation, and so on. Other concepts allude to the relatively unchanging features of social life. Examples of these are social statics, social order, social structure, social patterns and institutions. Theories differ in how much relative importance they attach to the static versus the dynamic. Some theorists assign an exclusive role to the relatively unchanging structures underlying human interaction. For example, Braudel's structuralist historical research discarded a mere history of events or 'science of contingency'. Instead, he explored the *longue durée*; that is, the relatively stable structures which affect and constrain people's actions (Braudel, 1966, preface; 1980). Other theorists underscore the temporal flux. For example, Herbert Blumer's symbolic interactionism dismissed the Durkheimian view of a social structure externally imposed upon people. Instead, he examined the interpretative procedures through which people make sense of and contribute to the making of the social. Even institutions or social patterns can be modified since they are contingent upon people's designation and interpretation (Blumer, 1969, esp. pp. 70–77).

A fourth dimension draws upon a distinction between several temporal levels. By temporal level I am referring to the type of duration that is the subject of sociological analysis. In this context scholars traditionally distinguish between historical time, institutional time, and *durée* of everyday interaction (see also Giddens, 1981, 19 ff.). 'Historical time' or '*longue durée*' refer to patterns of society over a prolonged period of time. The '*durée* of daily life' specifies remarkably short temporal spans that are linked to the routines of everyday interaction. Between these is the so-called 'institutional time', alluding to the persistence of specific properties at institutional level. I think these concepts are a trifle ambiguous, but they suffice to distinguish between social theories. Theories indeed differ in that some deal with historical or institutional time, others with the *durée* of daily life, and others again with the intersection between any of the three. For example, Marx's stages of historical development stretch over notably long periods of time. In contrast, Herbert Garfinkel's ethnomethodology investigates the skills and aptitudes involved in daily life, and different again is Anthony Giddens' attempt to explore the connection between the various temporal levels (Garfinkel, 1967; Giddens, 1984, pp. 34–37).

A fifth dimension alludes to a broader methodological issue. Whilst the social world is necessarily investigated from a present perspective, that which is being investigated has in some obvious way already happened. The object of empirical research might or might not have occurred a long time ago, it might even be part of a larger process which is still in operation, but that object is part of the past nevertheless. There are various ways in which theorists or practising historians have dealt with that discrepancy between past and present. Many believe that social scientists ought to make an effort to relive or rebuild the past, for example by capturing the context in which social practices took place or by reinvoking the anxieties, aspirations and fears underlying the practices. For example, Max Weber argues that one has to bridge the gap between past and present through *Verstehen* or interpretative understanding which allows the researcher to grasp the hopes, predilections and apprehensions of the people in the past (Weber, 1949). Others try to deny the discrepancy between their present perspective and the object of research by appealing to mechanisms which are allegedly independent of time, place and context. For example, nineteenth century positivist inclined sociologists, like Auguste Comte, referred to 'wheels of history' that are, so they maintain, unaffected by the various meanings attributed by people to the social world (Comte, 1969). Yet others, like Nietzsche or (more ambiguously) Foucault, embraced some form of 'perspectivism'. For them, it is impossible to erase the temporal locus of the social researcher: one can only present a narrative amongst many, and there are no clear-cut rules to judge between them (Foucault, 1977b).

11.2. History

Although I will not explicitly refer to the above list in what follows, it will serve as a frame of reference nevertheless. With these five components in mind, I am now able to explore the history of social theory in relationship to temporality. For the sake of clarity and brevity, the discussion will be confined to the nineteenth and twentieth century. Of course, significant developments in social theory emerged before this time. But I will mention earlier forms of social theory only in so far as they clarify my discussion.

11.2.1. Nineteenth century models; theory through history

Three types of theory formation in the nineteenth century may be distinguished. What follows is clearly not an exhaustive list, nor are these types, in practice, always viewed as mutually exclusive. There is a logic to the selection in that the three types of theory formation are central to the history of the discipline: each type is linked to the work of one or several major thinkers, and each is regarded as a 'theoretical tradition' or canon by most contemporary sociologists. All have in common that they study processes over long periods of time, and *mutatis mutandis* stress the importance of historical research.

The first type is what I would call 'pre-Darwinian unilinear evolutionism'. I associate this doctrine with for instance Saint-Simon or Auguste Comte's *System of Positive Philosophy*, and to some aspects of Karl Marx's *Communist Manifesto*. The intellectual roots of this type are twofold: one is non-naturalist and goes back to the Hegelian dialectical method (as in the case of the early Marx), the other to the positivist pursuit of historical lawlike generalisations (as with Comte's study of 'social dynamics'). Whether naturalist or not, these unilinear evolutionists have in common that they see every society as inevitably going through essentially the same stages or phases. All societies follow roughly the same path throughout history (some of course faster than others), but the path is one of progress: however brutal, ignorant or unjust earlier stages, the course of history eventually leads towards a desirable end. That path tends to be irreversible; once a certain stage has elapsed, it is difficult, undesirable and evidently unwise to resurrect that phase. The task of the social scientists is to uncover the path, partly because it allows people to facilitate or accelerate that which is inevitable. The relationship between present and anticipated future is a complex one. On the one hand, the present is at the brink of the desirable future; people are close to the completion or end of history. On the other, the prophesied final stage of society will be radically different from the present.

The second type is 'Darwinian evolutionism'. It is exemplified in Herbert Spencer's *Social Dynamics* and Durkheim's *Division of Labour*. These Darwinian evolutionists resemble the previous type, but in addition they draw heavily upon analogies with and metaphors of biological evolution. They assume that there is a direction to history in that every society becomes gradually more complex. In Spencer's or Durkheim's parlance: as time goes by, societies tend to become more differentiated, and this makes for a superior adjustment to their physical environment. One basic system might previously have fulfilled various functions, but that system has now been broken up into units, each of which specialises in one or a limited number of functions. Division of labour is the example *par excellence* of structural differentiation in an advanced society. That movement towards increasing complexity is accompanied by a momentary crisis. The different parts of the system need to be readjusted to a state of intensified complexity, and social scientific forms of knowledge are essential for the accomplishment of this readjustment. In short, as with pre-Darwinian unilinear evolutionists, Durkheim, Spencer and others assume that human history moves in a particular direction, that this path is common to all cultures and virtually irreversible. But added to these are the Darwinian notions of differentiation, adjustment and selection. Furthermore, these Darwinians perceive less discontinuity between the present and future. If anything, the radical shift in society (the move towards differentiation) occurred some time ago. In the mind of Durkheim and others, there does not seem to be much scope for the vision of a radically different, Utopian future. In a similar vein, the notion of progress, if present at all, becomes qualified. The path of history involves a distinct *possibility* for progress – not its necessity.

The third type of nineteenth century theorising is 'methodological individualism'. It is associated with the work of Max Weber (exemplified in his *Protestant Ethic*) and to some extent Alexis de Tocqueville (in particular, his *Democracy in America*). These authors distance themselves from both pre-Darwinian and Darwinian evolutionary conceptions of history. For Weber and fellow methodological individualists, it is intellectually inappropriate or even foolish to search for meta-narratives or all-transcending laws of history. It is simply not the case that the study of the past will enlighten one with incontestable knowledge regarding the precise future state of society. Furthermore, the magic of history lies in its unpredictability for all the people concerned, and those who study history ought not to depict it otherwise. So it is futile to be looking for those grand schemes of history that transcend human conduct and that necessarily unfold themselves through time. The starting point of sociological analysis should be to grasp people's motives and purposes (something which is achieved through *Verstehen* or 'interpretative understanding') since they are the causes of their actions. In turn the actions lead to outcomes, some of which are intended, some unintended and unforeseen. For sociology especially, the unanticipated effects of purposive action are of crucial importance. The task of sociology is to reconstruct the causal logic that links the goals and beliefs behind the actions, the actions themselves and their effects.

11.2.2. Early functionalism and structuralism; synchronic pursuit of the atemporal

At the beginning of the twentieth century, functionalism and structuralism were advanced as theoretical alternatives to both the evolutionary approach and methodological individualism in the study of society. The writings of a small number of social anthropologists became decisive in the initial development of both doctrines: whereas Malinowski and Radcliffe-Brown established a functionalist paradigm to assist their field work, Lévi-Strauss sought the help of structuralist linguistics (Malinowski, 1944; Radcliffe-Brown, 1952, 1958; Lévi-Straus 1993, 1994). Early functionalism set out to account for the persistence of social practices by demonstrating how they contribute, often unintentionally, to the cohesion or solidarity of the system in which they are embedded. Structuralism attempted to reveal hitherto hidden structures beneath the surface level of observed phenomena. Various structures can be identified, ranging from mental or cultural structures to modes of production.

Early functionalism and structuralism differed substantially from their nineteenth century predecessors, especially with regard to issues involving temporality. I will mention three main differences all involving some aspects of time. First, with some notable exceptions, early functionalists and structuralists were less pre-occupied with identifying transition points or stages, or with accounting for social transformations. Instead they became captivated with the question of how social order is brought about and how society maintains itself. Relatedly, they attributed epistemological or ontological priority to the deeper levels of reality that are presumed to hide beneath the temporal flux of immediately accessible data. Indeed, many searched for durable structures which are not readily available; these underlying structures are supposed to affect or attribute meaning to the surface level of observed phenomena. Second, rather than making broad conjectures regarding historical development, they opted for in-depth synchronic analyses. They considered sociological 'laws of history' to be based on highly unreliable data and thus 'pseudo-science'. Add to this their holistic perspective according to which the meaning of a cultural artefact or practice can be inferred from its inter-relationship with other items currently in use within that culture. It follows from this holistic stance that, in order to grasp the present meaning of an item, it is no longer necessary to trace back its lineage or geneaology. Third, none of them embraced the notion of inevitable progress which tended to underpin various writings of nineteenth century evolutionists. Instead, they were notably cautious, anxious to avoid the ethnocentricism and empirical inaccuracies that had guided the triumphalism of earlier days. As a matter of fact, structuralist writings tended to be highly critical of our present Western condition, and, especially in the case of Lévi-Stauss, articulated nostalgia for a lost spiritual sphere.

Although structuralism led many social researchers to attribute methodological primacy to synchronic analysis, not all historians remained indifferent or hostile to the new creed. Some historians, like the members of the *Annales* School, converted whole-heartedly. The *Annales* comprised a number of young scholars, mainly led by Fernand Braudel, who repudiated the *histoire Sorbonniste* which was the dominant way of history of the day (Braudel, 1980; Burke, 1990). This orthodoxy, so Braudel and others argued, relied upon a positivist epistemology, presenting a mere sequence of events, or a narrative regarding leading personalities who had moulded the past. Instead, the *Annales* school contended that history should take on board sociological insights. In his later writings Durkheim had shown that beyond the temporal flux of events and actions lies a deeper, more significant structural realm, and Braudel and his followers imported this structuralist methodology into history. Rather than remaining at the illusory and fast moving level of immediately accessible data, members of the Annales school tried to reveal latent, relatively durable, structures. Or in Braudel's terminology, as opposed to the *'courte durée'* of the *'histoire événementielle'*, they searched for the *'longue durée'* (Braudel, 1966, preface). In contrast with the frenzied, erratic nature of the immediately observable level, the *longue durée* refers to the more permanent features of social life: for instance, mental structures, and geographical and climatological constraints.

11.2.3. Normative integration and ethnomethodology; accounting for order

The banner of structural-functionalism was carried forward by the American Talcott Parsons. But it would be a mistake to conceive of Parsons' work as a unified entity; there are distinct stages to his intellectual development. I will now deal with the early Parsons, as exemplified in his *Theory of Social Action*; a discussion about other aspects of Parsons is postponed to a later stage. The early Parsons is especially relevant. His main question has a substantial bearing on the problem of time, and has influenced a significant part of American sociology, including ethnomethodology.

Like Malinowski and Radcliffe-Brown, the early Parsons showed little interest in social change or conflict. Inspired by Hobbes and Durkheim, Parsons' main concern is how social order is brought about (Parsons, 1949). How can one account for social order given that people pursue their own goals? Parsons goes through a variety of classical authors ranging from Weber to Vilfredo Pareto, but he basically provides a Durkheimian answer. In an attempt to avoid the pitfalls of idealism, positivism and utilitarianism, Durkheim referred to the internalisation of shared central values and norms of society within the need-dispositions of the personality structure. People generally do not adopt an instrumentalist attitude towards internalised goals. By pursuing their own goal, each individual unintentionally contributes to fulfilling the needs of a larger whole to which they belong. This 'normative integration model' dominated American sociology for some time, and it had a number of repercussions on the nature of empirical social research carried out. In line with the system theoretical notions of the day, the focus of attention was on how, faced with external and internal interferences, social systems manage to absorb those disruptions and thus maintain themselves. The answer was thought to reside in the complex functional relationship between various sub-systems, in particular the personality system and the social system. It would be unfair to say, as some commentators have done, that this functionalist frame of reference *necessarily* brackets out the lapse of time. But the social scientists at the time did assume that the relationships, balances and exchanges between subsystems *can* be studied synchronically, and they often proceeded on that basis.

Even those who went against the grain put forward questions remarkably similar to the normative integration model. One prime example is Harold Garfinkel's ethnomethodology (Garfinkel, 1967; see also Heritage, 1984). Once a pupil of Parsons, Garfinkel has often been regarded as the *enfant terrible* of American sociology in that he put forward a radically different research programme from the dominant one in the States. But a closer look shows a remarkable similarity between Garfinkel's proposal and Parsons' orthodoxy. Like Parsons, one of Garfinkel's central questions is how social order is brought about, and his research programme is primarily directed towards that issue. Nevertheless, Garfinkel's account of social order differs from Parsons and other functionalists in a number of ways. First, Garfinkel uses a narrow definition of social order: he mainly refers to the production and reinforcement of shared meaning. In contrast, Parsons' notion of order was a broad one: it also entails, for instance, a relative consensus regarding the allocation of scare goods and resources. Second, Garfinkel's answer to the problem of order is different. Compared to Parsons, socialisation and internalisation do not any longer occupy a pivotal role in Garfinkel's account of social order. Order needs to be seen as a skilful unintended accomplishment by individuals who continuously display and apply practical knowledge. People operate in 'real time': whilst looking back upon the past and anticipating the future, they unwittingly rely upon and contribute to the predictability of everyday life. Third, compared to functionalist inspired investigations, etnomethodological research into order unequivocally calls attention to processes over time. That is, shared meaning and the predictability of everyday life are understood as a practical achievement across time, and therefore encourage, if not necessitate, diachronic analysis. Finally, in contrast with functionalist attention to the mechanisms in operation at a systemic level, ethomethodologists investigate the intricacies at the level of the *durée* of daily interaction. Whereas Parsons attempts to fathom social order as a functional balancing act between the social and the personality system, order is now seen as accomplished in our everyday routine practices.

11.2.4. Pragmatism and symbolic interactionism; the philosophy of the present

Meanwhile, a very different theoretical tradition emerged in the States. At the beginning of the century, William James and John Dewey's pragmatism dominated American philosophy. This doctrine had a decisive impact on the emergence and growth of symbolic interactionism as a separate sociological school from the 1930s onwards. The Chicago-based philosopher George Herbert Mead formed the link between the two strands of thought. Together with Dewey he became one of the leading exponents of pragmatic philosophy of his time. He also advanced a number of views on social psychology, which were taken up by Herbert Blumer, one of his pupils. Blumer founded the school of symbolic interactionism, which was largely based on Mead's thoughts on the relationship between the individual and society.

Mead is especially known for his posthumous *Mind, Self and Society*, based on his lectures in social psychology (Mead, 1970). His starting point is that human beings have a self in that they are able to be both object and subject. That is, it is possible for individuals to look at themselves from an outsider's perspective. It is important to recognise the social nature of the self: by putting themselves into the perspective of *other* individuals, people are able to reflect upon alternative ways of expressing themselves, and eventually capable of choosing one of them and acting. The self is intrinsically linked to the 'generalised other': precisely because people are able to take up the attitude of the whole community or generalised other can they anticipate the meaning of their various gestures to others. The self implies shared symbols: an individual is able to put him or herself in the perspective of the generalised other because he or she and other members of the same community tend to attribute the same meaning to gestures.

Mead's later reflections on time are virtually unknown. However, they are crucial to the understanding of the temporal features of symbolic interactionism. Most of his ideas on time can be found in his *Philosophy of the Present* and *Philosophy of the Act* (Mead, 1959, 1972). Especially in the first book he put forward a general theory of time, partly influenced by Henri Bergson and Alfred Whitehead. His philosophy operates at various levels, ranging from the cosmological to the societal and individual realm, the latter two of which will receive attention in what follows. There are two ways in which Mead's proposal is a philosophy of the present: it is meant to have contemporary relevance in that it synthesises the process philosophies of the day, and it attributes special status to the present.

The *Philosophy of the Present* opens with the enigmatic statement that the 'locus of reality' lies in the present. This proposition has a dual meaning. First, the past and future are, in general, represented in the present. That is, people are able, in the present, to anticipate the effects of imaginary *future* lines of conduct by drawing upon *past* experience regarding shared meaning. Second, the meaning of past and future can be altered in the present. (see also Mead, 1929). There might be some novelty to the present which throws new light on what has happened and what is likely to happen. Influenced by evolutionary theory and opposing a Laplacean view of causality, Mead regards the novel or the emergent as unpredictable: even a complete account of what preceded the event would not allow it to be predicted. There are two ways in which emergence or novelty appear in the social realm. First, people regularly encounter unexpected or unanticipated events that possibly make them reflect differently upon the past and the future.

Second, although people necessarily draw upon shared rules and assumptions whilst considering alternative lines of conduct, their final decision or choice remains unpredictable.

None of the symbolic interactionists referred at length to Mead's work on time. Moreover, whilst several secondary sources on symbolic interactionism expound on its indebtedness to Mead's social psychology, they fail to mention its connection to Mead's thoughts on time. However, the symbolic interactionist research programme does not exclusively borrow from Mead's theory about society; it is also perfectly in line with his philosophy of the present. Reminiscent of Mead's process philosophy, Blumer and his followers are highly critical of theories that conceive of external structures simply imposed upon people (Blumer, 1969). There are no pre-existing and self-perpetuating forms of social life that are independent of people's interpretative procedures. Even the most repetitive of practices has continuously to be formed anew. In other words, this kind of social reproduction is contingent on the skilful accomplishment of individuals, their interpretations and designations. Likewise, there are no universal, intrinsic meanings to objects, neither does meaning reside in a person's mind. The meaning of an object for an individual derives from his or her tendency to act towards that object. So meanings are not fixed; they can change at any time.

11.2.5. Structuration theory and genetic structuralism; diachronic analysis of order

In the course of the 1970s several authors, among them Anthony Giddens, Pierre Bourdieu and Roy Bhaskar, saw it as their main task to bridge the gap between apparently rival perspectives: between structuralist and hermeneutic approaches, between so-called structure-orientated views and actor-orientated views, and so on. I think it is fair to treat Giddens structuration theory, Bourdieu's genetic structuralism and Bhaskar's critical realism as kindred spirits (Bourdieu, 1977; Bhaskar, 1979, Giddens, 1984). I call this way of thinking about social theory the 'new orthodoxy' - as opposed to what Giddens terms 'the old orthodoxy' of structural-functionalist social theory and positivist epistemology. The label 'orthodoxy' alludes to the fact that Giddens, Bourdieu and Bhaskar are not isolated thinkers. Indeed, they they have a significant number of followers.

The new orthodoxy differs from the old one in a number of ways. First, as mentioned earlier, Giddens and Bourdieu wish to transcend what they see as polarities between structure- and action-orientated approaches. Structure-orientated or objectivist approaches explore the extent to which unacknowledged structures constrain and determine people's actions and thoughts. Action-orientated or subjectivist approaches investigate the various ways in which people exercise agency, attribute meaning to their environment and act accordingly. Both theoretical angles have something to offer and deserve attention, but none is sufficient on its own. Second, people's ability to exercise agency ties in with tacit and practical knowledge. In daily interaction people continually apply sophisticated skills. Although individuals are not necessarily aware of this knowledge, they continuously exhibit it in their actions. Take the example of language. People know how to speak their language in that they know practically how to speak in accordance with the rules of grammar. They do not have to know their native language in any explicit manner in order to speak properly. Third, structure and action are mutually dependent in that structure is both medium and unintended output of social practices. That is, structure is the precondition for the emergence of agency, whilst dependent on the very same agency for its reproduction. For example, as they speak, people necessarily draw upon the syntactical rules of the English language, and these utterances help to reproduce the very same structural properties. Fourth, people are able to develop knowledge about their social surroundings, but that knowledge is nevertheless limited. The limitations are twofold: there are unacknowledged structures and unforeseen consequences. On the one hand, people are not necessarily aware of the structural conditions which affect or make possible their actions. On the other hand, people are not aware of all the possible effects that their actions might entail.

The new orthodoxy makes for a radical transformation with regard to temporality. First, the orthodox consensus is highly suspicious of some of the reifying tendencies in structural-functionalist reasoning. In particular, Parsons, Niklas Luhmann and other system theorists occasionally attribute to systems an inbuilt tendency for re-adjustment and equilibrium. Giddens, Bourdieu and others wish to distance themselves from those metaphors that suggest an inherent ability in social systems to maintain or restore stability. Second, the adherents of the new orthodoxy feel uncomfortable with several characterisics of the evolutionary paradigm. They are particularly hostile to what they perceive to be the historicist tendencies of evolutionary thinking: people constantly incorporate knowledge as they go along, so social life is inherently unpredictable and does not allow for lawlike generalisations. In a related way, they are highly distrustful of the applicability of the notion of adaptation to the social realm, and they tend to dispute the view that there is a structural homology between societal and individual development. Third, the advocates of the new orthodoxy conceive of social order as a skillful accomplishment across time. To study order is, therefore, to take a diachronic analysis of society. In structural-functionalist writings, time and change tend to be equated; they fail to recognise that the lapse of time also brings about order. The new protagonists assert that 'nothing will come of nothing' in that even the most radical transformations can only be brought about through the reproduction of prior structures. Fourth, Giddens and others intend to show the relationship between the various temporal levels. Some theories explore long temporal spans; other theories examine shorter durations. The new orthodoxy wants to go beyond these dichotomies: it aims at disclosing the extent to which the reproduction of structures at the level of the longue durée is contingent upon people's routine practices at the level of everyday life.

11.2.6. Post-structuralism; the history of the present

Post-structuralism is a general term covering the work of otherwise very different authors such as Roland Barthes, Jacques Derrida, Jacques Lacan and Michel Foucault. Most post-structuralists have been highly influential in literary theory, with the exception of Michel Foucault whose impact has been most significant in the area of social theory. In what follows I will explore Foucault's methodology of history which radically challenges how we think about the relationship between historical writings, the present and the past.

Foucault's notion of knowledge acquisition is quite distinctive. Knowledge in the social realm is generally seen as an attempt to gain access to the unfamiliar by drawing upon analogies with the familiar. Instead, Foucault tries to gain access to and undercut the familiar present with the help of the unfamiliar past. This explains why he calls his own history a 'history of the present' (Foucault, 1990, p. 36). That is, the main aim of his historical writings is to uncover the taken-for-granted present. In this respect, Foucault reacts against a commonplace notion according to which the main objective of history as an academic discipline is to uncover the past. For Foucault, uncovering the past is not an aim in itself; the past is a medium for access to the present.

There are, roughly speaking, two periods in Foucault's writings: his archaeological period, and subsequently his genealogy. I will deal with each in turn, and show how in both cases a history of the present is developed. His archaeological period, partly emanates from his work in the 1960s, partly emanates from structuralist historical research, especially the *Annales* School (e.g. Foucault, 1989a, 1989b). Remember that, according to the *Annales*, proper history should attempt to uncover the *longue durée*; that is, the unacknowledged, stable structures which stretch over a long period of time. Apart from the *Annales*, Foucault has also been inspired by French history of science. In particular, Bachelard and Canguilhem's notions of discontinuity or epistemological rupture catch Foucault's attention (Bachelard, 1984; Canguilhem, 1978). It follows that Foucault portrays history as a succession of relatively stable structures, each radically different from the others. Every *longue durée* eventually breaks up, and then a very different, though equally stable structure emerges. Foucault's archaeology is a history of the present in that the hitherto latent structures of the present are made visible once juxtaposed with dissimilar past structures. Archaeology also undercuts the present in that what appears to have a fixed meaning is shown to be variable through time.

In the course of the 1970s Foucault becomes increasingly indebted to Nietzsche, especially his anti-essentialism, his notion of power and his method of genealogical history (e.g. Foucault, 1977a, 1977b, pp. 140–164; 1980). Like Foucault's archaeology, his genealogy endeavours to demonstrate shifts in meaning. But unlike archaeology, it tries to demonstrate how these shifts are due to contingency and power struggles. As far as the history of the present is concerned, the method of genealogy resembles archaeology: the study of the past enables the latent present to be made manifest, and what appears invariable is shown to be changing. But there are two additional ways in which genealogy aims at elucidating and undercutting the present. First, genealogy reveals that what appears to be innocuous is not. Whilst moral and belief systems are often presented as innocuous or honourable, genealogy shows that their emergence in the past was tainted by power struggles. Second, genealogy discloses that what is *prima facie* necessary is not. Whilst moral and belief systems are sometimes assumed to be the necessary outcome of a historical process, genealogy shows that they are often the mere product of contingency.

11.2.7. Later functionalism; systems encountering unpredictability

Whereas earlier versions of functionalism exhibited a synchronic bias, further developments secured a closer link with the study of long-term change. It is possible to distinguish two developments in latter-day functionalism: the later work of Parsons, and so-called neo-functionalism. I will first discuss Parsons whose earlier writings had been criticised for ignoring historical change. Partly rebuffing this reproach, Parsons devised a 'paradigm of evolutionary change', based on resemblances to biological evolution. Parsons introduces a number of concepts, of which 'differentiation', 'adaptive upgrading', 'inclusion' and 'value generalisation' are the most important (Parsons, 1966, 1977). With time, differentiation takes place: whereas previously little specialisation occurred, functions now become increasingly fulfilled by sub-systems. Differentiation leads to adaptive upgrading: a differentiated system is better adjusted to its environment than a non-differentiated one. Differentiation implies a need for special skills, which can only be accommodated through inclusion in the economic sphere of previously excluded groups. A differentiated society also implies the need for value generalisation: values need to be pitched at a higher level so that they direct and coordinate activities and functions in the various sub-systems. With regard to temporality and social theory, Parsons' evolutionary paradigm is not radically different from Durkheim's or Spencer's. Parsons too attributes a clear direction to history: it moves from the simple to the complex. From this, Parsons infers that it is unlikely and undesirable that societies restore a less differentiated state. As with his nineteenth century predecessors, Parsons conceives of present problems as temporary, solvable and eventually to be overcome.

So-called neo-functionalists deviate from Parsons and mainstream functionalism in a number of ways (e.g. Alexander, 1985). The most obvious difference is that neo-functionalists try to make use of rival perspectives, and incorporate them into a functionalist framework. But other differences have substantial bearing on issues of temporality. Remember that functionalists regularly alluded to an inbuilt tendency in social systems towards harmony and towards mutual (re)adjustment between its various sub-systems. Instead, neo-functionalists are particularly sensitive to the malfunctioning and the various conflicts that might emerge between the sub-systems. Simlarly, whereas functionalism was inclined to treat social integration as given, neo-functionalism recognises how problematic it is in contemporary society. Another difference is that functionalism tended to disregard the significance of the *durée* of everyday life; sociology was supposed to tackle macro-sociological questions for which micro-studies offered no answer. In contrast, neo-functionalism acknowledges the importance of shorter temporal levels for understanding broader issues like social order. In addition, functionalists conceived of social change in terms of increasing differentiation across the board: all sectors of society seemed equally affected by an all-embracing increasing complexity. Neo-functionalists realise that there is uneven differentiation: some parts of society are more affected than others. Finally, functionalists saw differentiation as a quasi-irreversible process. Neo-functionalists realise that occasionally de-differentiation takes place: systems might become less complex.

Niklas Luhmann is one of the leading proponents of the neo-functionalist cause, and I will briefly discuss those tenets of his theory that have some bearing on notions of temporality, in particular issues concerning increased complexity and contingency. Luhmann employs insights from system theory and the theory of autopoiesis. Any system, so Luhmann argues, interacts with its environment. Environments can be very complex in that there might be a large number of actual or possible events. Systems tend to reduce that complexity by selecting events, and by reducing the number of ways in which the environment is tackled. Social systems are different from other systems in that complexity is reduced through communication of meaning or, more precisely, 'double contingency'. Double contingency refers to the process by which in interaction people have to take on board the perspective of others. From this Luhmann infers that social systems are 'autopoietic': the system allows for the interpretion of the environment so that the very same system is reproduced and its autonomy enhanced (Luhmann, 1990). Maturana and Varela define autopoietic systems as producing components that eventually lead to the reproduction of the very same system (Maturana and Varela, 1980), and Luhmann argues that social systems are autopoietic in that sense. As modernity implies increased openness of the future (growing contingency and complexity), more sophisticated mechanisms are needed to allow social systems further adjustment to their environments. Self-reflexive procedures and differentiation are examples of how modern social systems accomplish this (Luhmann, 1982, 1990). Self-reflexive procedures are those that can also be applied to themselves, and they allow for continuous adjustment to an increasingly unpredictable environment. Luhmann also distinguishes between various forms of social differentiation. Some forms of differentiation allow for even better adaptation to increased complexity.

11.3. Outline of a temporalised social theory

In my view the appeal for a temporalised social theory should be part of a broader attempt to become aware of the wide range of possible objectives of social research. Implicit in a significant number of contributions to social theory or methodology is the presupposition that social research aims solely at explaining a world-out-there. For instance, it is often taken for granted that the exclusive objective of historical research is to explain past phenomena. However, other aims might underlie social research, and it is unnecessarily restrictive to reduce the wide range of possible aims to only one. It is perfectly possible to conceive instead of social research that draws upon what I call 'a self-referential concept of knowledge acquisition': that is, it aims at making making manifest our underlying cultural presuppositions. In particular, the section on Foucault shows that, in the case of historical research, the past might then become a medium for access to the present. Indeed, the juxtaposition with past cultural forms makes visible the taken-for-granted nature of the present.

It now becomes clear that my perspective contrasts sharply with what I called the new orthodoxy. Although Bhaskar, Bourdieu, Giddens and others argue that temporality and historicity ought to be at the centre of social research, one is left wondering to what extent their methodology, in practice, differs from good old-fashioned historical sociology. They still operate roughly within the boundaries of a naturalist philosophy of science in that they tend to presume that the inquiry into the social and the natural realm share the same objective: explanation (possibly leading to prediction). My point is that other objectives are possible, one of which leads to the self-referential concept of knowledge acquisition mentioned earlier. To take time seriously is not simply to put forward the truism that social processes (both change and stability) occur through time, and therefore call for an in-depth study of the past. To take time seriously is also to be sensitive to what I call 'the methodogical interplay' between the different temporal modes. That is, whilst we cannot help but gain access to the past via the present, the former can also be seen as a path for elucidating the latter.

Even leaving the notion of a self-referential concept of knowledge acquisition aside, my argument for a temporalised social theory also differs from the new orthodoxy in other respects (see also Baert, 1992). For the sake of clarity, I will use some examples from the history of science and political science to illustrate my approach and to demonstrate how it contrasts with the new orthodoxy. As mentioned earlier, Bhaskar, Bourdieu and Giddens highlight the relationship between self-monitoring, tacit or practical knowledge, shared rules and the skillful production of order. My approach differs in that it attributes methodological significance to the Bergsonian and Meadian concepts of 'novelty' and 'emergence'. It accentuates that, once people face unexpected experiences (the emergent or the novel), they might become discursively aware of previously tacit rules and assumptions, and they might attempt to alter or maintain the very same rules and assumptions. What I call 'reflection of the second order' refers precisely to this process by which people become aware of and put discursively previously implicit rules and presuppositions. For example, it has been shown by historians of science that, in a period of scientific turmoil, the members of a scientific community become aware of previously tacit shared assumptions and publicly discuss these presuppositions. Or take the collapse of the Soviet Union, as a result of which several ethnic groups started to reflect upon their lost culture, language and political autonomy.

One of the core notions is the self. I borrow G.H. Mead's definition of the self as implying the ability of people to be both object and subject; that is, the ability to be object to themselves. But crucial to my temporalised social theory are two further distinctions. There is first and foremost the distinction between reflection of the second order and first order reflection. First order reflection comes into play when people look upon their real or imagined actions. For instance, whilst writing, one might reflect upon alternative ways of expressing oneself before actually writing. Second order reflection refers to people's inquiry into rules and assumptions underlying their actions, for example scientists' questioning of the assumptions of their theories. There is secondly the distinction between individual and collective reflection. Collective second order reflection refers to a public-collective exploration of previously tacit rules and presuppositions. It is vital for explaining both deliberate change or conscious maintenance of social structures. For example, a public debate in Parliament is bound to be more forceful than the private thoughts of a few politicians.

The unexpected or the emergent is central to my argument. People start to reflect upon a hitherto taken-for-granted world either because they are confronted with previously undisclosed cultural forms, or because they are faced with unforeseen effects of their previous actions. Taking the example of the history of science once more, scientists start questioning their paradigm once they are confronted with too many anomalous experimental results which are the *unintended* outcome of their attempts to articulate the very same paradigm. Likewise, to be confronted with the suboptimal collective effects of myopic self-interested actions (e.g. traffic congestion or global warming) might make people become aware of the limitations of our instrumental rational actions. This second order reflection might, especially when collectively acquired, lead to efforts to alter the previously seen-but-unnoticed rules and presumptions underlying people's practices. For instance, the scientists might eventually decide to *modify* the dominant presuppositions of their discipline. In the case of the fall of the

Soviet Union, several ethnic groups tried to *revitalise* their lost culture and language and opt for political autonomy. Or take public-collective awareness and discussion of the free-rider problem which can lead to people's attempts to alter the shared rules of the game according to which they have operated until now. Note that this reflection of the second order often goes hand in hand with a symbolic reconstruction of the past: people now attach importance to very different events in the past, and more significantly they attribute a different order or logic to the past. For instance, scientists often alter their paradigm by *rewriting* the history of their discipline: previously ignored experimental results suddenly acquire significance and validity, whilst other results, previously milestones, lose importance. Likewise, the past Russian presence is now seen in a different light, and local historical events have gained enormous significance. The examples demonstrate that the reconstruction of the past is an essential motivating factor for the making of collective and co-ordinated action.

Another crucial difference between my temporalised view and the new orthodoxy concerns evolutionary theory. I hold the belief that to put temporality and process at the centre of social theory necessitates taking into account the current developments in evolutionary biology. In contrast, Giddens and Bourdieu reject any form of evolutionary thinking. The reasons which they provide are vague. They tend to associate evolutionary sociology with its nineteenth century excesses, and fail to engage with the possibilities raised by new developments in the evolutionary paradigm (e.g. Jensen and Harre, 1981; Boyd and Richerson, 1985; Runciman, 1989, 1998; Cavalli-Sforza and Feldman, 1992; Reed and Harvey, 1992; Hodgson, 1993; Eve et al., 1997; Byrne, 1998; Blomley, 1998). They mistakenly believe that the differences between biological and socio-cultural development are so vast that it becomes pointless to use analogies with the former to elucidate the latter. It is, of course, important to acknowledge the differences, but I would like to emphasise that most are differences of degree rather than kind, and that Giddens and Bourdieu fail to realise that disanalogies can be informative as well. Seven apparent differences are worth mentioning, and give some indication of the wide variety of insights provided by evolutionary thinking. The most prominent dissimilarities will be mentioned first, but it will be shown that all seven are relative differences, and all stress the importance of second order reflection.

First, whereas biological evolution is clearly Darwinian and not Lamarckian, the case of socio-cultural evolution is more ambiguous. In a Darwinian view, mutations come into being independently of the environment; the environment (in the broadest sense) selects so that through time organisms with appropriate mutations are more likely to leave offspring. Lamarck and his followers misunderstood the selection process: they mistakenly believed that acquired characteristics are transmitted across generations and that mutations are 'adaptive' towards the environment. But in so far as one takes social practices as 'interactors' (the social analogue of organisms), rules and assumptions as 'replicators' (the social analogue of genes), and therefore changes to rules and assumptions as analogous to 'mutations', Lamarck's perspective appears relevant for socio-cultural evolution. Changes to rules and presuppositions *can* be (and occasionally are) the product of second order reflection, and it follows that the changes might well, at least to some extent, be adjusted to the environment. It is, of course, true that changes to rules and assumptions are often mere 'imperfect replications'. For example, the grammar and vocabulary of languages constantly change, but rarely as a result of a decision process. Instead languages tend to evolve in that every act of speaking might lead to minor alterations to grammar and vocabulary, to people replicating the incomplete replications, and so on. But my point is that people are able to reflect, able to act upon that knowledge, and that they regularly do so. There are plenty of examples in economics: changes to the corporate structure of a firm tend not to be random, but aimed precisely at superior adjustment to its environment. This quasi-Lamarckian mechanism even applies to the improbable case of language: the spelling of Dutch language was deliberately changed twice within a span of half a century, and there have been several attempts to discourage or ban the use of English words in the French language.

Second (and related), there is controversy within biological circles today as to the level at which selection operates. Some argue that the selection operates mainly at the level of the gene, some opt for selection at the level of the individual organism, and a few claim that entire groups can be selected. But there seems to be a consensus that 'selection pressure' operates at least largely on interactors (organisms). This contrasts sharply with the social realm. Selection pressure in the social world is different in that it can clearly affect directly the replicators. Take for example a firm that wants to implement new procedures which happen to be ineffective. The firm might make enormous losses and thus be selected out, in which case the selection hits at the institutional level. But it is also possible that the company directors perceive or foresee the problem and alter the procedures before it is too late or even before implementing them. In other words, the second order reflection of the people involved implies that selection can take place at the level of rules and assumptions. Note that, in this example, the rules and assumptions are indeed ineffective, and therefore institutional selection has been avoided due to timely cultural selection. This is not necessarily the case. It is indeed possible that people, for whatever reason, decide not to implement or to abort procedures that are actually considerably effective (so that, in the extreme case, selection at a cultural level might even lead to institutional selection). But in either case, people have the ability to imagine and anticipate alternative futures, and to act upon these thought processes. In either case, people's second order reflection (and its attendant foresight) not only makes for the vulnerability of customs and habit but also makes for the precarious and transient nature of its own (real or fictitious) creations.

Third, there is much controversy as to the nature and origins of speciation in biology. Traditionally, the emergence of a new species was defined as the product of geographical separation. Geographical separation leads to reproductive separation, and over time the accumulated variations become so vast that a new species develops. This view has now been challenged, and some biologists have pointed out that organisms might play a more active role in the process of speciation: for instance, reproductive separation might occur without geographical separation. But most biologists would agree that speciation is irreversible in that, once it has occurred, the new and the old species cannot successfully reproduce. The equivalent to speciation in the social realm is obvious. That is, rules and assumptions are replicated via practices through time; sometimes the institutions, in which the people find themselves, become seperated or differentiated, the result of which might be the gradual development of rules and assumptions in an ultimately very distinct direction. There are numerous examples of such speciation in the social realm: whether one is dealing with economic, religious or political organisations, differentiated sub-systems may evolve differently so that distinct cultures emerge. But one should not stretch the metaphor of biological speciation too far as there are differences as well. Whereas the degree to which organisms play an active role is a matter of dispute in biology, it is undoubtedly the case that institutional separation is often the product of (at least some) people's second order reflection and their power to act upon it. Furthermore, whereas biological speciation is irreversible in the way in which I described it, the social equivalent is not. Of course, institutional separation leads to the development of different cultures and thus makes fruitful communicative interaction less feasible, but it does not rule that interaction out all together. Whereas biological speciation resembles a branching tree, the social equivalent has branches that can always reconnect. Indeed, people's encounter with a different form of life may force them to confront themselves and their own culture.

Fourth, although biological mutations may eventually affect the environment, newly implemented rules and assumptions regularly influence it more rapidly and more significantly. In the social realm there is an 'evolutionary feedback mechanism' between innovation and selection in that cultural innovations become part of and alter the very same surroundings that make for the selection of culture. For example, changes to the corporate strategy of one firm might lead to further second order reflection in other firms, the result of which is overall change to the environment that selects. Two ideal types of such evolutionary loops can be distinguished. First, in the case of a 'homeostatic loop' (similar to what some economists have called 'strategic complementarity' and biologists might call 'positive feedback'), an innovation helps to create an environment that reinforces the very same innovation. For example, once a new framework is widely accepted by a scientific community, scientists will start looking at the world through that lens, and are then likely to be sympathetic to writings that exhibit that very same framework. Second, in the case of a 'self-defeating loop' (or what some economists term 'strategic substitutability' or what biologists might call 'negative feedback'), an innovation contributes to the making of surroundings that undermine the further spread of the innovation. Take the example of a new strategy in a game, which when first introduced takes the opponents by surprise, but consequently becomes less effective and thus less used because the opponents are aware and know how to anticipate it. In reality, one tends to find a combination of homeostatic and self-defeating loops. For example, take a new commercial strategy that gives one an advantage as long as not more than half of the competitors are adopting it. One would expect the commercial strategy first to enter a homeostatic loop, and then, once about half of the competitors have adopted it, a self-defeating one.

Fifth (and related to the previous point), various innovations in the social realm are immediately interrelated. Again, there are similar processes in biological ecosystems that show organisms to be interconnected (e.g. the phenomenon of the Red Queen), but it seems that pivotal innovations in the socio-cultural realm lead more easily and more swiftly to other vital innovations. Let me take the example of Darwin's contribution to science and philosophy. Darwin's framework was an innovative explanation of the mechanisms underlying biological evolution. But Darwinism did not only change the way people thought about biological processes. It also affected the theorising by sociologists and economists, and quickly challenged deep-rooted teleological and humanist assumptions. Darwinism also altered our notions of science and causality, shedding doubts, amongst other things, over the Laplacean notion of causality. Darwin's 'dangerous idea' had significant knock-on effects in aligned (and even in remote) fields of inquiry, and it took surprisingly little time for the new set of interrelated innovations to replace the old one. Likewise, numerous examples can be found in the area of technological innovations. Take the computer which, as we know, made possible new types of calculation, action and organisation. In addition, the computer made redundant basic clerical and mechanical tasks, and it altered the nature of the jobs that did survive. Together with communication technology, the computer made possible the separation of work and workplace, while allowing for more widespread surveillance and central control. It changed the way people think about how the mind works, and how organisations do and should function. Like Darwinism, the implementation of computer technology had important knock-on effects. In short, computers tie in with other innovations, and together they form a set of interrelated innovations. The emergence of a new set swiftly replaced the old one.

Sixth (and following on from the previous point), there is the phenomenon of lock-in and chreodic development. The notion of chreodic development is derived from evolutionary biology. It occurs when an evolving system tends to be locked into certain paths even at the expense of more efficient developments. This notion ties in with the well-known 'butterfly effect', according to which small changes in initial conditions can have enormous repercussions because they set a path. There is controversy as to whether chreodic development is so prominent in the biological world, but there is no doubt that it is of paramount importance in the social realm. There are various examples of technological innovations or economic decisions that form a chreodic development. They often lock us in precisely because of their knock-on effects. Take the car as a means of transport. Opting for the car (rather than a railway system, for instance) means investing in roads, in the technology of traffic lights, and so on. But the more this infrastructure is in place, the higher the costs of switching to other options compared to not switching. Also, people will act with that structure in mind so that the structure gets reproduced: for instance, more people will start living far from their workplace because they know they can rely upon the car, the result of which is more cars.

My seventh point sums up and follows from the previous ones. There is a growing interest in the extent to which, in the natural realm, small changes to initial conditions can have enormous ramifications. Perturbations in a far-from-equilibrium system can lead it to an unstable state of fluctuation. The system then arrives at a bifurcation point, in which it is impossible to predict the future direction of the system. Minute interferences might lead the system into one or another pathway. But once a pathway is 'chosen', the system becomes relatively stable until it confronts further disturbances that lead to bifurcation, and so on. One finds striking (and probably less ambiguous) examples in society of situations in which de facto unpredictable phenomena trigger off significant social and cultural shifts. I mentioned earlier that technological innovations, like the car, have significant knock-on effects, not just on other technological innovations (e.g. different kinds of fuel), but also on lifestyle (e.g. drive-in movie theatres), urban structure (e.g. the emergence of suburbs), international relations (e.g. the importance of the Middle East in international relations) and so on. The new set replaces the old. But once the new set is in place, path-dependency emerges, the development becomes chreodic, and a new period of relative stability has emerged. Note that, in these social analogies to non-equilibrium thermodynamics, the symmetry between explanation and prediction breaks down. It was virtually impossible for anyone to predict the exact nature of the invention of the motorcar before it was actually invented (several other technical options were available at the time). But one can, with some plausibility, reconstruct post hoc what happened as a result of that innovation.

Although the above argument for a temporalised social theory shows some continuity with the other theories, it also differs substantially from them, and it can be used to point out some of their weaknesses. Let me give some examples.

As in the intellectual project of many nineteenth century social theorists, what I have tried to develop underscores the importance of diachronic research in that the phenomena described refer to processes across time. But unlike some nineteenth century evolutionary schemes which treat the future as always already given, I have tried to draw the sociological consequences of a Bergsonian or Meadian world view in which novelty and emergence occupy a central role. I have done so by both exploiting the notion of unexpected experiences (for instance, unforeseen effects of purposive action, which might force people, individually or collectively, to reflect upon what was hitherto taken for granted) and innovations (the social analogue of mutations which, as I have tried to show, can have enormous knock-on effects and lock us in). Following Weber and Tocqueville, I have emphasised that people act purposefully though bringing about effects that are unintended and even unanticipated. Indeed, at critical points, people's reflection of the second order makes for noteworthy innovations the consequences of which far outwit the very same people who set the process in motion. But unlike Weber and Tocqueville, I have tried to take full advantage of the fact that people are often confronted with the very same unanticipated consequences of their actions (or of other people's actions), and that this confrontation might lead to second order reflection and ultimately to processes of significant restructuration. As with American pragmatism and symbolic interactionism, I have taken seriously the sociological opportunities of a process philosophy. As in Mead and Blumer's oeuvre, the concept of the self is central to my argument. But whereas symbolic interactionists focus exclusively on first order reflection, I have taken the juxtaposition with second order reflection to be of central importance, and likewise the distinction between individual and collective forms of reflection. As regards structuralism, ethnomethodology and structuration theory, I have recognised their pioneering work in exploring the various ways in which the underlying structures (rules and assumptions) make possible and restrict our social practices in everyday life settings. But I differ from them by highlighting that people are able to develop discursive knowledge of the very same structures, in other words, able to make the tacit manifest, and to put that reflection into practice. Instead of Giddens' general ontological claims regarding the constraining and enabling nature of social structures, it can be argued that lock-in phenomena and the mechanisms underlying their eventual collapse, as described earlier, are more powerful and insightful tools for capturing the solidity and vulnerability of societal constraints. In common with the functionalism of Merton and Parsons, I have recognised the significance of homeostatic loops, but only for particular and limited periods of time. As far as the dynamics of the social realm are concerned, the study of systems in equilibrium are of restricted use compared to dissipative and self-organising structures which operate at levels far from equilibrium.

References

Alexander, J.C. (ed.)., 1985, Neofunctionalism (Sage, Beverly Hills).

Bachelard, G., 1984, The New Scientific Spirit (Beacon Press, Boston).

Baert, P., 1992, Time, Self and Social Being (Ashgate, Aldershot).

Baert, P., 1998, Social Theory in the Twentieth Century (Polity Press, Cambridge).

Bhaskar, R., 1979, The Possibility of Naturalism: A Philosophical Critique of the Coantemporary Human Sciences (Harvester Press, Brighton).

Blomley, M., 1998, Complexity Theory and Social Sciences (Mimeo, University of Cambridge).

Blumer, H., 1969, Symbolic Interactionism; Perspective and Method (Prentice Hall, New York).

Bourdieu, P., 1977, Outline of a Theory of Practice (Cambridge University Press, Cambridge).

Boyd, R. and Richerson, P.J., 1985, Culture and the Evolutionary Process (University of Chicago Press, Chicago).

Braudel, F., 1966, La Méditerranée et le Monde Méditerranéen à l'époque de Philippe II (Armand Colin, Paris).

Braudel, F., 1980, On History (University of Chicago Press, Chicago).

Burke, P., 1990, The French Historical Revolution: The Annales School, 1929–89 (Polity Press, Cambridge).

Byrne, D., 1998, Complexity Theory and the Social Sciences; An Introduction (Routledge, London).

Canguilhem, G., 1978, On the Normal and the Pathological (Reidel, Dortrecht).

Cohen, I., 1989, Structuration Theory; Anthony Giddens and the Constitution of Social Life (Macmillan, Basingstoke).

Comte, A., 1969, System of Positive Philosophy; Volumes 1–4 (Burt Franklin, New York).

Durkheim, E., 1915, The Elementary Forms of Religious Life (Allen and Unwin, London).

Durkheim, E., 1963, Primitive Classification (University of Chicago Press, Chicago).

Elchardus, M., 1988, The rediscovery of Chronos: the new role of time in sociological theory. International Sociology **3**, 35–60.

Eve, R., Horsfall, S. and Lee, M. (eds.), 1997, Chaos, Complexity, and Sociology; Myths, Models, and Theories (Sage, London).

Foucault, M., 1977a, Discipline and Punish: The Birth of the Prison (Allen Lane, London).

Foucault, M., 1977b, Language, Counter-memory, Practice, ed. D.F. Bouchard (Cornell University Press, Ithaca).

Foucault, M., 1989a, The Archaeology of Knowledge (Routledge, London).

Foucault, M., 1989b, The Order of Things; An Archaeology of the Human Sciences (Routledge, London).

Foucault, M., 1990, Politics, Philosophy, Culture; Interviews and Other Writings 1977-1984 (Routledge, London).

Garfinkel, H., 1967, Studies in Ethnomethodology (Prentice Hall, Englewood Cliffs, NJ).

Giddens, A., 1981, A Contemporary Critique of Historical Materialism: Volume 1; Power, Property and the State (Macmillan, London).

Giddens, A., 1984, The Constitution of Society; Outline of the Theory of Structuration (Polity Press, Cambridge).

Heritage, J., 1984, Garfinkel and Ethnomethodology (Polity Press, Cambridge).

Hodgson, G.M., 1993, Economics and Evolution; Bringing Life Back into Economics (Polity, Cambridge).

Jensen, U.J. and Harre, R. (eds), 1981, The Philosophy of Evolution (Harvester Press, Brighton).

Lévi-Strauss, C., 1993, Structural Anthropology, Part 1 (Penguin, London).

Lévi-Strauss, C., 1994, Structural Anthropology, Part 2 (Penguin, London).

Luhmann, N., 1976, The future cannot begin: temporal structures in modern society. Social Research **43**, 130–152.

Luhmann, N., 1979, Time and action: a forgotten theory. Zeitschrift fur Soziologie **8**, 63–81.
Luhmann, N., 1982, The Differentiation of Society (Columbia University Press, New York).
Luhmann, N., 1990, Essays on Self-Reference (Columbia University Press, New York).
Luhmann, N., 1995, Social Systems (Stanford University Press, Stanford).
Malinowski, B., 1944, A Scientific Theory of Culture (University of Carolina Press, Chapel Hill, Carolina).
Martins, H., 1974, Time and theory in sociology. In: Approaches to Sociology: An Introduction to Major Trends in British Sociology, ed. J. Rex (Routledge and Kegan Paul, London), pp. 246–294.
Marx, K., 1990, Selected Writings (Oxford University Press, Oxford).
Maturana, H. and Varela, F. (eds.), 1980, Autopoiesis and Cognition: The Realisation of the Living (Reidel, Dortrecht).
Mead, G.H., 1929, The Nature of the Past. In: Essays in Honor of John Dewey (ed J. Coss) (Henry Holt, New York), pp. 235–242.
Mead, G.H., 1959, The Philosophy of the Act (University of Chicago Press, Chicago).
Mead, G.H., 1970, Mind, Self and Society (University of Chicago Press, Chicago).
Parsons, T., 1949, The Structure of Social Action (McGraw-Hill, New York).
Parsons, T., 1966, Societies: Evolutionary and Comparative Perspectives (Prentice-Hall, Englewood Cliffs, NJ).
Parsons, T., 1977, The Evolution of Societies (Prentice-Hall, Englewood Cliffs, NJ).
Popper, K., 1986, The Poverty of Historicism (Routledge, London).
Radcliffe-Brown, A.R., 1952, Structure and Function in Primitive Society (Cohen and West Limited, London).
Radcliffe-Brown, A.R., 1958, Method in Social Anthropology (Univerisity of Chicago Press, Chicago).
Reed, M. and Harvey, D.L., 1992, 'The New Science and the Old; Complexity and Realism in the Social Sciences.' Journal for the Theory of Social Behaviour. **22**(4), pp. 353–381.
Runciman, W.G., 1989, A Treatise on Social Theory, Volume 2: Substantive Social theory (Cambridge University Press, Cambridge).
Runciman, W.G., 1998, The Social Animal (Harper Collins, London).
Saussure, F., 1959, Course in General Linguistics (Peter Owen, London).
Weber, M., 1949, The Methodology of the Social Sciences (Free Press, New York).

chapter 12

POLITICAL THEORY AND TIME

MELISSA LANE
University of Cambridge

Time in Contemporary Intellectual Thought
Patrick Baert (Editor)

CONTENTS

12.1. Introduction

Peculiar to political theory's concern with time is the way that conceptions of time may be invoked to explain and legitimate, or delegitimate, structures of political authority. Indeed, one great division in the history of political theory is between those theories which invoke time to legitimate political structures, and those which reject temporal considerations in favour of timeless principles such as those of reason. The interplay between theories of these two kinds, the various refinements of and mergers between them, will structure our sketch of time in the history of political thought.

Past, present, and future are all dimensions of time relevant to political theory, and they are interrelated; conceptions of the past and judgements about the future are themselves integral to determining action and belief in the present. Conceptions of the past constitute history, and the shape of historical time – the question of whether history as a whole has developed in a discernible direction, or the philosophy of history[1] in short – has both transcended and informed the narrower history of political thought. Judgements about the future are crucial for an individual's engagement in practical reasoning: one's sense of one's own interests, values and goals, and which means to choose to pursue them, depends on the length of one's time-horizon and the risks and possibilities which one foresees. One argument of this chapter is that judgements about the future have not always been central in the history of political theory on a reflective level (though they have always had a role in the decisions of political agents) and that a shift in the relative balance between the weight of the past and the weight of the future is part of what constitutes modern political theory as such, though the proper balance will remain controversial.

The question of judgement imports time not only as experiential – as lived in the past or expected in the future – but also time as a dimension of the particularising of objects and actions. How political judgement may grasp just this object or action as what is appropriate to a given moment, and whether indeed political judgement can stably do so, has been a longstanding concern of political theorists. The relation of time to the legitimation of authority, on the one hand, and

[1] Manuel (1993) is a good introduction to this topic, distinguishing broadly between 'cyclical' and 'progressive' theories.

to the judgement of particulars, on the other, are the twin themes to be traced in this chapter.

The earliest sources of Western political thought include classical Greek culture and Hebrew culture – the one later mediated by a pagan, the other by a Christian, Rome. To draw fundamental distinctions between them is tempting: in the philosophy of history, it has become a cliché to contrast Greek notions of 'cycles' with Jewish (later Christian) notions of 'linearity' and 'progress'.[2] More relevant for the question of authority is a contrast which has been noted 'between the Hebrew sense of time as the dimension of divine action in the human world and the Hellenic sense of time as the dimension of human finitude and disorder.'[3] Yet this contrast, also, is too general. The real interest of political theory is in the way individual thinkers develop, rework, or dissent from, the conventional assumptions of their contemporaries and forebears both in intellectual and practical contexts. To see this in the Hellenic case, the best way to begin is also the traditional starting point of Western political theory, by introducing the exceedingly complex thought of Plato and its relation to some ways time was handled in the Athenian democracy in which he wrote and lived.

12.2. Plato and Athens: acting in time

Although Plato sometimes referred to the prevalent Greek belief in cycles of history,[4] most salient for him in the *Republic* was not renewal but corruption and decay. The *Republic* classically presents time as the arch-enemy of secure political authority because it is also the arch-enemy of the true knowledge on which such order depends.[5] Knowledge in the *Republic*, indeed the very possibility of explanation, is limited to the intelligible world of timeless and imperishable entities ('Forms'), in contrast to the sensible world of particulars which are perishable and contingent (475d-480a). The aspiration to establish a just city in order to enable all to have just souls, depends on the philosophers' ability to rule, making particular decisions about individual events and things, in light of their profound knowledge of timeless universals. But the principal speaker Socrates predicts that philosopher-rulers who possess such knowledge, and rule in virtue of possessing it, will eventually fail to master the perverse demands of the changing moment (*kairos*) and will thus initiate the downfall of their own regime (Book

[2] Examples of this stance are given and decisively criticised by Momigliano (1966).

[3] Pocock (1969) 297, summarising Gunnell (1968).

[4] E.g. *Laws* III 677aff.

[5] Wolin (1960) 40–50 gives a beautifully perceptive account of Plato's concern to establish an 'architectonic' political order (contrasting this on p. 216 with the 'manipulative' aims of Machiavelli). But he sees this atemporal urge in the *Statesman* and the *Laws* as well. It will be argued here that the *Statesman* involves a very different view of time and authority from the *Republic*.

VIII).[6] This claim that the demands of the moment will eventually prove their Achilles' heel shows that this relationship between knowledge – conceived as necessarily timeless – and judgement of particulars in time is precarious. If knowledge must prescind from time, time threatens to take its revenge by undermining the grasp of that knowledge on a changing reality, and so the authority of its rule.

The speculative magnificence of this view of knowledge, time, and authority, is such as to have led many students of Plato to grant him the posthumous triumph of identifying 'Greek' or 'Athenian' time-consciousness with his own. Yet it is crucial, and explicit in the dialogues, that Plato is setting out to provoke and counter – not to express – the prevailing winds of opinion. On looking further we can find some of those opinions in the legitimating views of time generated by the institutional structure and practice of the Athenian democracy[7] of which Plato was so vituperatively critical. Moreover, Plato himself seems to criticise one of these Athenian conceptions and appropriate another in assaying, elsewhere, a different relation between time and knowledge.

Athenian conceptions of the link between time and authority deploy at least two levels of authority: the memory of charismatic founders and their legacy of laws, and the day to day deliberations of the demos in Assembly and courts. The interplay between these two levels of authority is an exciting question in current research.[8] Suffice it here to point, on the one hand, to the authority accorded at least in speeches to those customs and laws (collectively, *nomoi*) ascribed to the great and near-legendary lawgiver Solon.[9] These *nomoi* enjoyed the patina of immemorial authority; in the respect for them, time as tradition evinced an authority of its own.

On the other hand, these same laws and customs had to be invoked in the everyday decision-making of the Assembly and Council, and the judging of the courts, in ways free from the bureaucratic accumulation of precedent and open instead to daily contestation and reversal.[10] Here time, as handled in the orators'

---

[6] *Republic* VIII describes the degeneration of the ideal city, ruled by philosophers, who will eventually fail to discern the proper time or season (*kairos*, 546d2) to order the consummation of sexual unions necessary to perpetuate the best offspring. (*Kairos* here is to be read temporally, *pace* Adam (1963) vol.2 208, who confuses it with the reference to numbers of marriages in V 459eff.).

[7] This brief discussion aims to sketch an answer to Pocock (1969) 299, who noted the usual focus on Greek philosophical accounts of time and suggested that it would be surprising if democratic institutions had not generated their own conceptions.

[8] Ober (1989), Cohen (1995), and Allen (1996) emphasize the role of demotic deliberation in Assembly and courts as sovereign over and above the laws whose interpretation and application was in their hands; Ober also argues convincingly for a method treating the orators' speeches as evidence of democratic ideology, which I have relied on below. Contrast Campbell (1986) who emphasizes instead citizen adherence and loyalty to the 'guardianship' of the laws.

[9] The late sixth-century aristocrat Kleisthenes may deserve more credit for democratisation of the Solonic regime, having reorganised the local basis and defined the Council and Assembly, but it was Solon whose name was revered as founder. See Hornblower (1993) and Dunn (1993b) 240–242.

[10] Allen (1996) 176–183 clarified my thinking on this point.

speeches which appeal to and contribute to democratic ideology, became conceived as the fleeting and unique context of circumstantial decision or judgement. The notion of the *kairos*, the correct and fitting moment for word or deed, came to exemplify the *demos'* temporal mastery on which its own survival would depend. The fourth-century orators Demosthenes and Isocrates applied the *kairos* to political decision-making, vying to persuade an audience that the *kairos* required this or that course of action.[11]

Plato's *Statesman* issues a radical challenge to the conventional authority of the *nomoi*. (Gill, 1995) But it does so not on the grounds of timeless knowledge used in the *Republic*, but rather by wresting the *kairos* from the orators to make it definitive instead of the true political knowledge of the Platonic 'statesman'. In contrast to the *Republic*'s sharp distinction between timeless knowledge and problematic temporal particulars, the *Statesman* defines political knowledge precisely as mastery of the temporal flux and the changing actions which this requires. (Lane, 1995, 1998) Thus one step was taken toward breaking down the idealist position that knowledge and time can never mix. Yet the genesis of such political knowledge is never investigated in the *Statesman*, in which it rather seems to spring full-blown like Athena from Zeus' head. On the one hand, the contribution of the past – the experience, habit, and consultation to which the orators ascribed their grasp of the *kairos* – are denied relevance to the Platonic statesman's political knowledge. On the other hand, the problems posed by the uncertainty and contingency of the future to the rightness of political choices are simply ignored. The claim to judge correctly what the moment requires – the claim that there is indeed one correct and required action – involves denying that calculation of the future and its influence on one's present judgements is inherently an imperfect art. Once the uncertainty of future events is admitted into one's calculations, the very notion that there could be an objective and absolutely correct action to take in the present – one whose correctness the future will not mock – falls into obscurity.

12.3. Cognition, common law, contract

One path out of this obscurity is to alter, or weaken, the claim made for the judgements about what action the moment requires. Aristotle altered Plato's idea of the *kairos* as knowledge by distinguishing between theoretical and practical knowledge; the latter, which deals with judgements in time, still claims to be absolutely correct but not to be theoretically deducible or reproducible. Machiavelli went further, assigning mastery of particulars, change, and action not to a strictly cognitive faculty but rather to the faculty of the will which could

[11] Jaeger (1963) 129–54 discusses the battle over the *kairos* between these two orators. On the Sophistic background in which *kairos* applied primarily to rhetorical success, see Cole (1991).

exhibit *virtù*, an inimitable combination of will, prudent perception, and daring. If the *kairos* matched action to moment by claiming to discern an objective fit, *virtù* was rather the quality of wrenching the moment to favour one's action.[12] In his concern with this quality, Machiavelli shared the Renaissance concern with the knowability of particulars and the consequent instability of political regimes, which was crucial to the revival of republicanism. Yet other Renaissance and, later, early modern theorists would find that Machiavelli had gone too far: 'Machiavel' was pilloried in part for this violent and willful indifference to 'truth' and 'knowledge'. Some sought to restore grasp of time to a cognitive faculty which could, however, renounce claims to demonstrability and certainty. The notion of 'experience' was developed as a kind of non-rational, but cognitive, faculty of learning in and from the passage of time. It was put to use most decisively in defence of the English common law, justifying the common law's cognitive superiority as due to its roots in immemorial tradition altered only slowly by the humble workings of experience, and so embodying a cumulative wisdom of many men superior to any rational prescription of a rash individual (Pocock, 1972, pp. 202–32).

The history of the common law was proudly parochial and inescapably institutional. But interest in the non-parochial history of mankind as a whole, and in the pre-institutional phase of that history in particular, was burgeoning throughout Europe, due perhaps most profoundly to the encounter with the New World and its epistemological, commercial, and political implications; the bloody wars of religion which divided Christendom further and raised urgent problems of authority and legitimation anew; and the development of the printing press which made historical knowledge newly collatable, correctable and accessible.[13] The new availability and apparent relevance of the 'natural' history of mankind (though almost always within some sort of providential framework) allowed various kinds of legitimation to be sought in the annals of past time, and so returned attention to the philosophy of history from the obsession with particulars. To be examined here is that school of natural law theorists who sought to reconstruct the story of the human entrance into political ('civil') society (Haakonsen, 1996), either through a gradual process or through a decisive moment of entering a 'social contract' which established legitimate political authority for the first time.

The 'timing' of the social contract has long vexed students of political theory: one is forced to ask whether a given theorist viewed the contract as an event

[12] Pocock (1975) gives a brilliant account of Machiavelli's thought as a response to the problem of knowing particulars in time. This is perhaps the place to record my indebtedness to Pocock generally for his probing and evocative work on time in political theory, which not only charts the Renaissance and early modern responses but also formulates general issues about institutional self-awareness as distinct from individual mortality and the philosophy of history.

[13] On the influence of the printing press in particular on conceptions of historical time, see Eisenstein (1966).

having occurred in real time. Those who did, did so not to nominate a specific past date, but to affirm that moral and political events wrought in the past continue to bear on the legitimacy of contemporary polities. 'Continue to bear', not 'exhaustively determine': even Locke, in whom the contract is dressed in the most historical of trappings, argued that authority might be so badly abused that an existing generation might be justified in setting it aside and starting afresh; not for him the binding of the present generation forever by the sins of the fathers. But neither would he have held, with Jefferson, that political authority should be reconstituted purely on principle every twenty years, so as to give each generation its right to start afresh whether suffering from abuses or not. Locke, like Grotius and Pufendorf, vindicated the moral relevance of certain decisions, actions, and institutions made by humans in an indefinite, but recognisably historical and reconstructed past; and Locke for one held that the history of these institutions gave them a *prima facie* though not unbreakable claim on the allegiance of their inheritors. A version of this vindication of the relevance of history will be discussed later in this paper when we come to the contemporary political theory of Robert Nozick.

For Thomas Hobbes, on the other hand, there was nothing primitive or irrecoverably historical in the idea of the 'state of nature.' This was rather an ever-present possibility in human affairs, the abyss waiting to open whenever effective political authority should collapse or be destroyed. Hobbes thus used the language of the 'state of nature' to make a very different point about the relation of history to authority. For Hobbes, the general history of humankind is inadequate to establish a natural fact about political authority: such authority is the supreme artifice and is dependent upon a rational decision and the subsequent power to enforce order and provide security. Equally, the history of a given regime – whether it be steeped in the patina of centuries as monarch, or newly arrived by violent usurpation – is irrelevant to the question of whether its authority should be accepted. What matters is not history but the rational judgement that the balance of security and self-preservation is overwhelmingly better served by submitting to political authority than by retaining the right to judge matters for oneself. The rationality of this judgement may be only fitfully apparent to such contentious, spiteful and proud creatures as are human beings, but Hobbes wrote to remind waverers of the mortal stakes and to justify the Leviathan's use of power and fear to keep them obeying as they would (almost) always have good reason to do. Thus time and history for Hobbes are swept aside in favour of present-tense judgments about the efficacy of given regimes and the rationality of obedience.

Indeed, Hobbes' scepticism about the power of time to bind or loose human obligation was even more profound than his rejection of any historical action as an unchallengeable source of political legitimacy. Whereas Locke seemed to found political obligation on the cross-temporal self-binding of the act of promising, Hobbes argued mordantly that the human moral sense itself, and indeed the very stability of our perception of time, rested on sufficient political stability for interaction and intersubjectivity to be secured. (Wolin, 1960, pp. 264–5) Arts, sciences, trade and the very 'account of Time' are impossible in the state of war which is by definition without effective political authority (*Leviathan*, ch.XIII). The ability to plan for the future is itself a precious artefact of political society. Thus, while sweeping aside the claims of past allegiance or legitimating acts, Hobbes invokes the need for humans to plan stably for the future, and justifies the alienation of rights to political authority as the choice humans would make were they able rationally to foresee the consequences of continuing to live in a 'state of war' with one another. If only present-tense ratiocinations of humans can, for him, establish legitimate authority, these present-tense judgements incorporate a rational assessment of the radical uncertainty of the future and the need for authority to lessen one's consequent vulnerability.

12.4. Apocolyptic futures

The future, for Hobbes, is a crucial ingredient in the rationality of judgements in the present, and enters those judgements as a claim of radical uncertainty. For other political theorists, however, authority has been linked not to a judgement of the uncertainty of the future, but rather to claims about what it must certainly bring. The Jewish account of a Messianic Age was absorbed into a Christian expectation of a Second Coming perhaps more structurally crucial for the meaning of the 'First Coming' than the Messianic Age ever was for the Jews. The pagan Romans among whom Christianity spread were themselves more filled with forebodings for the future, the stability of the Empire, than the Greeks had been (despite the musings of their philosophers). It was the future's fearful obscurity which the various charismatic sects of the two centuries on either side of Christ's birth sought to dispel with their visions of apocalypse and redemption. Augustine had later to attack those Christians who combined apocalypse with a belief in historical cycles, arguing that Christian doctrine lay in a linear view of history as opposed to the benighted pagan cycles (*City of God*, XII, 13). Cyclical or not,[14] however, apocalyptic visions continued to influence political, as well as purely religious, views of the future throughout the Christian era.

Such apocalyptic visions tended to be destructive of established authority, manifest in sectarian and enthusiast rejection of church or state based on the hopes of a millenarian prophetic faith. Another kind of vision of the future has tended to be kinder to the powers-that-be: that which views the future as continuous and evolutionary from the past. Eighteenth and nineteenth century thinkers interested

[14] It is worth remembering that an apocalypse can precede the inauguration of a new cycle (as in Greek catastrophism) as logically as it can be made to presage the ultimate and final realisation of a linear development of time. The expectation of apocalypse means that the preceding time can be described either as in decline (to the crisis) or as progress (to the subsequent redemption).

in the moral, and later the cultural, progress of mankind manifest this kind of interest in the future, an interest which was reinforced and transformed among some by the assimilation of Darwinian evolution in intellectual circles. (Manuel, 1993, pp. 70–135) The progress of human culture or morals could mean that whatever is, is best (for its time, or absolutely when viewed at the peak of the progress of Spirit: Hegel), or else that whatever is, will be deservedly superseded by something better in the fullness of the future (Marx).

In the twentieth century the Marxist philosophy of history, crucially mediated by Lenin's belief in the necessity of conspiracy and disciplined action to bring the promised future about, made the philosophy of history into a potent tool for political struggle. The Marxist philosophy of history guaranteed the objectivity of knowledge, while the Leninist conception of strategy and tactics secured that knowledge to judgement and action in the realm of the context-bound particulars. Here a link between time as history, and temporally bound judgement, was forged in the crucible of determinism and the authority of the revolutionary party.

Marxism having become the dominant form of the philosophy of history in twentieth-century political theory, its explanatory and political failings left the very idea of an overarching account of historical direction in bad odour. Liberal theorists of the Cold War era analysed the intrinsic unfalsifiability of 'historicist' thought and claimed to reject all such 'ideologies' as now otiose.[15] But the absence of sustained thought about the direction of history among most liberal thinkers of this century arguably has deeper roots as well. The urgency of concern with the effects of time, as we have seen, was linked in the history of political thought with urgent anxiety about the maintenance or establishment of political authority. Seeking to understand the course of history could offer resources for mastering it, not least by claiming that legitimacy flowed from some source in the past. As state authority became consolidated and legitimacy established, however, the questions of day-to-day use of such relatively secure state power were less obviously questions which an understanding of history was crucial to answer.

12.5. Rejecting the past? Normative politcal theory

The precociously stabilised polity of late eighteenth century England witnessed a clarion rejection of the relevance of the past to the political policy of the present: the utilitarian political philosophy of Jeremy Bentham. Bentham's clear rejection of the untrammelled authority of the past was arguably born in the revolutionary experiences of the eighteenth century. The firebrand Tom Paine prefigured

[15] Popper (1945, 1961) drew his work on the philosophy of science to argue that 'falsifiability' was a necessary criterion also for political theories. Bell (1960) diagnosed an 'end of ideology' in Western political argument. It is ironic, however, that triumphalist works such as Fukuyama (1992) in the wake of the breakup of the Soviet Union offered a liberal version of the 'end of history' structurally analogous to the communist ones which they claimed had been proved false.

Bentham's faith in the potential beneficence of a future freed from the benighted imposters of the past. Paine's *Age of Reason* provoked the great statement of the opposite case by Edmund Burke, whose *Reflections on the Revolution in France* pitted the custom, tradition and experience of the past against the claims of Paineite rationalism. If Burke provided a compelling statement in favour of the past's rightful disposition of the present in political affairs, however, the future by and large of political theory belonged to the heirs of Paine. The scorn which Bentham poured on the historical pretensions of natural rights theorists in his 'Anarchical Fallacies' (1824) demonstrated his rejection of the idea that any actions in the historical past could rightly constrain the rational determination of present actions by a legislator seeking to maximise utility.[16] Utilitarianism gives authority to judgement in the present, though this must be shaped by assessments of the future consequences of action. The authority of rational judgement invokes a tolerably though not necessarily absolutely predictable sense of the future, in order to banish the dead hand of the past.

The implications for political theory of this emancipation of present-tense rational choice from the obscure entrails of history have been profound. (They were reinforced by the growth of a scientific approach to society generally, a development beyond the scope of this chapter.[17]) Indeed, it is largely this concern for rational principles of normative choice which defines 'political theory' as a professional discipline today, allied to the philosophy of games or language more than to the philosophy of history. 'Political theory' in this sense is different from the history of ideas or of past political thought, although these may well be judged to be its indispensable resource;[18] it seeks legitimation for political authority in normative argument rather than in the appeals to tradition, covenant, or redemption which we have seen endemic in earlier political thought. It is conventional to date the 'rebirth' of political theory from the publication of John Rawls' *A Theory of Justice* in 1971.[19] But the move to reject the authority of history is older and broader than this, visible in the logical-positivist analyses of the 1950s and, as argued above with regard to Bentham, rooted in utilitarianism as well as in the Kantian tradition of separating an autonomous realm of morality to which Rawls acknowledges his debt. Rawls is exemplary but by no means

[16] Bentham did hold that the expectations generated by legal rules had to be respected, so that the legal past should in this limited way constrain future legislation. But the middle to long range of historical events were irrelevant in themselves to the choice of legislation now, except of course as they had shaped propensities to experience pleasure and pain.
[17] On the growth of scientific approaches to society, see Hughes (1977) 31–66.
[18] See Tuck (1993) on the relationship between the history of political thought and modern political theory.
[19] But see Pettit (1993), which argues for dating the rebirth from the publication of Brian Barry's *Political Argument* in 1965.

unique in conceiving the nature of political theory as entirely indifferent to any inquiry into the directional course of history.

Rawls revives the notion of a social contract not as an historical event but as a device of reason, used to illuminate the choice of basic principles of justice which it would be rational to make in circumstances in which such a choice would be fair. These circumstances must, in Rawls' view, eliminate all knowledge of one's individual talents, beliefs, or goals (indeed one's entire individual history) and model a respectful indifference toward the talents, beliefs, and goals of others. The principles chosen in such circumstances are rational principles which express a society's considered convictions about justice in a society of fair cooperation among free and equal citizens. Because their authority in the end is the authority of 'reflective equilibrium' between one's deepest convictions and the results of the social contract thought experiment, they do not claim to be authoritative for any and every society which might exist: they are not 'ahistorical' in the sense of claiming validity for all historical epochs. But they are 'unhistorical', in the sense that they are not valid in virtue of any historical acts or facts. Time's yoke is not binding on political legitimacy, which must yield rather to the dictates of reason.

Three sets of critics have challenged Rawls' attitude to history and to time more generally. The first group of libertarians was led by Robert Nozick, who revived against Rawls' erstwhile Kantianism a Lockean version of the social contract which reinscribed the authorising power of history. (Nozick, 1974) Nozick began from a Lockean starting point, assuming that humans have natural rights which legitimately constrain the actions of others, including the formation by oneself and others of a political authority. These rights govern the legitimacy of property acquisition and transfer, such that the just ownership of an item is determined solely by the history of just transactions involving it and is not available for any rationally based redistribution such as advocated by Rawls. Nozick's striking phrase was that commodities were not, as he thought Rawls had assumed, 'manna' from heaven, but artefacts already and inextricably tied to owners by the history of production, acquisition, and exchange. (Nozick, 1974, p. 198)

Criticisms can be made of Nozick's theory, in particular his failure to account for social cooperation in production, the legal and political framework which legitimates ownership in the first place, and the impossibility of determining or rectifying the ownership history of many commodities. But his work was significant in reviving the claim that the history of social interactions could determine their legitimacy. Nozick, however, still remained within the framework of reconstructive normative analysis: his historical chains of right-governed transactions were, ironically, even more ahistorical than Rawls' contract in the sense defined above, in that they claimed to apply naturally and so universally to any society whatsoever. This issue of a-historicity was raised against both Rawls and (by implication) Nozick by the second group of critics, those known as 'communitarians'.[20] These thinkers defend a range of positions; to be examined here is Michael Sandel's rejection of the claim that there can be principles of justice which are universal, independent of the particular political and social values and traditions of a given society (Sandel, 1982).

One confusing feature of this communitarian criticism is that Rawls himself, as noted above, acknowledged that the convictions modelled by his theory were those of a contemporary liberal democracy. According to the communitarians, however, Rawls' admission of a social and political context for his theory remains unacceptably general. In particular, charged Sandel, beliefs about justice, the good life, and the relation between them, beliefs shaped by membership in historically rooted communities, are deliberately excluded from the 'original position' in which principles of justice are to be chosen. How, the challenge goes, can any choice at all about justice be made in a vacuum of meaning and commitment? And even if it could, why should such hypothetical choices be binding on real individuals who hail from particular communities and adhere to particular values and beliefs?

Sandel's argument claims to defend the beliefs, values and practices of actually existing historical communities, though just how much authority, which beliefs, values and practices, and what size or definition of community are not well specified. Nevertheless, it is fair to say that the authority granted to these communities mixes a recognition of the historical roots of the present with an approval of some sort of principle of self-determination or local autonomy. That is, although the vicissitudes of the communal past are assumed to have moulded the outlook of the present generation, communitarianism in this form is really a granting of authority to the present-tense deliberations of members of a community to govern themselves as they judge best (in the light of their history) rather than an ascription of authority to any particular event or tradition from the past. A community will, on any sensible account, be the inheritor of many traditions, and communitarianism does not presume to validate one over the other; it gives authority to do that to living members of the community. In this sense communitarianism does not, despite what one might expect, develop a philosophy of history: it says nothing about why the past happened the way it did, only that the residues of that past in particular communal values and beliefs are valuable and should be accorded self-determination.

Unlike the communitarians, the third school of critics of Rawls, to be labelled here 'deliberative-democrats' do specify the kind of reflections in which particular communities should engage, if their judgements are to deserve political authority. Two kinds of argument have been advanced under this broad heading, of which the first can be construed as responding especially to the need to recover the

[20] 'Communitarian' has been applied by self or others to a diverse group of thinkers, including Amitai Etzioni, Michael Walzer, Charles Taylor and Alastair MacIntyre. I will focus on the work of Michael Sandel, the closest critic of Rawls among them.

passage of time in deliberation, while the second is especially concerned with the need for attention to temporally bound particularities.

Whereas Nozick and the communitarians faulted Rawls, in different ways, for ignoring the normative relevance of the past, the deliberative-democrats fault him in the first place for ignoring the normative relevance of political argument in real time. They object to the original position not because it models fair constraints on choice of principle (they accept the idea of procedural constraints on argument), but because it fashions these constraints so as to make plurality in deliberation redundant.

Rawls wrote in *A Theory of Justice* that 'the veil of ignorance ensures that everyone should reason in the same way' (Rawls, 1971, p. 564), and though he emphasized that plurality matters for reasons of ensuring public commitment and stability, he never required actual dialogue among the parallel (indeed, identical) reasoners in the original position. The deliberative-democrats reject such use of normative argument to simulate the outcome of debate. They hold instead that only debate ('democratic deliberation') can elicit the transformations in outlooks necessary to attain genuine consensus. Theory must stop trying to short-circuit the time needed for democracy to function; its role is only to outline the conditions for fair and open discussion of principles of political legitimacy or justice without seeking to preempt or predict the outcome.

This recovery of time as vital to the unfolding of democratic dialogue, and consequent legitimation of political authority, is linked in the deliberative-democrats' second concern to the recovery of particularity in ethical and political judgement. This revives the ancient link between time and knowledge in a distinct form: time is treated as a condition for the formation, through debate, of the knowledge relevant to politics. For the ancient Athenians, as we have seen, deliberation applied not just to general questions about values but to decisions to take particular actions; and the effort of thinking about the demands of the particular circumstances led as noted above to the notion of an objective *kairos*, matching moment to deed. The particularity of deliberation has been said to have been lost in Kantian ethical theory, though that charge has been rebutted well by contemporary Kantians concerned to explore the faculty of judgement.[21] It has also been under threat in sociological terms, from the clear modern differentiation between the legislature, responsible for deliberation but normally excluded from applying their deliberations to particular actions, and a bureaucratised executive branch meant to apply lawlike rules to such individual cases. Yet the stark clarity of this distinction between deliberation and application cannot really be maintained. The Constitution of the United States, a document which formalised this idea of separation of powers as an achievement of eighteenth-century political

[21] This is well argued by Herman (1993) and (1996); the latter argues that universal moral principles constrain but do not fully determine the rightness of context-bound judgements, which can be legitimately bound also by local values.

thought, nevertheless assigns to the deliberative legislative body the power to declare war: a power to make a particular decision in time.[22] Deliberation about particulars is revived, in deliberative democracy, as a condition for moral knowledge and political action. This revival relies on a conception of 'moral knowledge' as practical and, in embodiment, fallible, two features which preserve it from the perfectionist ambitions of the *kairos* and leave it dependent on the cultivation of good judgement about the future.[23]

We have seen, then, that modern political theory is increasingly alive to the ancient debate about universals, time-bound particularity, and the way in which political judgement must mediate between them. The other major theme of our survey of time in political thought – the relevance of history – has fared less well in contemporary theoretical debates. This is so despite popular claims that the end of the Cold War has 'unfrozen' the sources of nationalism, a potentially virulent form of historical memory. Nationalist and other ascriptions of authority to the putative memory of a common past are in dubious odour among political theorists. On the one hand, many, like the communitarians examined above, pay homage to the power of collective pasts to shape individual identity and values, and so conceptions of practical reason and political action. On the other hand, few are prepared to grant such wellsprings of memory – all tailored by the processes of national formation and invocation – untrammelled authority. The question of how much authority to grant the past has become increasingly controversial as purportedly 'rational' schemes of government – a mantle claimed by the United Nations no less than the fallen Communist regimes – have been challenged by the claims of memory and tradition[24] once again.

12.6. The timeliness of political emotions

One path forward is for political theory to consider not only the rationality of cognitive judgements, but also the rationality of emotions, in evaluating the claims of memory and tradition. To construe the standoff between tradition and justice as between emotion and reason is to fall into old, destructive, and erroneous habits of dualism which feminist research, and the philosophy of mind, have begun to expose and go beyond. Emotions such as regret, directed toward the past, and

[22] I am indebted to Elaine Scarry for her paper *Thinking in an Emergency*, which drew attention to this fact. She argued that the frequent circumvention of this power in the last half-century reveals a modern suspicion of the notion of deliberating in an emergency; in the context of the present discussion, it may also reveal a modern suspicion of the notion of deliberating in a temporal context about particulars at all.

[23] Dunn (1996) develops a notion of 'prudence' which embodies similar recognition of the fallibilist nature of judgement about the future, with more concern for the systematic obstacles to collective prudence than deliberative-democrats have tended to show.

[24] It is important to note, however, that virtually all nationalist movements owe at least part of their origins and 'memory' to assiduous inventors and purveyors of the national idea. Anderson (1983) is a seminal work on this topic.

hope,[25] directed toward the future, fuse feeling and judgement such that they colour perception and also constrain action. The cognitive component means that the emotions are, potentially, open to reflection and criticism,[26] as is the cognitive capacity of memory, although the criticism of memory, both individual and 'collective', is intellectually underdeveloped and politically controversial.[27] In the political sphere, memory and the temporal emotions fuse individual identity and memory with the common social experience of passage through the hazards of time. Political judgement and action are susceptible to their sway. Political theory needs to learn how to evalute their claims, without succumbing to their claims to blanket authority on the one hand, but without retreating to insulated principles of pure rationality on the other.

In evaluating political legitimacy, for example, one must recognise that adherence to minimal rational norms of justice is necessary but not sufficient as a litmus test. Rawls has tended to elide the difference between legitimacy and justice: though he recognises the special problems of stability, justice is for him the first virtue of politics and the sine qua non of political legitimacy.[28] It follows, as Stanley Cavell[29] has noted, that those challenging the principles of justice must be able to show just cause for their resentment, so that the society's liability to reproach itself is strictly limited. Yet as Cavell argues, self-reproach and shame cannot always be penned within purely rationally justifiable bounds. One may well feel ashamed of one's society's failings, whether past or present, without claiming to bear responsibility for them or being able to claim they involve clear public liability. Membership in a political society is more hapless and complicit than principles claiming to distribute and measure liability for blame can acknowledge. Questions of political legitimacy must acknowledge these feelings of complicity, whether complacent or embarrassed, feelings which a theory of justice can help to evaluate but cannot entirely overturn or constrain. And such questions will often involve not only the historicity of individual emotions, as it were, but also the historical memory of the commons as a whole: the question of responsibility for those acts perpetrated by ancestors and compatriots. The

---

[25] Hannah Arendt drew attention to the political significance of hope, and indeed to the significance of time-bound judgements and emotions generally, in her aptly titled collection *Between Past and Future* (1961).

[26] The path to a critical theory or practice of the emotions remains underdeveloped. The Stoic theory of the emotions has been revived and expounded by Nussbaum (1994).

[27] Recent controversies include the discussion of 'false memory syndrome' in individual psychotherapy, and the reliability of memories of the Holocaust. The latter is discussed under the rubric of 'testimony' in Felman and Laub (1992).

[28] Rawls (1971) 3: 'Justice is the first virtue of the social institions, as truth is of systems of thought. A theory however elegant and economical must be rejected or revised if it is untrue; likewise laws and institutions no matter how efficient and well-arranged must be reformed or abolished if they are unjust'.

[29] Cavell (1990). I am indebted to Stephen Mulhall's paper on *Promising, Consent and Citizenship* for drawing Cavell's treatment of Rawls to my attention.

significance of the collective past, the fidelity of its memory, are politically relevant and need to find a place in political theory which goes beyond simply enshrining or attacking them.

A political society's memory of its own past, what it has done and suffered, is inextricably linked to any assessment of its legitimacy in the present or prospects for the future. Recognition of, and willingness to accept responsibility for, one's memories of the past and one's expectations of the future, is at the heart of the nexus between the normative concerns of contemporary political theory and the history of political thought. To recover the significance of time in the history of political thought is to restore, also, a necessary depth to its contemporary manifestations and future prospects.[30]

[30] I am indebted to Monique Deveaux, Emile Perreau-Saussine, Soran Reader and Karl Thompson, each of whom offered very helpful comments on an earlier draft of this chapter.

References

Adam, J., 1963, The Republic of Plato (New York and Cambridge).

Allen, D., 1996, A Situation of Punishment: The Politics and Ideology of Athenian Punishment (unpublished Ph.D. dissertation, Cambridge).

Anderson, B., 1983, Imagined Communities: Reflections on the Origin and Spread of Nationalism (London).

Arendt, H., 1961, Between Past and Future (London).

Baynes, K., 1992, The Normative Grounds of Social Criticism: Kant, Rawls, and Habermas (Albany, New York).

Bell, D., 1960, The End of Ideology (Glencoe, Illinois).

Benhabib, S., 1992, Situating the Self: Gender, Community, and Postmodernism (Cambridge).

Benhabib, S., 1994, 'Deliberative Rationality and Models of Democratic Legitimacy' Deutsche Zeitschrift für Philosophie 43, 1: 3–29.

Campbell, B., 1986, 'Constitutionalism, Rights and Religion: the Athenian Example' History of Political Theory 7, 2: 239–273.

Cavell, S., 1990, Conditions Handsome and Unhandsome: the Constitution of Emersonian Perfectionism: the Carus lectures, 1988 (Chicago).

Cohen, D., 1991, Law, Sexuality, and Society: the Enforcement of Morals in Classical Athens (Cambridge).

Cohen, D., 1995, Law, Violence, and Community in Ancient Athens (Cambridge).

Cohen, J., 1989, 'Deliberation and Democratic Legitimacy'. In: A. Hamlin (ed), 1989, The Good Polity (Oxford).

Cole, T., 1991, The Origins of Rhetoric in Ancient Greece (Baltimore).

Connolly, W. ed., 1984, Legitimacy and the State (Oxford).

Dunn, J. ed, 1993a; c.1992, Democracy: The Unfinished Journey 508 BC to AD 1993 (Oxford).

Dunn, J. 1993b, 'Conclusion'. In: Dunn, 1993a, 239–266.

Dunn, J. 1996, The History of Political Theory and Other Essays (Cambridge).

Eisenstein, E. L., 1966, 'Clio and Chronos. An Essay on the Making and Breaking of History-book Time,' History and Theory 6: 36–64.

Felman, S., and Laub, D., 1992, Testimony:

Crises of Witnessing in Literature, Psychoanalysis, and History (New York and London).
Fukuyama, F., 1992, The End of History and the Last Man (London).
Gill, C., 1995, 'Rethinking Constitutionalism in Statesman 291–303'. In: Rowe, 1995, 292–305.
Goodin, R. and Pettit, P. (eds.), 1993, A Companion to Contemporary Political Philosophy (Oxford).
Gunnell, J.G., 1968, Time and Political Philosophy (Middletown, Connecticut).
Haakonssen, K., 1996, Natural Law and Moral Philosophy: From Grotius to the Scottish Enlightenment (Cambridge).
Herman, B., 1993, The Practice of Moral Judgment (Cambridge and London).
Herman, B., 1996, 'Pluralism and the Community of Moral Judgment'. In: D. Heyd (ed.), Toleration: An Elusive Virtue (Princeton, 60–80).
Hornblower, S., 1993, 'Creation and Development of Democratic Institutions in Ancient Greece'. In: Dunn, 1993a, 1–16.
Hughes, H.S., 1977; c.1958, Consciousness and Society: The Reorientation of European Social Thought 1890–1930 (New York).
Jaeger, W., 1963, Demosthenes: the Origins and Growth of his Policy, tr. E.S. Robinson (New York).
Lane, M., 1995, 'A New Angle on Utopia: The Political Theory of the Statesman'. In: Rowe, 1995, 276–291.
Lane, M., 1998, Method and Politics in Plato's Statesman (Cambridge).
Manuel, F.E., 1993; c.1965, Shapes of Philosophical History (Aldershot, Hampshire).
Momigliano, A., 1966, 'Time in Ancient Historiography,'. In: History and the Concept of Time: History and Theory, Studies in the Philosophy of History 6 (Middletown, CT): 1–23.
Mulhall, S., 1997, 'Promising, Consent, and Citizenship. Rawls and Cavell on morality and politics' Political Theory 25, 2: 171–192.
Nozick, R., 1974, Anarchy, State, and Utopia (New York).
Nussbaum, M., 1994, The Therapy of Desire: Theory and Practice in Hellenistic Ethics (Princeton).
Ober, J., 1989, Mass and Elite in Democratic Athens: Rhetoric, Ideology, and the Power of the People (Princeton).
Pettit, P., 1993, 'The contribution of analytical philosophy'. In: Goodin and Pettit, 1993, 7–38.
Pocock, J.G.A., 1969, Review of Gunnell, History and Theory 8: 295–301.
Pocock, J.G.A., 1972, Politics, Language and Time (London).
Pocock, J.G.A., 1975, The Machiavellian Moment: Florentine Political Thought and the Atlantic Republican Tradition (Princeton).
Popper, K., 1957, The Poverty of Historicism (London).
Popper, K., 1966, The Open Society and Its Enemies, 2 vols. (London).
Rawls, J., 1971, A Theory of Justice (Cambridge, Massachusetts).
Rowe, C.J. (ed.), 1995, Reading the Statesman: Proceedings of the III Symposium Platonicum (Sankt Augustin, Germany).
Sandel, M., 1982, Liberalism and the Limits of Justice (Cambridge).
Tuck, R., 1993, 'The Contribution of History'. In: Goodin and Pettit, 1993, 72–89.
Wolin, S., 1960, Politics and Vision: Continuity and Innovation in Western Political Thought (Boston and Toronto).

chapter 13

TIME AND SOCIAL ANTHROPOLOGY

ALFRED GELL
London School of Economics

Time in Contemporary Intellectual Thought
Patrick Baert (Editor)
2000 Elsevier Science B.V.

CONTENTS

13.1. Introduction

Anthropology sometimes appears the most ambitious of the human sciences, claiming, not only the capacity to represent the whole world, but in addition the ability to understand how these representations come into being as social facts. Social (biological) reproduction, and material and ideological production all fall within its ample scope, and it is often sufficient to say, that if a thing exists or may be conceived, then there is an 'anthropology' of it, i.e. an account of how it plays a part in the humanly constituted world. Naturally, anthropologists, who are conditioned to accept the claims of their chosen discipline, have not failed to conclude that there must also be an anthropology of time. No living creature is immune to time in the form of environmental rhythms and as a biologically inherent constraint on organic life; and how much more true this must be of human beings, who construct their ambience conceptually, and who recognise their mortality, their time limitations? Anthropologists have therefore been accustomed to speak of time as falling within the cultural domain, as something shaped and conditioned by society, by interaction, by practical and symbolic dispositions.

I am one of these anthropologists, and I plead guilty to having written a whole book on the *Anthropology of Time* (Gell, 1992). But the experience I gathered while researching this book – which I promised to my colleagues and myself before I had the least notion as to its contents – has rendered me more skeptical than I was at the outset of my project. Without meaning to, I lost my assurance as to the the merits of the anthropological case regarding 'social time', and this chapter expresses a more cautious view than the one which I think is still characteristic of the discipline as a whole. I came to believe that anthropologists allow themselves too much liberty in asserting that this or that fundamental aspect of the world is 'socially constructed' and therefore belongs to them, as of right, and they are insufficiently critical, sometimes, of their relativist assumptions.

13.2. Social construction of time: Durkheim and Evans-Pritchard

I will outline my reasons for skepticism in due course, but first it is necessary to trace the history of the notion of time which has become installed within *forms of the Religious Life* (1915, pp. 9–11) proposed the social origins of human temporal awareness in the following terms:

> What philosophers call the categories of the understanding, the ideas of time, space, class, number, cause, personality . . . correspond to the most universal properties of things, . . . they are like the solid frame surrounding all thought [which] does not seem to be able to separate itself from them without destroying itself, for it seems that we cannot think of objects which are not in time and space, which have no number etc. . . . Now when primitive beliefs are systematically analysed, the principal categories are naturally found. They are born in religion and of religion, they are a product of religious thought. . . . Religious representations are collective representations which express collective realities . . . so if the categories are of religious origin . . . it is allowable to suppose that they are rich in social elements.
>
> . . . we cannot conceive of time except by distinguishing its different moments [i.e. periodicities] . What is the origin of this differentiation? . . . observation proves that these indispensable guidelines are taken from social life. The division into days, weeks months, years, etc. correspond to the periodical recurrence of feasts and public ceremonies. A calendar expresses the rhythm of collective activities, while at the same time its function is to assure their regularity . . .. what the category of time expresses is the time common to the group a social time, so to speak.

What is perhaps most surprising about this sociological hi-jacking of the Kantian categories is the extent to which it was, and still is accepted as valid by social scientists. Durkheim appears to be saying, and really is saying that, but for 'religion', human beings would not know whether it was day or night, summer or winter, or whether the moon was waxing or waning. Such an evidently absurd collection of propositions survives because certain questions are never asked, being obscured by an over-riding disciplinary interest in demonstrating the ubiquity of 'social' motives. (Which is linked to a moral position emphasising linkages between ethical virtue and social solidarity).

Durkheim's thesis, i.e. that human time cognition and time concepts were socially determined, found favour with a later generation of British social anthropologists, who responded both to Durkheim's intellectualism (his focus on 'collective representations') and his functionalism – representations could be explained on the basis of their contribution to the 'organic' life of society. Among these British post-Durkheimians, the most notable were Evans-Pritchard and Leach. In 'The Nuer' (1940) Evans-Pritchard made a sensible distinction between 'oecological time' and 'structural time' which enabled him to preserve the more useful features of the Durkheimian theory[1] while silently abandoning its more grandiose Kantian claims. Essentially, what Evans-Pritchard did was to separate out 'practical' (environmental, oecological) time from ideological or 'social' time, which was only relevant within certain symbolic frames of reference, notably religion and politics. The time-frame of (oral) genealogical reckoning, with its conventionally tidied-up generations, and its insistence on maintaining a fixed number of generations between founding ancestors and living descendants is the

[1] To which he continued to pay lip-service, Perceptions of time, in our opinion, are functions of time reckoning and hence are socially determined (1939, p. 209).

paradigm case of 'structural time' . Thus, time concepts are socially determined only in contexts of intrinsically sociological types of discourse, otherwise they are oecological. Evans-Pritchard here anticipates Bloch's (1977) contrast between ideology and cognition.

Predictably, Evans-Pritchard's prestige as an anthropological theorist owes more to his analyses of the fictions of 'structural' time than to his account of the hard facts of oecological time, and his school (based at the Institute of Anthropology, Oxford) produced a series of illuminating studies of temporal fictions, notably the work on African age-grade systems (Spencer, 1965; Hallpike, 1972; Stewart 1977) and 'generations' (Needham, 1974).

Meanwhile, the leading British anthropologist of the ensuing generation, Edmund Leach, remained more firmly within the Durkheimian orbit, at least so far as the anthropology of time was concerned. In 1961 Leach republished two essays written in the 50s which proclaimed 'The idea of Time, like the idea of God, is one of those categories we find necessary because we are social animals, rather than because of anything empirical in our objective experience of the world' (1961, p. 125). According to Leach, our category of 'time' conflates two 'basic' experiences (1) that certain processes are repetitive, or cyclic, and (2) that human life consists of irreversible changes, beginning with birth and ending with death. 'Religion' (society) creates 'time' only in order to fool us into thinking that life → death (change) is actually only a 'phase' of recurrence, (repetition) life → death → life, etc. Time is humanity's answer to entropy.

13.3. Directionality, cyclicity and circularity

Edmund Leach, in his capacity as 'social animal' and Provost of King's College Cambridge, attended innumerable divine services, but one is entitled to doubt whether he really borrowed his notions of time (or God) from traditional Christian eschatology. He would first have had to exclude himself from his account of cognitive time in order to find an external point of vantage from which to diagnose its fundamental illogic; yet in the passage quoted, he seems to include himself in that sweeping 'we'. This is a characteristic anthropological stance, ambiguously poised between acceptance of belief systems at face value, in the name of seeing the world 'from the native's point of view', yet obliged to discount any genuine possibility that the natives might have got things right, in the name of science and reason. In fact, Leach's evocation of archaic temporality as a pendulum swinging between now and not-now, never able to free itself from the repetition of what was before and will be hereafter (forever and ever) is wildly off the mark when placed against even the most schematic ethnography of the time concepts of non-western peoples.

Barnes (1974) made a necessary distinction, systematically elided by Leach, when he pointed out that societies which operate with 'cyclical' notions of time do not imagine that time literally goes round and round, or back and forth. In fact, such societies may have no visual metaphors of time at all (he is writing of Kedang in Indonesia, but the point applies very generally). They simply have collective representations of time which consist of a schedule of repeatable events which they regularly anticipate, and around which they plan their individual and group activities. Having a repetitive/cyclical schedule for events does not mean having an idea that time itself is repetitive.

We can make this clearer by means of the type-token distinction. The Kedang have a social time-schedule which consists of a temporal ordering of event-'types' (astronomical indices, seasons, agricultural operations, ritual festivals, etc.) 'tokens' of which recur periodically in unrepealable, oriented, time. Only tokens are real events, with real consequences, and the purpose of cyclical time-schedules is not to allow for a denial of time's irreversibility via the metaphysics of recurrence, but precisely to anticipate real events (tokens) advantageously. 'Be prepared' is their motto. The Kedang, just to confirm this point, are in fact very anxious to ensure, not that the dead are reborn in 'cyclic' time, but that they stay safely dead in irreversible time, and do not return to plague their relatives in the land of the living. Here 'religion' is devoted to keeping time linear, against the countervailing pull towards cyclicity which is built into peasant calendrical lore.

The Durkheimian equation of social schedules and time itself ('category' time) has misled many anthropologists and sociologists, like Leach, into drawing a broad distinction between archaic 'cyclical' time and modern linear time (Gurvich, 1961). In fact, were time actually cyclic, events could not 'recur' at a period of once per cycle. If all of time were one week, say, then we would not have 'repeated' tokens of Wednesday, but just one instance of Wednesday in the whole of time. The repeatedness of successive Wednesdays depends precisely on the (local) linearity and orientedness of time which alone allows us to say, 'Wednesday – *again*'.

Unfortunately, anthropologists, confronted with religious rituals apparently designed to bring the year – or the life cycle – back to its point of origin, have reasoned that such behaviour would be inexplicable unless people believed that time itself were cyclic. This move is really only a way of dealing with the problem of other peoples apparently irrational beliefs (which have to be accepted as 'true') by constructing metaphysical scenarios within which they might actually be rationally true. Thus, *if* time were cyclical, then the world would return to its point of origin. If we observe a ritual which is designed to renew the world, ergo, the parties to it believe that 'time is cyclical', and because they believe it, it is (locally) true.

The attribution of the idea of 'cyclic time' to the ethnographic Other arises because the anthropologist needs to rationalise the Other's behaviour, and one way of doing this is to attribute to this Other an imaginary metaphysic of recurrence. Actually, if time were circular, there would be no point to the world-renewing ritual, since the world would be renewed anyway, with or without the ritual, so the construction of this scenario is self-defeating and explains nothing about the ritual practices that provoked it. But anthropologists like to attribute non-standard metaphysics to the Other, and indeed to claim that the Other lives in a totally differently constituted 'reality' than our own.

13.4. Temporal cultural relativism and its discontents

This position is known as 'cultural relativism' The thesis that different cultures 'live' in different temporal frameworks can be called 'temporal cultural relativism' and is widespread, deriving not just from Durkheim, but also from the neo-Kantian traditions within American Cultural anthropology from Boas onwards. Cultural relativism is not necessarily a mistake; indeed, unless belonging to a different culture and experiencing the world in the light of a particular set of cultural premises made a difference, there would be nothing for anthropologists to study, describe, and analyse. The difficulty is in defining the scope and import of the relativity of cultures, not in admitting the existence of the phenomenon.

The position I take is that there is almost no limit to the substantive beliefs (representations) people may entertain as to this world, and other worlds; what beings are to be found therein, what spirits, influences and powers, by what means they may be influenced and controlled, and so on. But (like Kant) I believe that the formal properties of (cognitively accessible) time derive from the bare possibility of having representations of the world, independently of their content, which may be indefinitely variable. In other words, the world is a process which goes on in time; different cultures may posit entirely different pictures of this process, and in that sense, occupy different culturally-constituted 'worlds' – but this leaves unaffected the schema of time per se, which is logically prior to any specific concept of the world-process which is understood to transpire in time (and space). Temporal cultural relativity is not a sub-species of cultural relativity in general (as religious cultural relativity, or gastronomic cultural relativity, or judicial cultural relativity might be considered sub-species of cultural relativity). Cultural relativity is only possible because 'categories' (time, space, number and other logical parameters) allow for diverse representations of the world, i.e. because these categories are not culturally relative at all, but logico-cognitive universals.

These categorical universals have to be sharply distinguished from empirical facts. I do not at all want to say that all cultures must see the world in the same way (fundamentally) because the factual make-up of the world obliges them to. There are infinitely many sustainable interpretations of the world as factually constituted because it is always possible to sustain false premises on the basis of true factual conclusions. (Thus, 'if sorcery is true, those who have many enemies will be ensorcelled' – my uncle has many enemies, and there he lies, coughing, so sorcery is clearly to blame). The position I take is entirely permissive in this regard, which is all that is necessary to make the anthropological investigation of cultural relativity a feasible enterprise.

13.4.1. Geertz and the Balinese

In the literature on temporal cultural relativity the debate has become polarised between relativists and anti-relativists arguing at cross purposes. Let us take the most hotly debated example, the Balinese, as described by Geertz (1973). Geertz describes the Balinese as 'detemporalised' . Their lives are enacted within 'a motionless present, a vectorless now'. (ibid. 404) How does Geertz come to this assessment? Geertz has two main lines of evidence. Firstly, he cites the cyclical character of the Balinese kinship and naming systems, in which living individuals are strongly socially identified with their same-sex ascendants of the grand-parental generation, to the extent of being treated socially as these very individuals reborn. Secondly, he draws attention to the proliferation and complexity of Balinese calendars. The most 'Balinese' of these calendars is the ritual-permutational one, which combines five- six- and seven-day cycles to produce a 210-day 'year' each day of which is specifiable via a unique trinomial expression. Each individual day, says Geertz, has its own specific character (of auspiciousness for this or that purpose, inauspiciousness for others) so that time is read qualitatively and non-progressively rather than quantitatively and progressively.

Geertz has been criticised by a number of writers, some, like Howe (1981) only seeking to moderate his rather sharp polarisation between Balinese time-keeping and our own, others, like Bloch (1977) denouncing him in stronger terms for confusing ritual ideology and practical cognition. It seems certain that the Balinese are just as adept in using their calendar(s) to plan and predict as we are in using our own; and though the Balinese calendar looks unnatural to us, that is only because we are habituated to a calendar which has oddities of its own, e.g. so-called 'months' which ignore the phases of the moon, the weekly cycle, and which may have 28, 29, 30, or 31 days – features which seem quite perplexing to non-users of the Gregorian calendar.

But to focus on such practical issues is to miss the point which Geertz is really driving at, which is to suggest that the Balinese may (in some, if not all contexts) conceptualise the world/process in terms different from our own. Their 'de-temoporalizing' of genealogical succession, such that each living individual is so to speak a 'token' of an immortal person (type) who reappears in each alternate generation is a feasible reading of how persons come to be in the world, but not a theory about cyclic time; equally, their assumption concerning the calendrical determination of lucky and unlucky days (which has echoes in western notions about Friday the 13th) is a feasible reading of how contingency operates, rather than a theory about time as such. Geertz is trying to put us into the shoes of a Balinese for whom certain potential aspects of the world/process are much more salient than they are for us; but in order to do so he often sounds like a more radical relativist who is trying to demonstrate that the Balinese live in a world whose time dimension is differently articulated than in the west, which is hardly so.

Bloch (1977) however, does not see things in this way. He sees Geertz as a proponent of Durkheim-derived relativism in its full-blown, rather than 'interpretative' form. If Balinese time is so different, he asks, how is it that anthropologists are not the ultimate authorities on time, dictating to physicists and others the concepts of time they must employ, if these are 'cultural' products? Bloch proposes a basic division between 'cognitive' time and 'ideological' time, and he accuses Geertz, and other anthropologists, of mistaking ideological fictions, designed to legitimate authority, for 'reality', when it is precisely in order to mask reality that these fictions have been invented. The de-temporalizing character of the Balinese ritual-permutational calendar immobilises time in order to obviate the possibility of questioning authority and inducing social change. It serves a sectional interest, the interest of the elite and the patrons of ritual demonstrations of symbolic power. But ideological representations of time do not abolish the capacity of human beings to cognise temporal relations practically, and thus undercut the fictions which sustain the politico-religious status quo. In effect, says Bloch, Geertz is the willing victim of the propaganda of the lords and their pundits, who want to circumscribe reality in an unmoving frame; but we should not be beguiled.

Bloch contrasts practical time to ideological time in a number of ways. Practical time is linear and progressive, ideological time is cyclic/immobile; practical time comes from experience, ideological time from ritual dogma and performance. Practical time has biological roots in innate schemata necessary for learning both to act and to speak (Piaget, 1971), ideological time is arbitrary, and so forth. While it is obviously useful to draw a distinction between the kind of elaborate cosmological schemes which are sometimes enacted in ritual, and the schedules and time-handling schemes which are deployed in practical contexts, it is not so easy to divide up 'practical' and 'ideological' time in quite the cut-and-dried manner Bloch suggests.

The Balinese calendar, for instance, does not just determine ritual life, but also commercial, domestic, and political life, because in all these domains, one needs 'luck' in order to succeed. From the Balinese perspective, finding an auspicious day is the most 'practical' of all considerations, limited only by lack of knowledge and conflict among authorities (see Davis, 1976; Tannenbaum, 1988, for similar considerations relating to Thailand). Nor are 'cyclic' time-schemes necessarily ideological rather than practical – after all, agriculture everywhere is cyclical and repetitive, year after year. Astrological almanacs are resorted to by peasant farmers not because they are the gullible dupes of ideology, but because they need a framework for practical planning, and the seemingly arbitrary advice they obtain often incorporates tried and tested traditional agronomic principles. Different environments and productive technologies can motivate a variety of calendrical devices adapted to specific circumstances, and it would be hard to extract from this diverse mass of cultural schemes a few which could be said to be 'basic' temporal- cognitive universals.

In effect, the anthropology of time, once it abandons the Durkheimian notion that 'time' is socially determined, becomes the anthropology of time-use, and time-talk (both of which are very culturally variable) rather than the anthropology of time as such. The study of time-use is one which anthropology shares with social geography, which has been productive of much research and theory in this field (Parkes and Thrift, 1980; Carlstein, 1985), the study of time-talk with comparative linguistics (Comrie, 1976, 1985). But anthropology has something special to offer in the study of the organisation of temporal relations in social life, and the description and analysis of contrasted temporal regimes. One writer who has contributed to this field greatly is Bourdieu, and it is to his treatment of the subject that I will now turn.

13.4.2. Bourdier and the Kaybele

Bourdieu's work (1963, 1977) on the anthropology of time is based on his experience among the Kaybele of Algeria, but it can also be taken more generally as representative of temporal attitudes in societies outside the orbit of modern capitalist production. The Kaybele *fellah* (peasant farmer) lives according to a temporal rhythm determined by the divisions of the ritual calendar and the cycle of agricultural operations. These schedule-bound technical activities are not understood abstractly, but are constructed according to schemata embodied in a rich accumulation of traditional attitudes; practical life is 'mythology in action'. Man lives by nature's grace, but only by violating nature with ploughs and with fire. These necessary liberties must be recompensed by sacrifices and the maintenance of ritual respect towards the earth. It is important not to be too greedy or to attempt to hurry things along.

It is useless to pursue the world
No one will ever overtake it

The Kaybele are immersed in nature and are part of it. They do not abstract time and set it apart from the flux of interlocking and culturally preordained events which carries them along. Times are not specified by the hour or the minute, but by social conventions (we will meet 'at the next market'). This lack of specificity about time is adequate because if an event is not already an inevitable element in the working out of the preordained flow of socially expectable happenings, then there is no point in making special provisions to bring it about; indeed, to do so borders on sacrilege, disrespect towards the established order of things,

The Kaybele know nothing of the standardised, metered, objectified duration which rules the lives of modern city folk.

> The intervals of subjective duration are not equal and uniform. The effective points of reference in the continual flux of time's passage are qualitative nuances read upon the surface of things. . . . Temporal points of reference are just so many experiences. One must avoid seeing here points of division, which would presuppose the notion of regular measured intervals, that is to say, a spatial conception of the temporal. The islands of time which are defined by these landmarks are not apprehended as segments of a continuous line, but rather as so many enclosed units . . . the lapse of time which constitutes the present is the whole of an action seen in the unity of a perception embracing both the retained past and the anticipated future (1963, pp. 59–60).

Because Bourdieu is trying here to evoke a species of temporal experience and attitude which is by definition 'not' that available to his (metropolitan, educated) readers, he is obliged to use rather indirect language, and anybody would be forgiven for finding his words baffling. Perhaps a more extended parable may help. I think that most of my readers will at some stage in their lives have been on a roller-coaster (or 'Big Dipper') ride. Going on a ride like this is, first of all, an intense experience, or sequence of experiences, of 'presentness'. While the ride is in progress, the ride constitutes the whole available world, as a single, whizzing, plunging, swerving, trajectory. The ride is a temporal whole focused around a 'now' which continually transforms while remaining continuous with itself, undivided into discrete blocs or periods. And one notices another thing. As the car plunges down one apparent cliff-face, one is already anticipating – and indeed already living through – the racketing crunch which will occur as one hits the bottom of the slope. And moreover, when that crunch comes and is followed by the next dizzying ascent, the violence of this transition is signally increased by the fact that so far as one's stomach is concerned, the plummeting descent is still going on, so that one's entrails seem to be going one one way and the rest of one's shattered body another. This exhilarating sequence of temporal dislocations – experiences of anticipating an imminent future ('crunch') which is as present as the present itself (if not more so) and of a past which seems, likewise, to live on into its own future as a visceral wrench – is something urbanites will happily pay good money for. It is a certain experience of vivid temporality unbounded by divisions and schemes, which is simply lived through. It is a time of past-present-future all rolled into one.

Now imagine the whole of life as a ride on a very, very slow roller-coaster, and you have Kaybele time, according to Bourdieu. And it could be so. It could be that the anticipated future could emerge out of its present, not on the time scale of seconds, as in a roller-coaster ride, but on a time-scale of days, weeks, years and whole lifetimes. It could be that the past lives on into the present as a visceral inertia, experienced as a rootedness in established routines and rhythms. We can only have access to this kind of continuous, unbounded self-transforming time in the artificial setting of a funfair, but that is because the regimentation of time has become an imperative of capitalist production (Le Goff, Thompson) – but for others, not so constrained, why should time be anything other than a self-generating 'present' coming out of a past and oriented towards a future which are both included within it it, as lags and anticipations?

Bourdieu's emphasis on the 'experience' of 'pastness in the present' and 'futurity in the present' as diagnostic features of time outside the orbit of capitalist production, represents a signal advance on earlier attempts to distinguish between 'primitive' cyclic time and non-primitive 'linear' time as distinct temporal topologies. Bourdieu overcomes such structuralist schematics by concentrating in the different ways in which time can be 'lived through' rather than 'classified out'. And I am inclined to agree, on the basis of my own field experience with peasant and tribal societies (in India and in New Guinea) that there is much to recommend the idea that quite different temporal attitudes prevail there, as compared to the west. It really is as if one were immersed in an evolving present, rather than passing from period to period according to an abstract scheme, imposed on all by the rigorous scheduling constraints of technologically-dominated life.

So much is true: but as Bourdieu himself recognises, there is another side to temporal experience, even where peasants and tribesmen are concerned. The dense, evolving past-present-future of the Kaybele is articulated by a calendrical scheme, by a series of feasts and fast, periods of intensified labour and periods of relative relaxation; and though it is true that these feasts, fasts, etc. seem to loom up like changes in the landscape viewed from the windows of a moving train, it is not true that they cannot be contemplated except in the vividness of 'presence'. They can also be regarded synoptically, as a schedule, which everybody has internalised and can recount, at least to some degree (i.e. the ordering of months, market-days, ritual seasons and so forth). There may not be perfect agreement among all informants as to what, precisely, the community wide schedule or calendar is; but all are equally sure that such a schedule exists, and it is used, practically, to coordinate action. It is just not such a matter of dominant, even obsessive, concern as it is here.

13.5. Re-universalising time

We are not alone in having to calculate in time; and against Bourdieu's occasionally rather rhapsodic evocation of peasant time, it is necessary to consider other material which suggests that 'technical' attitudes towards time are by no means a monopoly of our own, though the particular guises in which technical mastery of time may appear, often seem strange to us.

Let me cite two kinds of instance, both of which are dealt with by Bourdieu himself. These are, the timing of exchanges, and the use of calendrical lore.

One of Bourdieu's most admired analyses concerns the precise moment at which recipients of gifts feel obliged to repay them with a counter-gift. A large proportion of anthropological literature concerns gifts and counter-gifts – of food, of valuables, of livestock, of brides and grooms – because pre-capitalist non-commodity 'embedded' economies are linked with kinship, marriage, politics and religion via 'gift' exchange institutions. We may take it that in most of the societies in which Bourdieu's past-present-future time prevails, gifts are given as part of the essential fabric of social life. But when should a gift be returned? Not according to a contract-date, like a commercial loan. But if not by an agreed due date, then when? To repay too quickly is to seem to despise the initial gift (and by extension, the giver) and it robs the giver of his enjoyment of the temporary ascendancy a gift-giver has over a gift-recipient. But to be tardy is equally to seem to despise the giver and the gift, because it suggests that the gift is too insignificant to need reciprocation, and the ascendancy of the giver so trivial as not to need to be reversed. There is, however, a moment between impolite haste and excessive laggardliness, which is 'just right' (like the temperature of the middle bear's porridge) – and this is the moment when the counter-gift should be returned. This moment is not (according to Bourdieu) known by calculation, but by a gut feeling, a feeling of rightness, which comes from lifelong habitual absorption in the rhythm of community affairs. That moment looms up as part of the unfolding present-in-being, and the gift-returnee just acquiesces in the flow of events, without having to ask what motivates his action.

This account has undeniable verisimilitude, but it has to be set against the actual testimony of real participants in gift-economies, as opposed to the 'typifications' of non-western practices reconstructed by anthropologists. Occasionally this testimony is disconcerting, as witness the following crystal-clear account of the timing of exchanges given by one Kisian of Tewara. He is describing his strategy in the famous 'Kula' exchanges which circulate ceremonial valuables in and among the islands of the Massim district, New Guinea.

> Suppose I, Kisian of Tewara go [north] to the Trobriands and secure a [famous, prestigious] arm-shell called Monitor Lizard. Then I go [south] to Sanaroa and in four different places secure four different armshells[2] promising each man who gives me a shell necklace, Monitor Lizard in return, later. I, Kisian, do not have to be very specific in my promise. It will be conveyed by implication and assumption for the most part. Later, when four men appear at my home at Tewara, each expecting Monitor Lizard, only one will get it. The other three are not defrauded permanently however. They are furious, it is true, and their exchange is blocked for a year. Next year, when I, Kisian, go again to the Trobriands, I shall represent that I have four necklaces at home waiting for those who will give me four armshells. I obtain more armshells than I did previously, and pay my debts a year late. . . . I have become a great man by enlarging my exchanges at the expense of blocking [the exchanges of others] for a year. I cannot afford to block their exchanges for too long, or my exchanges will never be trusted again. I am honest in the final issue (Fortune, 1932, p. 215.)

[2] Armshells are always exchanged for necklaces and vice-versa.

This passage shows that a 'calculative' notion of time is perfectly compatible with the Bourdieu notion of pragmatic time. Kisian shows that he is perfectly able to grasp the abstract 'temporal' principle which governs cash-flow in contemporary business organisations (i.e. that growth is dependent on ensuring that debts owed to the firm by customers are cleared marginally more rapidly than debts owed by the firm to its suppliers). At the same time Bourdieu is vindicated to the extent that Kisian has to rely on 'gut feeling' in determining just how long he can hold out against his angry suppliers before his credit collapses completely. But once again, that predicament is one shared by all too many modern businessmen, so the contrast between 'pre-capitalist' and 'modern' temporal attitudes proves less clearly drawn than at first sight.

It seems to me that while there may be a different balance between Bourdieuian 'evolving' time versus abstract, calculated and manipulated time as between peasant/tribal temporal regimes and our own, it cannot be said that either has absolute preponderance anywhere, because each requires, and calls forth, the other. The Kaybele have a calendar and a series of work schedules which require some degree of formal calculation to be implemented, while we also have, though in more muted form, the experience of social life as an imminent flow of events. The real problem for the anthropology of time is to understand how abstractly-represented time and 'lived' time are interrelated. So far, anthropology has not really come to grips with this problem, because the emphasis has always been placed, rather too heavily I think, on showing how much 'the Other' differs from 'Us'. But I will conclude this rather selective account of the anthropology of time, with one more example, which might point towards a way of making more progress towards this objective.

13.6. Calendars, commensurability and calculation: the Mursi

Numerous non-technological societies in the world have 'lunar' calendars, i.e. 'months' are lunations, and successive lunations throughout the (solar) year have names. Given that ordinary day-to-day work in many of these societies (and in particular the Mursi of Ethiopia, whom I am about to discuss) is not organised on a prescriptive calendrical schedule, it is something of a puzzle to know why these month-naming systems occur so widely. The mystery deepens when one reflects on the fact that these month-naming systems must encounter the intellectual challenge presented by the non-coordination of the lunar cycle and the solar year. The solar year is on average 11 days longer than 12 successive lunations of between 29 and 30 days each. A set of 12 month-names is too few to keep up with the solar (and meteorological) year, 13 is too many. Whatever use they may be, lunar calendars are going to come unstuck unless some extra machinery for 'intercalation' exist to cover the 11-day deficit.

Let us consider what happens among the Mursi, who garden and herd cattle along the escarpment overlooking the Omo river in southern Ethiopia. They have 12 'named' months (Bergu) each associated with a particular activity, and a thirteenth 'unnamed' month, which is associated with the annual flooding of the Omo river. I will not give the actual month-names, replacing them with numbers, for convenience's sake. The 'activity' calendar looks like this:

Bergu 1: Omo river subsides, move to riverside gardens
Bergu 2: Clearing riverside gardens
Bergu 3: Planting sorghum
Bergu 4: Planting maize, weeding
Bergu 5: Harvesting sorghum, bird-scaring
Bergu 6: Harvesting, firing gardens
Bergu 7: Store Crop, plant bush gardens
Bergu 8: Heavy rain, plant bush gardens
Bergu 9: Weeding young plants
Bergu 10: Weeding, bird-scaring
Bergu 11: Harvest bush crop, collecting honey, duelling
Bergu 12: Store bush crop, drinking, duelling
(Bergu 13: the Omo floods)

(Turton and Ruggles 1978)

The Mursi think they know, more or less which Bergu it is, as the year goes by, not by counting, but because they can easily see what activities they and their immediate neighbours are engaged in. But when pressed by a curious anthropologist to be totally specific on this point, they disclaim knowledge and/or cite the (unavailable) authority of dead or distant 'experts' as the following dialogue brings out. (This conversation is also worth citing because it gives a rare sidelight onto the actual process of anthropological investigation).

Anthropologist: What number is the Bergu now?
Mursi: Don't ask me.
A: Don't you know then?
M: Not me, I just listen to what people say about the Bergu.
A: Well, what do people say at the moment then?
M: Some say it's 5 and some say it's 6.
A: What do you think it is?
M: I told you, I just listen to what they say. I'm not an expert on the Bergu.
A: Who is, then?
M: Well, there's . . . [pause for thought] there's that Gongwi man who died the other day . . . what's his name . . . Chuah; he was a real expert on the Bergu. If he were alive now he would be able to tell you.

A: Is there anybody who *is* alive who could tell me?
M: Well there's . . . Girimalori [*a man living 65 miles distant*].
(Turton and Ruggles, 1978, p. 588)

What the Mursi have is not so much a calendar as a calendrical debate, an running argument about what month it it is, to which Turton and Ruggles' informant claims he only listens – though one suspects he may put in his two-pennies' worth when the anthropologists are out of earshot; and, from one point of view, it is just as well that this permanent state of collective indecision exists about what 'month' it is. This ambiguity allows the Mursi to 'intercalate' the necessary extra days to align the sequence of lunations and the solar year without making calculations at all. They can do this because the ambiguity is regularly ironed out by the unnamed 'Omo flooding' month which all agree marks the end of the year. Once this month arrives everybody can silently re-assess the assertions they may have been making about the month in the period immediately prior to the arrival of the floods.

Take Bergu 11 and 12, for instance:

11	12	13
Honey collecting	Drinking duelling	No activity

————————-[the Omo floods]—>

If we reconstruct the picture during the second quarter of lunation 12, it is likely that Mursi opinion on the Bergu will be divided into two parties, the 'leaders' who have finished honey collecting and hold that it is Bergu 12, and the 'followers' who have not, and hold that it is Bergu 11. Once the Omo floods, there is no reason to resist the consensus that the ensuing lunation is Bergu 13 as the 'leaders' were claiming. The 'followers' will have to keep their peace; but when next year comes round, they may be proved right, as the 'leaders' come to the end of (their) lunation 12, and the Omo fails to flood, and then it will be the 'followers' turn to feel smug.

This solves the intercalation problem, but one is still entitled to wonder why the Mursi bother with all this. Especially as the Mursi also have some surprisingly sophisticated means of cross-checking their running count of lunations with the progress of the solar year. The Mursi monitor the solar year by looking out, from observation posts atop the cliffs overlooking the Omo valley, for the winter solstice when, as they say, 'the sun goes into his house'. However, although the solstice is monitored in this way, the Mursi hold that the winter solstice does not necessarily fall in the same Bergu each year. Some years the sun goes into his house in Bergu 5, and in other years in Bergu 6. If the solstice comes in Bergu 5, or more precisely, in what any particular Mursi takes to be Bergu 5, then that portends a poor rainy season, come Bergu 8. Another Mursi, who thinks that it is Bergu 6, not 5, will not draw the same inference.

In other words, the sun is observed, like the weather, to provide partial clues as to which Bergu it is, but is not the source of evidence which is in any way stronger than the evidence provided by the weather, the progress of the agricultural year, or attempts to count lunations. It is all a matter of the flux of village opinion, there being no single knock-down argument – other than the annual arrival of the Omo floods – to identify the Bergu once and for all. Each may interpret the available evidence as he wishes. Turton and Ruggles rightly compare Mursi time reckoning with divination, a similarly tentative procedure, equally swayed by the currents of public opinion.

13.7. Conclusion: a shared constraint

The point is, that this continuous and unavailing effort to align the abstract scheme of lunar months against the flow of quotidian life, is productive of socially very useful knowledge. Charting the months means attending to life in an organized, structured, way, measuring the progress of daily affairs against an abstract scheme which produces continuous feedback and feedforward in day-to-day decision making. The babble of dissenting voices debating the identity of the 'current' month are channelling information about the progress of gardens and harvests into the common pool. Trying to keep track of time is part of the more general process of trying to keep up with events, seeking to anticipate marginal changes in conditions, the onset of a prosperous or particularly difficult season. Watching the sun go into his house and worrying – or not worrying – about the still distant rainy season is also part of the process of keeping up with time. The Mursi calendar shows how the mere existence of a classificatory scheme applied to time generates the very types of day to day observations of 'how we are getting on' which the tribal subsistence farmer, no less than the businessman poring over his sales charts and spreadsheets has to engage in in order to keep up. Bourdieu's concept of inertial, visceral, time only tells half the story; the other side of temporal awareness the calculative, detached, intellectualising side of temporal awareness also has its rightful role, among people who seem, but only seem, much less pressurised by time constraints than we are ourselves.

References

Barnes, R., 1974, Kedang (Clarendon, Oxford).

Bloch, M., 1977, The past in the present and the past, Man (n.s.12) 278–92.

Bourdieu, P., 1963, 'The attitude of the Algerian peasant towards Time'. In: J. Pitt-Rivers (ed), Mediterranean Countrymen. Paris Rechereches Mediterraneenees.

Bourdieu, P., 1977, Towards a Theory of practice (Cambridge University Press, Cambridge and New York).
Carlstein, T., 1985, Time resources, Ecology and Society (Allen and Unwin, London).
Comrie, B., 1976 (Cambridge University Press, Cambridge and New York).
Comrie, B., 1985 (Cambridge University Press, Cambridge and New York).
Davis, R., 1976, 'The Northern Thai calendar and its uses' Anthropos 71: 3-32.
Durkheim, E., 1915, The elementary Forms of the Regious Life (George Allen and Unwin, London).
Evans-Pritchard, E., 1939, 'Nuer Time Reckoning' Africa **12**, 189–216.
Evans-Pritchard, E., 1940, The Nuer (Clarendon, Oxford).
Fortune, R., 1932, Sorcerers of Dobu (Routledge, London).
Geertz, C., 1973, The Interpretation of Culture (Basic Books, New York).
Gell, A., 1992, The Anthropology of Time (Berg, Oxford).
Gurvich, G., 1961, The Spectrum of Social Time (Reidel, Dordrecht).
Hallpike, C., 1972, The Konso of Ethiopia (Clarendon, Oxford).
Howe, L., 1981, 'The Social determination of Knowledge; Maurice Bloch and Balinese Time' MAn (n.s.) **16**, 220–34.
Leach, E., 1961, Rethinking Anthropology (Athlone, London).
Needham, R., 1974, Remarks and Inventions (Tavistock, London).
Parkes, R and Thrift, N., 1980, Times, Spaces, Places (Wiley, Chichester).
Piaget, J., 1971, Structuralism (Routledge, London).
Spencer, P., 1965, Samburu (Routledge, London).
Stewart, F., 1977, Fundamentals of Age Group Systems (Academic Books, New York).
Tannenbaum, N., 1988, 'The Shan Calendrical System and its uses' Mankind **18**, 14–26.
Turton D, and Ruggles, C., 1978, 'Agreeing to Disagree: the measurement of Duration in a Southwestern Ethiopian community'. Current Anthropology **19**, 585–93.

chapter 14

POSTMODERNIST THEORIES AND THE QUESTION OF TIME

MARISSA QUIE
University of Cambridge

Time in Contemporary Intellectual Thought
Patrick Baert (Editor)
2000 Elsevier Science B.V.

CONTENTS

14.1. Introduction

> "Time provides not only a setting which permits the state of one moment to be compared historically with that of another moment. Time is also a constitutive property of society. Society is only conceivable as a system of varying states occurring at moments in time. Society displays its characteristic features not at a single moment in time but in various phases assuming various but related shapes at different and consecutive moments of time."
>
> (Shils, 1975, p. xiii)

All societies and cultures conceptualise themselves through time. All peoples have *some* perspective(s) on the past, present and future. There are now new experiences of time and place; new tensions between globalism and localism, socialism and capitalism, reality and spectacle; new instabilities of value, only meaning and identity – a dialectic between past, present and future? How are we to understand these changes? Postmodernists have offered radical reformulations across a broad range of disciplines and practices in response to economic, technological, political and sociocultural change. These reformulations challenge our ways of seeing ourselves, our communities and our world(s) within time and place.

An understanding of time needs to be at the very core of any perspective on our society and time cannot simply be conflated with social change. As Shils says: "Time is . . . a constitutive property of society" (Shils, 1975, p. xiii). Yet what I want to focus on here is precisely the diachronic element in time and the implications of its disavowal in postmodern theory. Postmodernism annihilates any real potential for change. It posits an empire of signs, whose powers of social control over individuals and societies is so absolute that no movement is possible. It stipulates an end to history and a death to the subject whose individual and collective action and whose self understanding(s) within time has made meaningful transformation and the contemplation of better future(s) possible.

The first question posed in laying the groundwork for my argument (although it can never be fully answered) is: "What is postmodernism?" Lyotard, a seminal figure in contemporary postmodernist writing devoted an essay to: "Answering the question" but hardly succeeded in resolving the matter. One of the reasons for such difficulties is the myriad of disciplines and practices which have been influenced by postmodernism. These include such diverse areas as literature, science, theology, philosophy and architecture. The commitment of virtually all of its practitioners to a rejection of any all-encompassing philosophy and their perpetual celebration of difference makes the task of defining the movement all the more difficult.

In Section 14.2, I focus on the way in which postmodernism is a response to both the impact and the failures of the Enlightenment project, a project which has its roots in the writings of the seventeenth century, pre-Enlightenment philosopher, Descartes. The Enlightenment advanced rationalism, universalism and transcendental forms of thinking. Postmodernism conceptualises itself as a challenge to such totalising discourses. The enormity of the challenge is difficult to comprehend. This is partly because its success would remove so much that is familiar in our (modernist) ways of thinking and partly because the transformation, in practice could not be as dramatic, as stark and, I would argue as implausible as the contrasts between the two 'isms' would have us believe.

The problems associated with the defining postmodernism are exacerbated by the fact that its definition (as we shall see in Section 14.3) is predicated on the meaning of modernism which is itself a disputed territory. Indeed, theorists like Alexander Callinicos (1989) see much of what is central to postmodernism's conception of itself within orthodox definitions of the modern. Postmodernists themselves demonstrate an awareness of this problem through their development of strategies of circumvention like Jencks's notion of 'double-coding' which allow for some vague form of interpenetration of periods.

Strategies of this kind are shared in part by writers like Lyotard who ironically enough was also instrumental in developing the idea that we are living in a new postmodern epoch. Lyotard argues that treating the 'post-' in the term 'postmodernist' . . . in the sense of a simple succession, of a diachrony of periods, each of them clearly identifiable is "totally modern . . . since we are beginning something completely new, we have to re-set the hands of the clock at zero. However, the idea of a total break with tradition is, rather, a manner of forgetting or repressing the past. That is to say of repeating it. Not overcoming it" (Lyotard, 1985, p. 6). If postmodernism lacks any clarity about its own position within time, how can it be a rescue of the past, of the histories of what it views as the crucial differences and identities which it says have been suppressed by the universalising metanarrative of modernity? This is the subject of Section 14.4.

Of course, postmodernists must perforce reject any concrete identification of a period style or the idea that socioeconomic, political or cultural phenomena coalesce at various times in a distinctive manner to produce an identifiable historical phase. Periodisations of this type are an anathema because they imply a simplicity and a closure which connects with distortions and misrepresentations of time. What is forgotten in such extreme characterisations is modernism's own potential for multiplicity. To use Wittgenstein's metaphor, "modernism is like a spun thread which gains its strength not from a continuous strand that runs its entire length, but from the overlapping and entwining of many separate fibres" (cited in Thier 1984, p. 201). Postmodernists fail to see how temporal categories can be used as aids for understanding – something which has been appreciated by writers as diverse as Weber, Baudelaire, Joyce, Harvey and Berman.

Postmodernists see themselves as distinct from their modernist forebears in their insistence on our inability to experience time as a coherent and integrated totality. Yet, paradoxically they also insist on their capacity to recover the past in ways which were lost by their modernist counterparts. As we shall see in Section 14.4, this 'recovery' of the past most often takes the form of montage, pastiche and unstructured eclecticism in which 'historical' accounts lose any semblance of structure and become an assemblage of fragments drawn from discourses or other cultural media.

In contradistinction to this, theorists like Taylor, Giddens and Gellner show us how historical categories need *not* imply a grounding in absolute truth. They demonstrate how an ordered chronology yields manifold rewards. Sequence clarifies, places things in context, underscores the uniqueness and/or the parallels between present and past events. It also allows for evaluation and for movement beyond the status quo. Postmodernists fail to recognise how their own insistence on incoherence and our supposed inability to capture even our own place in time or our contemporality can do violence. Such dogmatism has the effect of rooting us in the status quo and so undermines the potential for change that might be gained through critical perspectives on our past(s). Indeed for critics like Berman, the postmodernist refusal to take a temporal perspective has ultimately sinister implications. Berman stresses how the 'no future' attitude taken by writers like Baudrillard link ominously with the escape from freedom, from self-responsibility in the flow of time that was the motivation for fascism (Berman, 1992, pp. 33–58).

Another way in which postmodernists distinguish themselves from modernists is in their reassertion of the primacy of space or place over time. For them, the successes which are integral to modernity result in a severance of space from time and ultimately in the compression of time and space thereby negating the progressive potentialities of space. Their focus on space or place takes different forms – depending on how the category is defined. For some it is the recovery of an actual physical place and the generation of a kind of rootedness which helps to mitigate against the alienating effects of contemporary change. It is almost as if we could 'freeze' time for a moment and thereby recover some re-worked notion of the Heideggerian concept of 'dwelling' which might be appropriate to the postmodern condition. For others, place is a state of mind which in the contemporary epoch is open to continual reformulation so that we can choose from 'symbolic repertories' and try out new identities in accordance with our ever-changing needs. Sometimes these potentially progressive spaces or places have a link with time. Yet, as we shall see in Section 14.6 the link is never adequately specified. In addition, the progressive potentialities of space over time in these theories remain problematic.

As a movement against the excesses of modernism, postmodernism is important. Certainly elements of dogmatism which have the power to do violence to our understanding(s) of multiplicity and difference *can* be identified in modernism and should be resisted. For Newton, Locke, Pascal and Descartes, reason, rationality and science became the foundation of human progress and advancement. In Rousseau, progress and advancement became intertwined not only with increases in knowledge but also with the potential for human emancipation and equality. Fundamental to these ideas was the belief that underlying the superficial chaos of our world, there exists a rationality, a basic truth that can be identified and utilised for the improvement of the human condition.

Twentieth century theorists like Weber discerned the double-edged nature of rationality and the way in which it could stifle freedom in the modern world. His ideas have been extended and refined over the past several decades by the German social thinker, Jurgen Habermas. He seeks to separate out different forms of rationality and their different effects on our world.

For Habermas, our potential for freedom lies in language which raises the possibility of what he calls the 'ideal speech situation'. This in turn implies a particular form of communicative community. An ideal speech situation provides a paradigm against which the reality of power divided or exploitative social systems can be evaluated. He sees further hope in groups and movements like Greens or other ecological and feminist organisations which are critical of the 'productivism' of modern society. Such groups tend to reassert the primacy of social solidarity and communication. They therefore stimulate a demand for radical changes in the existing socioeconomic order which have real emancipatory potential.

Contemporary theorists such as Habermas continued therefore to defend the project of modernity. Yet for postmodernists like Lyotard and Baudrillard there can be no independent perspective through which history can be 'corrected' or 'set on the right track'. They reject the kind of critical engagement with society that a temporal perspective can give because of their fear and rejection of metanarratives.

We do not need, nor could we satisfactorily defend, the various versions of absolute truth and universality so central to the intellectual and political projects of the Enlightenment and its modernist progeny. Yet, as I try to show in the conclusion, we *can* adopt a more moderate perspective which eschews the extreme relativism of postmodernism without discarding the notion of truth altogether.

Modernity itself should not be understood in terms of the notions of fixity and stasis employed by postmodernists to render it antithetical to the unpredictability which characterises the 'postmodern condition'. In opposition to this, we need to see modernism as a movement which embraces many segments, directions and inner disputes. In this way, we can understand postmodernism and postmodernity *not* as a new and distinct type of social condition but as a part of the very diversity which is characteristic of both modernism and modernity.

An acceptance of the difference and contingency so celebrated by postmodernist thinkers need not signify a disavowal of the need for systematic social reflection and moral critique. We can only achieve this through clearly situating ourselves both within time and space. This is something which all societies and cultures have done throughout history. As Giddens says: "There is no society in which individuals do not have a sense of the future, present and past" (Giddens, 1991, p. 16). Accepting that history cannot be seen as a unified whole, or even as reflective of certain unifying principles of organisation and transformation does not mean that we need to sink into a kind of temporal chaos in which there are only incommensurable and idiosyncratic 'histories'. In all that we do, we confront our temporality. It is not that humans alone have a time-ridden existence, but that our memory and imagination enable us to encompass time and save us from being embedded in it in a way which is unreflective. We not only reach out and conceptualise ourselves within the past and the present, we also speculate about and strive for between futures. In other words, we *can* and *do* make comparisons and evaluations over time and it is the very capacity to do this which gives us the potential for progress and hope.

14.2. What is postmodernism?

Edward Shils says "Modern culture is . . . a titanic and deliberate effort to undo by technology, rationality and governmental policy the givenness of what came down from the past" (Shils, 1981, p. 197). Our precursors identified with a unitary antiquity whose fragmented vestiges became models for their own creations. With the advent of theories like postmodernism, our own more numerous and exotic pasts are divested of the iconographic meanings they once possessed. As Boorstin says, "we are flooded with disposable memoranda . . . but . . . we are tragically inept at receiving messages from our ancestors" (Boorstin, 1975, p. 788).

The Renaissance hinged on a reverence for classical antiquity which fostered rather than inhibited creativity. It generated a confidence that moderns might even surpass ancient greatness. The past was a source of inspiration which gave sustenance to humanists from Dante to Petrarch to Erasmus and Montaigne.

In contrast, the seventeenth century saw a polarization between the Ancients and the Moderns. It was at this point that distinctions began to be drawn between science's cumulative achievements and the isolated creations of art. There was a certain ambivalence among French Enlightenment philosphers who believed that the moral and artistic decline brought about by the ascendency of science and the general advance in knowledge and civilization impoverished the imagination (Manuel, 1965, pp. 67–8). Yet, cumulative progress ultimately became the hallmark of the Enlightenment. According to Hazard, by the end of the Seventeenth century "The past with all its mighty dead was set at nought" (Hazard, 1935, p. 30).

The *philosophes* sought to create a universal epistemology, replacing religion and metaphysics with science as the medium of truth. In the first instance, these ideas developed within the context of the study of nature. Later, the scientific paradigm was extended to human nature and society. By the end of the eighteenth century, there was a concerted effort to achieve conceptual and analytical unity by reducing the human and social domain to a limited number of general laws. In France and England, this took the shape of empiricist epistemology. Enlightenment philosophers believed this would destroy prejudice and undermine ignorance by revealing truths about human nature, society and history. Science would therefore make possible humanity's rational self-creation.

In the nineteenth and early twentieth centuries this project was further developed by the major thinkers who laid the groundwork for the study of society as a discipline: Comte, Marx and Durkheim. They extended the Enlightenment's privileging of science. They saw science as a mechanism for the attainment of objective knowledge. Beyond this, Comte believed his 'positive philosophy' and 'positive polity' would be a guide to social reconstruction in postrevolutionary France. Marx viewed his critique of political philosophy as a concrete contribution to political emancipation.

Postmodernism, in part, constitutes a radical deconstruction of the positivist philosophy which grew out of the Enlightenment tradition. Postmodernism takes issue with the core of the modernist dogma, that science itself is a privileged form of reason or medium of truth. It disputes the positivist claim that only scientific knowledge can be securely grounded. It rejects the idea that social theory has as its central role the securing of conceptual grounds for social research. Finally, postmodernism criticises the modernist view that science is, or should be, value-neutral. In contrast, it emphasises the practical and moral meaning of science.

In this challenge to classic ideas, it is not novel as Skinner says in his book *The Return of Grand Theory in the Human Sciences*. Rather, ". . . the empiricist citadels of English-speaking social philosophy have been threatened and undermined by successive waves of hermeneuticists, structuralists, post-empiricists, deconstructionists and other invading hordes" (Skinner, 1985, pp. 5–6). What all of these groups share in common is a 'strong dislike of ". . . all overarching theories and singular schemes of explanation."' (ibid). Hence, Lyotard's famous cliché, postmodernism is the "decline of grand narrative" (Lyotard, 1984).

Yet, postmodernism is broader than the conceptual formulations which make it part of cultural theory. This movement was echoed in philosophy by a group of

French theorists who came to be labelled as 'Post-structuralists'. In particular, Gilles Deleuze, Jacques Derrida and Michel Foucault. Despite the diversity of their thought, all three emphasised the fragmentary, heterogeneous and plural character of reality. They denied the capacity to arrive at any objective account of reality and reduced the self to an incoherent kaleidoscope of drives and desires. Homogeneous accounts of historical time are rejected in favour of differential accounts of temporality in which episodes or eras are not only discontinuous from each other but also heterogeneous within themselves.

In addition, postmodernism can be seen as a kind of ecumenical umbrella which embraces social, political and stylistic movements as well as views about changes in economy and society. It includes movements in the arts, feminism, ecology, theology, politics and even the sciences (see Prigogine and Stengers, 1984; and Gleick, 1987; Pagels, 1988). In terms of ideas about economic change, postmodernism spills over into theories of the post-industrial society. Such ideas refer (among other things) to the increase in communication, the growth of knowledge, the consequences of the consumer society, the rise of leisure and what Baudrillard calls the Disneyland *simulacra*. Other theories focus on the flowering of post-Fordism (with the corresponding insecurity of the workers); the emergence of a new world order (where neither pure capitalism nor pure socialism dominate and superpowers cease to exist); the decline of the nation-state and the emergence of supranational entities like the European Community, NAFTA and the global economy.

All of these diverse strands embrace some from of dialogic or heteroglossic style which eschews presentation in terms of unique facts and substitutes in their place multiple voices (see Rabinow, 1986, p. 245). The emphasis is on subjectivity, heterogeneity and disruption – a rejection of the universalizing themes of modernity.

14.3. Postmodernism and the problem of temporality

Perhaps the best illustration of the problem postmodernism has with time is its ambivalence about its own position within history. The most basic question we can ask is: when did postmodernism begin? This opens a Pandora's box of different responses. Jencks traces the initial usage of the concept back to the 1870s when an English Salon painter employs it to distinguish his work from that of modern French painting (Jencks, 1992, p. 10). Some theorists use the phrase to denote a cultural movement that precedes the modern while others view it as constituting a decisive break with the past. This definition is complicated by the fact that there is still no consensus about what is meant by the modern period. Michael Koehler, for example, concludes his survey of the term 'post-modernism' with the statement that it shows there is still little agreement over what can be counted as post-modern because of our lack of clarity about 'die Neuzit' (literally 'the new age', although it is usually translated as 'the modern period' or 'modern times'). Koehler himself refers to 'die Neuzit' as the time since the European Renaissance from about 1500 (Koehler in *Amerikastudien*, 22, 1, 1977, 16ff). Or, he says, it can be used to designate the most recent historical period from around 1900 and is then concurrent with the central concept of modernism.

The cultural concept of modernism can be itself divided into different periods and movements including Abstractionism, Expressionism, Surrealism or any other such modern (-ism) selected by the theorist. Further to this, several other meanings of the term modern could be added, for instance those attributed to it in the phrase describing the 'battle of the ancients and the moderns' from the late seventeenth century onwards (see: Jencks in *Architectural Design* 58, nos 1/2, 1988, p. 24). Koehler also suggests that insofar as the 'post' in 'postmodern' does not just imply a temporal relation, it designates a break with the preceding style.

Huyssen, who makes a systematic effort to periodise the concept (Huyssen in *New German Critique*, 33, Autumn, 1984, pp. 5–52) says that it goes back as far as the 1950s when it was used by Howe and Levin to lament the levelling off of the modernist movement. Postmodernism was first used in a positive sense, Huyssen believes, in the 1960s by literary critics such as Fiedler and Hassan who held widely different views of what postmodern literature was. It was only in the 1970s that the term gained a much wider currency encompassing for the first time architecture, dance, theatre, painting, film and music.

The decade of the 1970s is indeed crucial to postmodernism because of the publication of Jean-François Lyotard's *La condition postmoderne* in 1979. This work has been tremendously influential because, as Jameson suggests in his introduction to the English translation, it is a 'crossroads' in which different debates in different areas such as politics, economics and aesthetics intersect. Lyotard sometimes conceptualises postmodernity as part of the modern (Lyotard, 1984a, p. 79). Postmodernists are modern in their search for the new and their suspicion with all that has been received:

> "A work {of art} can become modern only if it is first postmodern. Postmodernism thus understood is not modernism at its end but in its nascent state, and this state is constant" (ibid).

Lyotard acknowledges that in nearly all material ways the modern tendencies continue: centralisation of power, demand for economic productivity. Even the exploding of once more integrated cultural systems or communities is not 'after' modern. It is, rather, something modernity has done throughout its existence. Yet, in other passages, Lyotard *does* tend to treat postmodernity as an historical period and postmodernism as a separable project.

In order to circumvent the problem of periodisation, interesting strategies have been developed by theorists like Charles Jencks and Umberto Eco. Jencks says that the juxtaposition of the modern against the postmodern is a methodology which is unlikely to yield anything of value. Such techniques are practised by writers like Ihab Hassan and David Harvey who construct tables composed of binary oppositions (for example: purpose versus play; centring versus dispersal; signified versus signifier and so on) (see Harvey, 1989, p. 43). In contrast to this Jencks says that the shift from one to the next ought not to be seen as a reversal or opposition. Rather it is a "hybridisation, a complexification of modern elements with other ones, that is double-coding" (Jencks, 1992, p. 12).

The term 'double-coding' is used by Jencks from 1978 onwards (in the 2nd edition of *The Language of Post-Modern Architecture*) and describes how the postmodern building may use both a 'modern' code (described as a code because it, like other architectural styles, is understood as sending out messages to its users in a similar way to other codes on speech acts) and at least one other code or style (such as, for example, the classical codes used in the Stirling Wilford and Associates' New Stuttgart State Gallery). Jencks distinguishes the postmodern in both art and architecture from the modern through the association of double-coding only with that which is postmodern. He further argues that theories of the postmodern which do not see it as aiming to double-code the modern with another code, but as simply 'deconstructing' the modern should themselves be seen as being 'late modern'. Jencks asserts that the idea of double-coding means that the pre-modern, modern and post-modern systems could all continue indefinitely in the future (Jencks, 1992, p. 15). This notion of interpenetration goes beyond Lyotard's failure to address the ambiguities in terms of a periodisation of one concept in his work. Yet it remains rather vague and cannot be reconciled with Jencks' central claim that postmodernism means both a continuation of Modernism *and* its transcendence (ibid: 26).

Another strategy of this type is the one employed by Umberto Eco in his *Postscript to the Name of the Rose*. Eco argues that postmodernism is not a trend to be defined. Conversely, it should be seen as a *Kuntswollen*, a way of operating: "we could say that every period has its own postmodernism, just as every period would have its own mannerism" (ibid., p. 73). In this way, postmodernism again means both the perpetuation of modernism and a progression beyond it.

At the extreme end of the spectrum, we find the notion that postmodernism is dead and that we are now in a 'Post-Post±Modern era'. Obituaries of this type began appearing in American architectural magazines and theoretical journals circa 1979. Postmodernists themselves repudiate this transformation and contend that such assumptions and their own lack of consensus about a periodization of the concept (both in its theoretical form and in substantive applications to changes in economy, polity and society) derive from the inherent pluralism of postmodernism and hence, are something to be celebrated.

Postmodernists suffer from great difficulties conceptualising their own project within time and yet they still assert that in contradistinction to modernism they reintegrate temporality into their work to gain new insights on the present. Central to this claim is the contention that postmodernism constitutes a rescue of the past.

14.4. Postmodernism: a rescue of the past?

What modernity shares with postmodernity is a desire to forget elements of the past. The most extreme formulation of this is found in Nietzsche who says "every past is worth condemning" (Nietzsche, 1957, p. 21). The past threatens us, diminishes us and endangers the present.

In *Literary History and Literary Modernity* Paul de Man focuses on a particular type of forgetting as the core experience of modernity (De Man, 1970, pp. 384–404). He suggests we consider "the idea of modernity" as consisting in a "desire to wipe out whatever came earlier in the hope of reaching at last a point that would be called a true present, a point of origin that marks a new departure. This combined interplay of deliberate forgetting with an action that is also a new origin reaches the full power of the idea of modernity" (ibid, pp. 388–9).

De Man deliberately emphasizes the paradox inherent in this, i.e., the more violent the rejection of anything that came before the greater the dependence on the past. Two stages might be discerned in this strategy of rejection. In the avant-garde, it took the form of a rhetoric of forgetting. In postmodernism it appears as a rhetoric of what Jameson calls *pastiche*. The attack on the avant-garde was oriented mainly towards the warehouse of collective memory: museums, libraries and academies. The appeal to forget reached its peak in the manifestos of the Futurists, who denounced intellectuals as the slaves of antiquated rites, museums as cemeteries, and libraries as burial chambers. However, the Futurists were not alone. As we have seen, the notion of a *tabula rasa* had already been trumpeted by Nietzsche and it recurred in a practical context in the work of avant-garde architects and town-planners like Le Corbusier and Mies van der Rohe. In postmodernism this perspective is often superseded by the omnipresence of pastiche in which the past is conceptualised as a vast collection of images – all styles of the past being potentially open to allusion.

"Modern life and art are removing the oppression of the past" argued Mondrian (quoted in Tafuri, 1980, p. 38). However, the breathtaking new world the modernists had to substitute for the old now strikes many as inhuman and sterile. Postmodernism is but one of the many reactions which have emerged against the avant-garde amnesia including historical eclecticism in the arts and traditionalism in literature.

In 1980 Paolo Portoghesi organised the architecture section of the Venice Biennale. It was entitled: *The Presence of the Past* and sought to highlight the element of historicism within postmodernism – something which Portoghesi and others saw as the most conspicuous trend within the movement. The avant-garde destroys and defaces the past. "The postmodern reply to the modern consists of recognising that the past, since it cannot really be destroyed, because its destruction leads to silence, must be revisited: but with *irony*, [my emphasis] not innocently" (quoted in Jencks, 1992, p. 73). Jencks refers to irony as *essential* and indispensable to any attempt to use classical vocabulary today.

Postmodernism views itself as a 'return to history', to what Eco calls "parody with critical distance" (ibid., p. 83). This, he says, allows for the ironic signalling of difference at the very heart of similarity. Eco's concept of parody is reminiscent of Jencks' notion of double-coding in that parody paradoxically enacts both change and cultural continuity. Jencks refers to the Greek prefix *para* which can mean both 'counter' or 'against' and 'near' or 'beside' (ibid). Jameson rejects this repudiation of the charge of randomness in use of the past. He argues that in postmodernism 'parody finds itself without a vocation' (Jameson, 1984, p. 65). Instead, postmodernists resort to pastiche: the past as an indecipherable jumble of images and styles with no inherent meaning or message.

Postmodernists stress the need for eclecticism when employing the past. They invoke the oxymoron of 'disharmonious harmony', 'dissonant beauty' and 'syncopated proportions'. In other words, their focus tends to be on simple ratios which are set up and then fragmented or violated. They celebrate what they see as the advantages of fragmented pluralism in both the political and cultural spheres. They make a virtue out of schizophrenia.

A polemical example of this is Dotty Attie's erotic narrative, *A Dream of Love* (1973) which incorporates 144 small drawings arranged on a grid. This is an idea which derives from the Pop Artist's use of the tackboard. Attie appropriates images from Gainsborough and Ingres, from cigarette advertisements and pornography. This culminates in a checkerboard of sexual fantasy that reads partly like a novel and partly like a classical painting. What is significant here and perhaps quite emblematic of postmodernism generally is the plurality and incoherence of both technique and reference.

The mixing of categories, genres and references is typical of postmodernism where the past is consumed, revived, plundered and often ridiculed. The movement illustrates both the enduring pull of the past and the difficulty of resuming a constructive intimacy with it after the modernist breach.

Huxtable says (with reference to architecture – but we can see how this statement has more general applicability): "They may be rediscovering the past, but their knowledge of it is still so spotty, their enthusiasm so arbitrary and episodic, that a lot of what we are getting is do-it-yourself history" (Huxtable, quoted in *New York Review of Books*, 1 May 1980, p. 26).

Postmodernism displays an unease with the past and established traditions. This manifests itself in an arbitrary revival of details rather than a real attempt to convey the spirit of previous times. Charles Moore's stainless steel and neon Piazza d'Italia in New Orleans, for example, provides a stunning collage and a dazzling spectrum of techniques but it lacks the sureness of reference necessary to a valid evocation of the past.

In a world characterised by what Henri Lefebvre has called the primacy of the 'neo', the past as a referent is gradually effaced (Lefebvre, 1971). Postmodernists employ an impressive range of techniques to convey the past but appear disinterested in the essential meanings of what they use. Treating the past as a spare-parts warehouse, in Moshe Safdie's view, they select historical motifs out of context and in ignorance and defiance of their origins and relationships (Safdie, *Atlantic Monthly* 248, December 1981, p. 68). In this way, the past becomes trivialized, it takes on the form of fragmentary rubbish rather than being a mechanism of enrichment for both the present and our powers of creativity.

However, this is clearly not the intention of those at the forefront of postmodernism. Writers like John Barth, for instance, outline a very different programme: "my ideal postmodernist author neither merely repudiates nor merely imitates either his twentieth century modernist parents or his nineteenth century pre-modernist grandparents. He has the first half of the twentieth century under his belt . . . but not on his back" (Barth, *The Atlantic* January 1980, p. 70). For Barth, there is no way of going back in time, no way of eradicating Nietzsche, Einstein or Lévi-Strauss. But postmodernists can use the past as both a source of enrichment and creativity.

The practice of postmodernists in their use of the past tends to contradict Barth's rhetoric. They revel in eclecticism and a lack of clear reference or meaning. Cigarette packets from the 1950s attract the same degree of interest as the relics of ancient Athens "who appear to be living" in Loren Eiseley's words, "amidst a meaningless mosaic of fragments. From ape skull to Mayan temple we contemplate the miscellaneous debris of time like sightseers to whom these mighty fragments, fallen gateways, and sunken galleys convey no present instruction" (Eiseley 1969, p. 6). In this way, the postmodernist rescue of the past is both gutless and vacuous. An indecisive attempt to use history as a source of new insights into our present situation.

Despite the fact that many postmodernists themselves in opposition to their modernist counterparts as 'rescuers of the past', disturbing parallels can be found between the ideas of the Futurists and central figures of the postmodernist movement such as Lyotard. His deconstructionist approach to the 'totalised metanarratives' of modernist theories ignores the potential for more moderate perspectives which allow a comparison across time and yet also accept discontinuity and internal difference.

The canonical narrative, the evil in history, appears in two almost equally sinister guises for Lyotard: in both Marxist and capitalist variants. Both share a totalitarian orientation and both demand absolute belief. This is how he describes the pragmatics of those narratives:

> If you are a citizen of one of these regimes, you are regarded as both the co-author of its narrative and a privileged listener, and . . .
>
> . . . you have to be word-perfect when you are told to act out an episode. You are officially assigned to three positions in the master narrative, and in every aspect of your own life . . . The money-narrative is its [i.e. capitalism's] canonical story because it . . . tells us that we can tell any stories we like, but it also tells us that authors must reap the profits on their narratives . . . The money narrative presupposes a first narrator, an author, an entrepreneur and a subject . . . it precludes the possibility of moving from narratee – narrated to that of narrator . . . So there is an element of religion in capitalism: the exclusive worship of the narrative entrepreneur.'
>
> (Lyotard 1989, p. 131, 149-50)

The establishment of metanarratives imposes silence upon those who seek to use them as tools for emancipation. Their final recourse is to narrate that there is nothing more to narrate, that there are no more listeners "and that there is no more reality to invent" (Lyotard, 1989, p. 132). This is Lyotard at his most extreme – suggesting a narrative of the destruction of narration. In its place, he proposes what Rorty sees as a philosophy of neo-tribalism: the erosion of master-narratives by "thousands of uncomfortable little stories" (Lyotard, 1989, p. 127).

Lyotard traces his farewell to history, and to the subject to the intellectual 'awakening' experienced as a result of his reading of Solzhenitsyn's *Gulag Archipelago*. This led him to the conclusion "that history consists of a swarm of narratives . . .; the people does not exist as a subject, it is a mass of thousands of little stories that are at once futile and serious . . . This succession of little stories is admirably common place, and it implies no recurrence and no return" (Lyotard, 1989, pp. 134–5).

Historical metanarratives are inextricably linked with the pragmatics of power. They therefore have no emancipatory potential. He rejects theory on the basis of its piety of remembering and its need to struggle against oblivion (Lyotard, 1989, p. 145). He counterposes those who have no names, and peoples and children' who like stories, because they are a language form in which it is possible to love time for its power to forget . . . That is why they do not set themselves up as subjects and retain almost nothing of the so-called lessons of history: cultures are made up of stories that are assembled in a series' (Lyotard, 1989, p. 145). These ideas are extended into a broader attack on theory because of its commitment to remembering and its struggle against all forgetfulness which becomes a polemic against memory.

Capitalism is a system which can make use of anything and this includes memory. Memory is manipulated, for Lyotard, into an incessant process of the museification of culture. Only the radical destruction of memory can halt this process. The attack must take the form of the destruction of everything which gives substance to it and the past: "Detruire l'oeuvre, mais detruire aussi l'oeuvre des oeuvres et des non-oeuvres, le kapitalisme comme musée, memoire de tout ce qui est possible. Démémoriser comme l'inconscient" (Lyotard, 1973, p. 303).

Here, the parallels with the Futurist attack on all elements of the past is ominous – both propose, in place of historical memory, a philosophy of forgetting. Hence, postmodernism returns, full circle as it were, to the worst excesses of the modernism it purports to transcend. This leads critics like Christa Burger to characterise the French philosopher's project not as postmodernist but *hyper-modernist* (Bürger, 1992, p. 13) a trend which reaches its peak in the work of Jean Baudrillard on the implosion of time

14.5. The implosion of time

> "The end of the dialectic signifier/signified which permitted an accumulation of knowledge and meaning . . . the end of . . . capital accumulation and social production. The end of linear discourse. The end of the classic era of the sign. The end of the era of Production . . .
>
> Power is no longer present except to conceal the fact that there is none . . . Illusion is no longer possible because reality is no longer possible" (Baudrillard, 1984). (Quoted in *Modernity and Identity*, eds. S. Lash and J. Friedman, 1992, p. 45.)

For Baudrillard we live in an age in which signs are no longer required to have any verifiable context with the world they allegedly represent. Representation has now passed on to the condition of pure simulation. Contemporary culture is a regime of simulation, characterised by the incessant production of images with no attempt to ground them in reality.

He believes our response to the fading out of the real is an attempt to manufacture it. Yet, our attempts to revitalise immediate experience, raw and intense reality are doomed to failure. We cannot repudiate the regime of what he calls the *simulacrum*. His example of the Philippine treatment of the Tasaday Indians in which they are returned to the jungle by the Marcos government shows how the freezing of time only extends the regime of the *simulacra:*

> "Of course these particular savages are posthumous: frozen, cryogenised, sterilised, protected to *death*, they have become referential *simulacra*, and the science itself of pure simulation" (Baudrillard, 1983, p. 15).

Baudrillard extrapolates from this type of example to move the generalisation that all of contemporary life has become a kind of facsimile in which we create manufactured objects and experiences which attempt to be more real than reality itself. This is what Baudrillard calls *hyperreality.* The most vivid overstatement of this idea comes in his later work where he contends the Gulf War could not {or indeed did not} happen. Conversely, it was a creation of media hype: part of the endless play of simulacra which in our information world have replaced reality.

What Baudrillard fails to acknowledge is the fact that symbolism, substitutes and what he calls *simulacra* begin with and are a part of language which is central to our very construction and critical perception of the social world. Although it is in some ways correct to say the media do not reflect realities but to some degree construct them, it cannot be said that reality has therefore ceased to exist and has been supplanted by an autonomous sphere of hyperreality where sign or image is everything.

In the *simulacra*-filled world Baudrillard describes, there can be no critical distance. All difference, including that of time – which might give us a critical perspective – are imploded. Opposites collapse into each other producing "a floating causality where positivity and negativity engender and overlap with one another, where there is no longer any active or passive" in which "every act terminates at the end of a cycle having benefitted everyone and scattered in all directions" (Baudrillard, 1983, pp. 30-31).

The logical extension of Baudrillard's notion of 'implosion' is the transformation of theory itself into the condition of simulation it theorizes. While most postmodernists focus on the complex question of the hermeneutic entanglement of theory with the social reality it describes, Baudrillard's work is extreme in its collapse of the distinction between theory and its object.

A similar problem is found in the work of Foucault who eschews any real conception of a critical historical consciousness. In his early works he stresses profound ruptures between historical periods and directs his attention towards the birth of modern power in the reformation of the institutions of, for example, carceral control in the seventeenth and eighteenth centuries in *Discipline and Punish* (Foucault, 1977). Yet his later work on sexuality abandons this perspective and implies that the mutuality of power and knowledge is universal, not distinctive of modernity, and that parallel analyses can be developed for all cultures and historical periods.

Foucault's later work abandons the attempt to give normative direction or to provide avenues of resistance. He rejects historical specificity as a tool for confronting different or incommensurable practices and resorts instead to incomplete and decontextualised borrowings from the past. Like other postmodernists he can, from this position, advocate only partialised resistance as opposed to emancipation. No viable direction can be given for the future because a critical comparison of past and present is not possible.

14.6. Progress through place over time?

Within the extreme world of Baudrillard's astral empire of signs there can be no vision of progress. The postmodern takes place through de-differentiation, but *not* in the direction of the particular, the local or some quasi-positive movement. Instead, implosion is into a further realm of abstraction, the implosion of an already abstract social realm, into an even more abstract semio-scape. For Baudrillard, exchange value, already a simulacrum in the modern, is imploded into sign-value, a second order simulacrum in the postmodern. The same could be said for Deleuze and Guttari, for whom the modern deterritorialisation of oedipus identity does not stretch far enough. For them, only a fully abstract, anarchic deterritorialisation of postmodern libidinal flows is acceptable.

Against this negativity, vision(s) of progress *can* be discerned in postmodernism and these are, often visions which see more potential in place than in time. Place or space is defined in a myriad of ways and is captured in terms like milieu, community, locality, location, locale, neighbourhood, region, territory and so on. Place also has an extraordinary range of metaphorical meanings. The place of women in society for instance, our place in the cosmos or a place where the oppressed can speak freely. The term is open to multi-layered interpretations and meanings. The emphasis on place/space is viewed as a deconstructive challenge. Because this category introduces innumerable contingencies, specificities and forms of difference, it is conceived by many postmodernists as being fundamentally undermining to the metanarratives which they see as characteristic of modernism. Time is seen as a category which is determinate and therefore more amenable to theories of determination. In contrast to this, postmodernists argue we *do* have a range of choices as to spatial configuration or more abstract forms of place or space. This allows for openness, unassimilation and toleration of difference. Place also has an extraordinary range of metaphorical meanings for example, the place of women in society, our place in the cosmos, a place where the oppressed can speak freely. The term is open to multi-layered interpretations and meanings.

Much of what is written about place/space and postmodern times emphasises a new phase in what Marx called the 'annihilation of space by time'. This process (according to many postmodernists) has now gained new momentum so that in the words of David Harvey (1989) we now face the threat of 'time-space compression' and the collapse of all forms of spatial identities. Indeed, for Harvey and writers like Sharon Zukin (1991) Postmodernism is fully the conquest of space by time. This process is also referred to in the literature in terms of 'globalisation', 'overcoming spatial barriers' and 'time-/space distanciation'.

Time-space compression is a term which alludes to movement and communication across space. It is a phenomenon which implies the geographical stretching-out of social relations referred to by Giddens (1984) as time-space distanciation. This can best be understood by taking account of a distinction which Giddens makes between 'social integration' and 'system integration'. The former applies to relations of reciprocity between persons who are in each other's presence. The latter refers to reciprocal relations between persons and collectivities who are not in circumstances of co-presence. Tribal and traditional small scale agricultural societies are characterised by what Giddens terms 'high presence-availability'; systemness is guaranteed or achieved through social integration. There is therefore very little 'stretching' of systemness in time-space. In contrast, civilizations of the class-divided type are able to stretch well beyond the immediacy of co-presence.

However, those who argue that we are undergoing a new phase of accelerated time-space compression usually do so from a very particular view of its determination. For Jameson and Harvey, these things are determined overwhelmingly by the actions of capital (Jameson, 1984; Harvey, 1989). This includes the process of commodification, market exchange, mass production and the spread of technological rationalisation. These processes result in a loss of spatial boundaries and a corresponding increase in alienation, insecurity and vulnerability. What Harvey and Jameson fail to take into account is the way in which different social groups might be affected in different ways by these processes so that some will be more disempowered than others. This is a point made by both Massey (1993, p. 61) and Giddens (1984).

Lyotard's vision of progress emphatically privileges space over time. He invokes a kind of neo-tribal paradise in which a set of spatially set forms of life carry on experiments, each in their own culture. In this vision, however, communication is impossible between tribes. Instead we are given a notion of 'minority affirmations' to which his *Instructions Païennes* (1977) are intended to serve as incitements. The overwhelming spatiality of these groups seems to annihilate the temporal dimension. Subject-less signifiers remain fixed in the absence of any real vision of the future in which some kind of commensurability and communication are possible.

Featherstone's (Featherstone, 1992, pp. 265–290) vision of progress (in S. Lash and J. Friedman (eds) 1992, pp. 265–290) also begins from a spatialised rather than a temporal liminality. He begins with 'Place' (i.e. in local markets and local fairs) and broadens out into the cosmopolitan, with the admission of travellers from the outside, of gypsies and exotic peoples. Here he constructs a notion of the 'carnivalesque'. This is meant to undermine the structured rule-boundedness of locality because space if flooded with unfamiliar signs symbolic inversions become the rule as the signifiers of the 'grotesque body' is reclaimed against that of the 'classical body' is reclaimed against that of the 'classical body'. From the standpoint of space, liminality is conceptualised as 'anti-structure'. Featherstone also believes his vision offers time/space choices of aesthetic and expressive reflexivity. In this vision, agents are free to choose their identities and to try on 'new masks'. Reference to a real temporal dimension are problematic in Featherstone's theory and history again is treated as a kind of pastiche which acts as a background for his ideas. Also absent is a credible treatment of the question of power and constraint and the ways in which individuals might be hampered (in both space and time) in their selection of new identities or even unaware of the variety of choices available.

Harvey believes that the conquest of space by time is accelerated in postmodernity. In the midst of such dynamic change what we need is 'Place construction' through a dialectical interplay of experiences, perception and imagination. His method would he says "find some sort of bridge between the concerns expressed in the Marxian and Heideggerian approaches" (Harvey, 1993, p. 17). It would also provide a framework across which social relations of class, gender, community, ethnicity and race could overrate. This he asserts would shed light on the "production of a spatially differentiated otherness as well as upon the chimerical ideals of isolationist communitarian politics and the dilemmas of a non-exclusionary and hence universal emancipatory politics" (ibid., p. 25).

Critics like Massey (Massey, 1991, pp. 267–81) argue that 'place' is conceptualised by Harvey as a source of stability and unproblematic identity. This renders his concept reactionary in that he focuses on single essential identities. Harvey's conception of place also seems to require a drawing of boundaries which ultimately have the exclusionary potential he so wants to avoid. These two problems derive from the influence of Heidegger on his work.

Harvey speaks of the integration of some notion of time with the recovery of place. However, his ideas in this context remain unclear. He evokes Heidegger when he speaks of the way in which place experiences should be time-deepened and memory qualified. He also speaks of the need to 'rewrite metatheory', to "specify dialectical processes in time-space rather than abandon the whole project" (Harvey, 1993, p. 5). Yet precisely how time should be conceived, remains unspecified. On a broader scale, the issue of preserving the particular within a non-exclusionary emancipatory politics has progressive potential but is too abstract within Harvey's schema to generate any plausible vision(s) of the future.

Against Harvey, Doreen Massey argues for a 'progressive sense of place'. This entails a linkage of place to places beyond. Places do not have to have boundaries, in the sense of divisions which frame simple enclosures. Conversely, definition can come in part through the particularity of linkage *to* the outside which allows for a notion of places as social relations. For Massey, places do not have single, unique 'identities'. Rather, they are full of internal differences and conflicts (Massey, 1991, pp. 267–81).

This multi-layered notion of place rejects a strict dichotomization of time and space. Here, place is not a static reality but an active, generative force which moves us beyond a simple dichotomization of the diachronic and the synchronic. What Massey hopes for is a progressive sense of place in which there is a "global sense of the local and a global sense of place" (Massey, 1993, p. 68).

Once again, the vision of progress is hopeful but nebulous. Exactly how we might understand a time-place linkage remains unspecified. The internal differences of place which are always defined in relation to wider global contexts are not dealt with in detail. The reconciliation of apparently incommensurable differences is never tackled.

Another theorist who finds hope in place is Marshall Berman. In the 1980s he, together with thinkers like Habermas made a stand against the excesses of postmodernism. Yet, unlike Habermas, he shares with the postmodernists a disdain for metanarratives. The title of his 1982 book: *All That Is Solid Melts Into Air: The Experience of Modernity* clearly conveys this point. All that is solid includes blueprints for the future and these must melt into air. Instead, Berman, pace Benjamin offers a *figural* vision of the *flaneur* looking out onto the streets of New York. He therefore begins from a localism of place and speaks not of disembodied ideal speech situations but of 'the signs in the street', of New York's upper West Side, of L.A.'s Compton of Berlin's Kreuzberg. This is what he calls (in opposition to postmodernism) 'low modernism' – a 'modernism in the streets'.

Berman disputes *post*modernism's rejection of history. This is central to his conceptualisation of modernity in *All that Is Solid Melts Into Air.* Yet his conception of temporality diverges from more orthodox interpretations. It is the self constructed time of Benjamin's *flaneur* who wanders through the Paris streets. It might also refer to the time of a winding medieval street. This is a conception of time which stresses irregularities, disruption and unpredictability. It reminds us of Henri Lefebvre's aestheticization of daily life and his notion of spatialization. While it is a disparagement of teleology, it does not reject visions of the future, historical consciousness and an ethics of responsibility. Berman wants to work towards an ethics – but an ethics which eschews grand blueprints. He seeks universalism – but one which fosters cosmopolitanism. He hopes for progress – but rejects our potential for complete emancipation.

What the analysis of postmodernist theories which stress the progress nature of place over time demonstrates is how the exclusion of time from social analysis on the privileging of categories like place/space seriously distorts our understanding of the ways social reality is or has been constituted. Berman's work (in the 'low modernist vogue') points the way (but does not go far enough) towards the kind of theories which are grounded in time and space and which allow for some form of evaluation of our past, present and future conditions.

14.7. Conclusions

Postmodernist theories allow for and indeed encourage the recognition of a multitude of voices in both time and space. As such, they can be conceived as an important effort to recover and come to terms with the facts of difference, fragmentation and otherness which have in many ways been suppressed by the legacy of Enlightenment universalism. However, it is this very preoccupation with fragmentation, combined with an inability to move beyond it, which undermines postmodernism's claim to be a liberatory and deconstructive force. The emphasis on fragmentation leads to a loss of contact with other forms of social practice which have much to teach. Conversely, we need, as Eric Wolf says, to "take cognizance of processes that transcend separable cases, moving through and beyond them and transforming them as they proceed" (Wolf, 1982, p. 17). this implies taking account of multidimensionality and historical variation but in addition making critical evaluations on the basis of the information provided by such analyses.

Fragmentation is an important feature of late modernity. However, postmodernists wrongly conclude that this marks the beginning of a new phase of development . . . a postmodern era. In opposition to this, theorists like Habermas, Giddens, Taylor and Gellner stress the way in which the unifying features of the present epoch are more significant than those which are disaggregating.

Giddens argues that in contrast to late modernity, the pre-modern world formed a genuinely fragmented array of human societies. The contemporary world is vastly different in that it produces a situation where there are problems and opportunities which no one can op out of. Examples of this include nuclear warfare and environmental disaster.

The construction of a kind of global 'we' also derives from the separation of time and space. In most pre-modern societies, time and space were fundamentally interconnected through place: "when markers were not just connected to the 'where' markets of social conduct but to the substance of that conduct itself" (Giddens, 1991, p. 16). In late modernity, place is pulled away from time and space. An important part of the beginning of this process is the diffusion of the mechanical clock. This results in a world which used a universal dating system and globally standardised time zones. The universalisation of time produces a standardised past and a globally applicable 'future'. Hence, a date such as the 'year 2000' becomes a recognisable market for the whole of humanity.

We have seen the way postmodernism has difficulty conceptualising itself within time and how it fails in its diffuse attempts to 'rescue the past'. An equally disturbing feature is its problems in dealing with the future or in conceptualising the future at all. The spatial replaces the temporal, while the fragmentation of experience underlines the contingency of the world. Postmodernist theorists are therefore unable to identify points of intervention in increasingly complex socioeconomic structures or to identify agents as bearers of social transformation. This perspective greatly undermines the potential for resistance, change or commitment to an alternative future or futures. In place of this, the emphasis on fragmentation leads to a fissiparity of political action thereby reducing the opportunities for a politics of emancipation. Thus, the status quo is unintentionally reinforced. This is a point which is implicit in the work of both Jameson and Harvey who see postmodernity as a product of late capitalism.

At the most fundamental level, postmodernism lacks any real conception of what is involved in the notion of a responsible human agent. It rejects the view that we are self-evaluating subjects who have the capacity to evaluate the world around us. Theorists like Habermas, Giddens, Taylor and Gellner believe we can come to an understanding of ourselves within time and space. Yet, such an understanding cannot rest on the kinds of claims to absolute truth put forward by some Enlightenment *philosophes.*

According to Taylor, theory must be a polyphonic discourse rather than a monological statement (Taylor, 1985, Chapter 10). It is, therefore, something which is always partial and incomplete. The objective of a polyphonic approach is to shed light on a problem from different perspectives. For Bakhtin (who first formulated the term), polyphonic meant an internal play of voices. Freud extended this idea to include the internal complexity of the individual as well as individuals. In Taylor's view, best theories are those in which there is some play of different voices, a dialectic which does not view the world in terms of a reducible whole. This perspective leaves open the potential for speculation about possible worlds and better futures in that it does not accept what exists at any one point in being all that exists or the limit of what is possible. Rather, it rejects closure in favour of internally contending possibilities.

Self-understanding can be extended to an evaluation of our own social context and beyond this to the contexts of those who inhabit times and places different from our own. In this vein, Taylor advocates the use of a language of perspicuous contrast – a kind of dialogue which enhances our understanding of ourselves and others. He says "we are always in danger of seeing our ways of acting and thinking as the only conceivable ones. That is exactly what ethnocentricity consists in. Understanding other societies ought to wrench us out of this; it ought to alter our self-understanding. It is the merit of the interpretive view that it explains how this comes about when it does" (Taylor, 1985, vol. 2, p. 129).

Postmodernists wallow in the oral and intellectual malaise of debilitating relativism. The ultimate postmodernist, according to Rabinod is "blind to her own situation as situatedness because, for postmodernist, she is so committed to a doctrine of partiality and flux for which even such things as one's own situation are so unstable, so without identity, that they cannot serve as objects of sustained reflection" (Rabinow, 1986, p. 252). Such extreme relativism, says Gellner, leads to both moral and cognitive nihilism which is simply false (Gellner, 1992, p. 71). This perspective denies the way we actually understand societies and cultures. It also negates or obscures vast differences in cognition and technical power. For Gellner, Taylor and Calhoun taking seriously other forms of life across time and space entails an evaluation of both technical efficiency and normative rightness – particularly when these forms of life are incommensurable with our own. The understanding and evaluation of these kinds of differences rests on a capacity to locate oneself within time and this understanding is crucial to an assessment of both present and future developments within human society. Gellner argues that a ". . . vision which obscures that which matters most cannot be sound" (Gellner, 1992, p. 72).

Understanding is therefore inseparable from criticism or evaluation, but this in turn is inseparable from self-criticism. The potentiality of this perspective is incalculable. Understanding other societies within time makes us challenge our own perspectives. Situating ourselves within the past and the present, acknowledging common meanings and evaluating different ones extends our knowledge of human possibilities and our visions of possible futures.

References

Baudrillard, J.-F., 1983, In the Shadow of the Silent Majorities (Semiotext (e), New York).

Barth, J., 1980, The Literature of Replenishment, Postmodernist Fiction, The Atlantic, pp. 65–71.

Benjamin, A., 1989, The Lyotard Reader (Blackwell, Oxford).

Berman, M., 1982, All That is Solid Melts into Air, The Experience of Modernity (Simon and Schuster, New York).

Berman, M., 1992, Why Modernism Still Matters. In: Scott Lash (ed.), pp. 33–58.

Boorstin, D., 1975, The Enlarged Contemporary (Reith Lectures, 1975), The Listener, pp. 786–9.

Bürger, C., 1992, Modernity and Postmodernity: Lyotard. In: Scott Lash (ed), pp. 73–93.

Callinicos, A., 1989, Against Postmodernism (Polity Press, Cambridge).

de Man, P., 1970, Literary History and Literary Modernity, Daedalus, **99**, pp. 384–404.

Eco, U., 1984, Postscript to the Name of the Rose (Harcourt, Brace Jovanovich, London).

Eiseley, L., 1975, All the Strange Hours: The Excavation of a Life (Scribner's, New York).

Featherstone, M., 1992, Postmodernism and the Aestheticization of Everyday Life. In: S. Lash (ed.), pp. 265–290.

Foucault, M., 1977, Discipline and Punish: The Birth of the Prison (Pantheon, New York).

Gellner, E., 1992, Postmodernism, Reason and Religion (Routledge, London).

Giddens, A., 1984, The Constitution of Society: The Outline of a Theory of Structuration (Polity Press, Cambridge).

Giddens, A., 1991, Modernity and Self Identity: Self and Society in the Late-Modern Age (Polity Press, Cambridge).

Gelick, J., 1987, Chaos, Making a New Science (Viking, New York).

Harvey, D., 1989, The Condition of Modernity: An Enquiry into the Origins of Cultural Change (Basil Blackwell, Oxford).

Hazard, P., 1963, The European Mind (1680–1715 (World, Meridian (1935), New York).

Huxtable, A.L., 1981, Is Modern Architecture Dead?, New York Review of Books, 16, pp. 17–22.

Jameson, F., 1984, The Cultural Logic of Capital, New Left Review 146, pp. 53–93.

Jencks, C., 1982, Free-style Classicism, Architectural Design 1/2, pp. 5–21, 117–20.

Jencks, C., 1991, The Language of Post-Modern Architecture (Academy, London).

Jencks, C., 1992, (ed), The Post-Modern Reader (Academy Editions, London).

Koehler, M., 1977, Amerikastudien **22**, 1.

Lash, S. and Jonathan F. (eds), 1992, Modernity and Identity (Blackwell, Oxford).

Lefebvre, H., 1971, Everyday Life in the Modern World (trans. S. Rabinovitch) (London).

Lyotard, J.-F., 1977, Instructions Païennes (Galilée, Paris).

Lyotard, J.-F., 1984, The Postmodern Condition: A Report on Knowledge (1979) (Trans;. Geoff Bennington and Brian Massami) (Manchester University Press, Manchester).

Lyotard, J.-F., 1989, Lessons in Paganism. In: A. Benjamin.

Manuel, F., 1965, Shapes of Philosophical History (Standford Univesity Press, Stanford).

Massey, D., 1991, The Political Place of Locality Studies, Environment and Planning A, **23**, 267–281.

Massey, D., 1993, Power-geometry and the Progressive Sense of Place in Mapping the Futures, ed. Jon Bird, (Routledge, London), pp. 59–69.

Nietzsche, F., 1957, The Use and Abuse of History (1873–6) (Bobbs-Merrill, Indianapolis).

Pagels, H., 1988, The Dreams of Reason, The Computer and The Sciences of Complexity (Bantam, New York).

Prigogine, I. and Isabelle, S., 1984, Order Out of Chaos, Man's New Dialogue With Nature (Bantam, New York).

Safdie, M., 1981, Private Jokes in Public Places' Atlantic Monthly, **248**, pp. 62–8.

Shils, E., 1975, Center and Periphery (Chicago University Press, Chicago).

Skinner, Q., 1985, The Return of Grand Theory in the Modern Sciences (Polity Press, Cambridge).

Rainbow, P., 1986, Representations are Social Facts: Modernity and Postmodernity in Anthropology. In: James Clifford and George E. Marcus (eds.), Writing Culture: The Poetics and Politics of Ethnography (University of California Press, Berkeley), pp. 234–261.

Tafuri, M., 1980, Theories of Architecture (Granada, London).

Taylor, C., 1985, Philosophy and the Human Sciences: Philosophical Papers II (Cambridge University Press, Cambridge).

Taylor, C., 1985, Human Agency and Language: Philosophical Papers I (Cambridge University Press, Cambridge).

Thier, A.,1984, Words in Reflection: Modern Language Theory and Postmodern Fiction (University of Chicago Press, London).

Wolf, E., 1982, Europe and the People Without History (University of California Press, Berkeley).

Zukin, S., 1991, Landscapes of Power: From Detroit to Disney World (University of California Press, Berkeley; Los Angeles).

chapter 15

TIME IN PSYCHOLOGY

WILLIAM J. FRIEDMAN
Oberlin College, Ohio

Time in Contemporary Intellectual Thought
Patrick Baert (Editor)

CONTENTS

15.1. Introduction

The psychology of time is an attempt to explain humans' experience of time and the ways in which humans and animals learn to adapt to a temporally structured world. Psychologists' interest in temporal experience is shared with many philosophers and writers, but psychological approaches are distinctive in the use of experiments and correlational studies to analyze underlying processes. The boundaries between psychological approaches and those in biology and neuroscience are less clear, and contemporary research on time perception and memory includes studies of brain mechanisms and biologically based cycles. Other psychologists focus on social and cultural influences on temporal perspective and behavior and work at the intersection between psychology and anthropology.

The experimental study of temporal experience dates to the second half of the nineteenth century, when researchers, mainly working in laboratories in Germany, conducted controlled studies of humans' judgments of brief durations (Fraisse, 1963). As part of a larger body of work in the field of psychophysics, they sought lawful relations between the objective physical characteristics of stimuli, in this case the amount of time that a stimulus is presented, and subjective experience. The approach has continued through most of the twentieth century and has supported the general conclusion that for intervals of time on the order of seconds to minutes, subjective time is a reasonably accurate reflection of true elapsed time (Eisler, 1976). Although the early psychophysical research was concentrated on the basic function relating subjective and objective time, later researchers began to study the influence of different types of time judgment methods, of specific influences, such as drugs, and of variation in different peoples' time judgments (e.g. differences associated with age, personality, or mental illness). Another change was prompted by research in the early to mid-twentieth century. Studies of animal learning revealed that pigeons and rats can learn to anticipate the end of some interval that the experimenter repeatedly interposes before food reinforcement is available, with animals withholding most responses until about the time that the reward is normally available. The discovery that non-human species can adapt to arbitrary intervals of time led to a body of research, mainly from the

1970s onward, designed to determine the nature of the mechanisms that underlie this ability (e.g. Church, 1984).

Other research, especially since the 1970s, has focused on the way in which cognitive processes, such as attention and memory, contribute to systematic distortions in perceived time in humans. Where psychophysics researchers mainly used 'empty' intervals in their tasks, the new studies were designed to examine the effects of the events that filled an interval or of other non-temporal tasks that subjects were asked to perform during the interval to be judged. During this same period, cognitive psychologists began to study how humans are able to judge the times when past events occurred and their ability to represent recurrent patterns of time. Finally, following the work of Piaget (1937/1954, 1946/1969), developmental psychologists have studied the origins and development of children's understanding of time. Psychological research on time is more diverse than this brief historical summary or the following sections can illustrate. The reader may wish to consult the following books for other approaches: Block (1990), Friedman (1990), Levin and Zakay (1989), Macar et al. (1992), and McGrath (1988).

15.2. Time perception

The term time perception is a metaphor; there is no sensory organ that receives temporal stimulation in the way that our eyes and ears transduce light and sound. Nonetheless, it is customary to refer to impressions of the duration of brief intervals of time, on the order of seconds to about a few minutes, as time perception. Psychologists who study time perception attempt to describe the basic properties of time judgments and the processes that contribute to these judgments. In studies with humans, researchers present stimuli (either a delimited, empty interval or an interval filled with carefully controlled events) for specific amounts of time and ask subjects to judge the amount of time that has elapsed. These judgments usually take the form of *estimation* in conventional units (e.g. seconds) or *reproduction* of the interval (e.g. having the subject depress a button for an amount of time he or she believes is equivalent to the stimulus interval). In a third method, *production*, subjects are asked to indicate the beginning and end of an amount of time that is specified in conventional units (e.g. 5 seconds). Time perception can be studied in animals by examining the subject's ability to learn to produce some behavior at a fixed amount of time after a stimulus, as in the example described earlier. In research with humans, an important distinction is drawn between *prospective* and *retrospective* time judgments. In the prospective paradigm, subjects are informed in advance that they will be judging an interval of time. This method is thought to measure impressions of quantities of time as they pass. In the retrospective paradigm, subjects do not know that they will be judging an interval of time until it is over. Retrospective judgments reflect impressions of the duration of intervals of time that are remembered.

Among the findings in this large body of research, four phenomena are particularly noteworthy and must be explained in an adequate theory of time perception. The first was referred to earlier: When humans are deprived of clocks and asked not to count, they are still able to judge intervals of time from less than a second to one or several minutes with considerable accuracy. Their performance can be described by plotting subjective time on the Y axis and objective time on the X axis for a variety of stimulus durations, and computing the exponent of the power function that best predicts Y from X. In a review of more than 100 studies, Eisler (1976) found an average exponent of 0.9. This value corresponds to a nearly linear function close to a 1 : 1 relation between subjective and objective time, although there is a reliable tendency to judge standard intervals greater than about 5 to 10 seconds as shorter than they really are. Clearly some source of information other than direct measurement or counting provides humans with impressions of quantities of time, impressions that approximately correspond to true duration.

Second, animals are also capable of judging the duration of intervals of time with some precision (e.g. Church, 1984; Gibbon and Church, 1990; Leak and Gibbon, 1995). In one procedure rats receive many repetitions of trials in which they are reinforced with food only after *N* seconds have elapsed between the illumination of a light and their pressing a lever. On some trials the light remains on for a long time but no reward is given. These trials can be used to measure when the rat 'expects' the food to be available. It has been found that maximum rates of lever pressing on these trials occur near the time when the reward has been available in the past, whether the delay that has been learned is 10 or 180 seconds or somewhere in-between. It is clear from this body of research that accurate timing is not restricted to humans and cannot depend solely on higher cognitive processes, such as language or the symbolic representation of time.

A third important phenomenon is that humans' and animals' time judgments are systematically affected by drugs (Meck and Angell, 1992; Block and Zakay, 1996). For example, when human volunteers take amphetamines, they *estimate* a given interval to be longer (or *produce* intervals that are shorter) than subjects given a placebo; objective time seems to pass too slowly by one's own standard and internal time too rapidly by a clock standard (Frankenhaeuser, 1959; Roberts, 1983; Hicks, 1992). Related effects are found in animals' performance in tasks such as the one described above (Meck, 1983; Church, 1984). Other drugs, such as alcohol and tranquilizers, have the opposite influence: External time seems to pass too rapidly and internal time too slowly (Frankenhaeuser, 1959; Fischer, 1966; Tinklenberg et al., 1976; Hicks, 1992). Many of the drugs that exert these speeding and slowing effects seem to influence the function of brain structures associated with the neurotransmitter dopamine. Other drugs that affect cholinergic transmitters seem to influence how long or short an interval of time is remembered to be (Meck, 1983; Meck and Angell, 1992). The specificity of drug effects suggest that certain brain structures play a special role in measuring and recording brief intervals of time.

A fourth research finding is consistent with everyday experience: our attitudes and activities and the things that happen during an interval can influence how much time seems to have elapsed. We all know that a boring interval spent waiting can seem endless, and an absorbing activity seems to pass quickly. Experimental studies of temporal illusions have revealed a number of reliable phenomena (Friedman, 1990). For several of these phenomena, the distinction between prospective and retrospective methods is important. Indeed, one finding is that estimates of a given amount of time are longer when subjects know in advance that it is to be judged than when they are unexpectedly asked for a judgment after the interval has ended (e.g. Brown, 1985; Zakay, 1992). Another finding is that in prospective tasks, intervals are judged to be shorter when subjects are required to engage in cognitively demanding tasks than when they perform tasks that make few demands (e.g. Zakay, 1989; and see Macar et al., 1994). A third example is that in studies using the retrospective paradigm, an interval will be remembered as longer if it was broken up into more distinct segments than if it was more temporally-uniform (e.g. Poynter, 1989; Zakay and Feldman, 1993; Zakay et al., 1994). These illusions show that drug manipulations are not the only factors that influence time perception. Cognitive factors, such as attention and how an interval is remembered, also affect subjective duration.

Several different kinds of theories have been proposed to explain time perception (Zakay, 1993; Church et al., 1994; Block and Zakay, 1996). None can account for all of the phenomena described above, and, as is common in psychological theories, all assume hypothetical mechanisms whose specific physiological basis remains unknown. A first theory, the scalar timing model, was designed to explain the second and third phenomena, animals' ability to learn to anticipate arbitrary but repeated intervals and the influence of drugs on their performance. The scalar timing theory assumes three main mechanisms: a clock, a memory store, and a comparator. The clock is analogous to a stopwatch in that it consists of a pacemaker that generates a stream of regular pulses, a switch that can begin the accumulation of pulses, and a readout of the current number accumulated. The memory store is needed to retain the number of pulses which on past trials corresponded to the delay before food became available. A comparator is needed to determine whether the current number of pulses is close enough to the value in the memory store to warrant attempting to obtain food.

The power of scalar timing theory lies in its ability to explain a number of specific findings from the animal studies. First, animal timing is flexible in that intervals of arbitrary duration (in the range of seconds to minutes) can be learned. This is explained by the capacity of the memory store to retain whatever number of pulses has been associated with the delay to food availability in the past. Second, the effects of drugs such as amphetamines and tranquilizers on performance are attributed to the drugs speeding or slowing the pacemaker. For example, if amphetamines cause the pacemaker to run more rapidly than usual, subjects should *estimate* a given interval as longer than under normal conditions because more than the usual number of pulses has accumulated in that time. Human's time *productions* of *N* seconds should be shorter than they otherwise would be, and animals should begin to respond earlier than they normally would on learned-interval tasks. Other drugs (e.g. pyrithiamine) which lead to consistently late responding are presumed to affect storage of accumulator values in the memory store. Third, by making quantitative assumptions about the mechanisms, the theory can explain some of the detailed temporal properties of animals' performance. For example, the times when rats' rate of lever pressing increase and then decrease on test trials has been found to be a constant function of the *proportion* of the optimal time that has been learned, whether the interval is, say, 30 or 50 seconds. The explanation is that the comparator uses a ratio rule: respond when the pulse count in the accumulator is some proportion of the value in reference memory.

However, scalar timing theory may not be an adequate general model of time perception because it does not easily explain the fourth set of phenomena, illusions in human's time judgments. For example, why should the subjective duration of an interval be longer in prospective than retrospective conditions or shorter when subjects engage in a cognitively demanding task? Or why, in retrospective paradigms, should dividing a task into distinct segments increase its subjective duration. Of course, one can make additional assumptions about the effects of attention on the switch that starts the timing (Allan, 1992) or the accumulation process (Block and Zakay, 1996), assumptions that make this model similar to the one to be described next. A more serious problem is that the scalar timing model seems ill-suited to explain retrospective time judgments: How does the system know to turn on the switch when an interval is not even defined until it is over? In general, the scalar timing model seems to be most applicable to the restricted set of situations in which a given interval of time is encountered repeatedly, and the system gradually builds a representation of the interval. This is true in the animal studies and in some studies of human time perception, where subjects must repeatedly discriminate particular durations (e.g. Allan and Gibbon, 1991). Indeed, it may be in such paradigms that the properties of human's time judgments most resemble those of animals (see Church, 1993). Scalar timing theory may also provide a good account of our everyday experience in situations in which we produce a response after some constant interval has elapsed. For example, with little attention to time, we seem to notice that something is wrong when a familiar traffic light takes longer than it should to change or a photocopying machine fails to produce a copy in the normal amount of time.

According to a second theory, attention-capacity theory, prospective time judgments are based on the output of a pacemaker, similar to the one assumed in scalar timing theory. However, in this model the number of pulses that are registered depends on how much of one's limited amount of attention is devoted to the passage of time. When subjects must perform a cognitively demanding task in addition to the time judgment task itself, fewer pulses are registered, and the interval is judged to be shorter than it otherwise would be. Conversely, when attention is focused primarily on the passage of time, as in situations in which subjects' primary task is waiting for the end of an interval, more pulses are registered, and the interval seems unusually long. The attention-capacity theory is successful in explaining the laboratory findings and common experience that impressions of amounts of time are inversely related to the amount of attention focused on time itself. It does not easily explain the influence of how segmented an interval is on retrospective time judgments.

In a third kind of theory, no special internal clock is assumed. Instead, humans judge amounts of time by the number of mental and environmental changes that are noticed. The greater the number of thoughts and feelings that are noticed and the more external events that are attended to, the longer an interval of time will appear to be. In prospective time judgment tasks where subjects simultaneously perform another cognitively demanding task, few thoughts and environmental events are noticed, so the interval seems relatively brief. Boring tasks, such as waiting, allow subjects to notice the numerous small changes that occur in the environment and the numerous thoughts they have, so the interval seems long. In the retrospective paradigm, dividing an interval into discrete segments increases the number of changes that can be remembered, thus lengthening one's estimate of its duration.

The main strength of change theories is their ability to explain illusions found in both prospective and retrospective time perception tasks. Another clear strength is their ability to explain performance in retrospective tasks. By basing time perception on cognitive processes that are always occurring, one avoids the problem that scalar timing and attention-capacity models have in explaining performance in retrospective tasks, where the start of an interval is not even defined until the interval is completed. Rather than making the implausible assumption that a pulse accumulator is switched on each time a new activity begins, change theories explain retrospective judgments in terms of the ordinary contents of memory.

Attention-capacity and change theories were not intended to explain animals' timing performance. Because attention-capacity theory assumes the existence of a pacemaker, it can be extended to animal timing. It is less clear how change theories, with their emphasis on noticing and remembering events, would apply to animals' ability to learn temporal intervals. However, drug effects can be explained by these two theories. Drugs which are assumed to speed or slow an internal pacemaker in scalar timing theory might be supposed to act in a similar way in attention-capacity theory, or they might exert their effects by directing attention to or away from the passage of time. Change theories explain the effects of drugs on subjective time by assuming that the drugs cause humans, and perhaps animals, to experience more or fewer changes in an interval of time than in a normal state.

No current theory seems to provide a comprehensive explanation of the variety of phenomena that have been discovered in studies of time perception. Because the ability to time intervals of arbitrary length is shared by quite different species, it seems clear that specific biological timing mechanisms are involved. Scalar timing theory provides a good description of the sorts of components that must contribute to this ability. However, it is by no means clear that specialized interval timing mechanisms play a major role in humans' experience of time. These mechanisms probably contribute to humans' ability to anticipate the end of repeated intervals of constant length, but they are unlikely to explain performance in retrospective tasks, where intervals are not even defined until they are over. Time perception in retrospective situations seems to depend on one's memory for the contents of the interval, and change models offer a useful way to describe the influence of these contents on the subjective duration of remembered intervals.

It is not clear how best to explain prospective time judgments, especially when only one or a few repeated judgments are required. There does not seem to be any way to tell whether humans are able to switch on the accumulator of some biological pacemaker at the start of an interval they have been asked to judge, or whether they base their impressions of amounts of time on the mental and other events that fill the interval. The fact that drugs influence judgments in such situations is not decisive support for biological timers, because drugs that shorten or lengthen subjective time may exert their influence either by speeding or slowing a pacemaker or by altering the number of mental and external events that are registered. It is clear that attention is an important determinant of the subjective duration in prospective time perception tasks, but again the findings do not tell us which theory best explains these effects. Attention towards or away from the passage of time may influence the registration of pulses from a pacemaker or it may influence the registration of mental and environmental events. It may be that attention to changes explains humans' performance when only one or a few judgments of a particular interval are required, and that biological timers play an increasingly important role as a given interval is repeated. But this suggestion is little more than speculation at present.

15.3. Memory for the times of past events

In addition to our impressions of the lengths of brief intervals of time, adult humans have what one might call a chronological sense of the past. Not only do we remember events that have occurred to us, but we have the impression that these events belong to particular times in our past, even if the times cannot always be determined with precision. This ability to remember the times of past events is an important topic in an area of considerable contemporary interest in cognitive psychology, the study of autobiographical memory (e.g. Rubin, 1996).

Psychologists have studied memory for the times of past events in laboratory studies, mainly by asking subjects to judge when particular words from a long list of words were presented (e.g. in which tenth of a list a word had been displayed). The times that are discriminated here range from the past few seconds to about the past half-hour. Other researchers have studied memory for time on much longer time scales by asking subjects to judge the times of news events or events the subject had recorded in a diary. Here the events come from the past weeks, months, or years before the testing session. Participants in these studies are usually asked to estimate the dates of the events (e.g. the month and year when an event occurred). Taken together, the laboratory studies and studies of memory for life events show that humans can distinguish the times of past events on both short and long time scales, but their judgments tend to be approximately, not exactly, correct. A number of other findings help us understand the psychological processes that underlie this ability (see Friedman, 1993, for a review of theories and findings).

The first of these findings is that time judgments are more accurate for recent events than for events from longer ago, whether the time scale is question is the last several minutes or the last several years. Clearly, some of the information that is used to discriminate past times is lost with the passage of time, and this process begins immediately and can continue over many years. A second finding from both laboratory studies and studies of memory for life events is that time judgments are more accurate for events that are better recalled than for those that are more poorly recalled. This suggests that general information remembered about an event contributes to the process of establishing when it occurred. A third finding is that participants in laboratory studies judge the times of stimuli more accurately when the word lists are divided into meaningful clusters (e.g. animal names, then flowers, etc.). A fourth is that subjects are especially accurate in judging the temporal position of items presented near the beginning of a list. The third and fourth findings show that participants in such studies notice and use the temporal structure of events to reconstruct when a particular event must have occurred.

Several other findings also indicate that humans use information that is associated with events in memory to reconstruct when the events must have occurred. In studies of memory for personal and news events, participants are often asked to explain how they arrived at their estimates. The most commonly reported methods involve using remembered aspects of the events to reconstruct when they must have happened. For example, in one study subjects who recalled talking about an earthquake during lunch correctly recalled that the earthquake had occurred just before noon (Friedman, 1987). This same example can be used to illustrate another finding consistent with the reconstruction of past times. It has been shown that people are often more accurate in judging the time of an event on a fine time scale than on a grosser time scale. In the earthquake study, many participants were very accurate in judging the time of day but inaccurate in recalling the day of the week or month of the year. This finding is compatible with reconstruction, assuming that information relevant to inferring the time of day could be recalled but not information relevant to inferring the month.

The results that have been summarized so far are all consistent with one of the main theories of memory for the times of past events. According to reconstructive theories, memory for time is a matter of relating whatever information we happen to remember about an event to our general knowledge of personal, conventional, and natural time patterns. In the earthquake example, lunch is known to occur in the middle of the day, so the approximate hour of the earthquake can be inferred. In other studies participants relate news stories to information about their lives, such as where they lived when they learned of the event, and use this information to constrain the year. Conventional time markers are not available in laboratory tests, but even here the evidence shows that participants notice and use temporal locations, such as the beginning of a new list or the part of the list when all the animal names were presented together. In reconstructive theories there is no special temporal code attached to events as they occur, nor is memory intrinsically organized by the order of events' occurrence. Instead, our memory for the context of an event (who was present, where the event occurred, what we were doing, what we saw, etc.) is the information used to determine the times of past events.

Another kind of theory explains memory for time in a quite different way. In distance theories some global property of the memory, such as its strength or vividness, declines with the passage of time. The strength or vividness of a memory when we recall it is used to determine how far in the past it must have occurred. This morning's news story might be judged to be quite recent because it is remembered clearly, but a news event from last year might be judged to be old because our memory for the event is quite dim. In most distance theories, strength or vividness are assumed to decline rapidly for some time after an event occurs but decline at an ever-slowing rate. If this is true, one could explain one of the main findings on memory for time, subjects' greater accuracy in judging the times of recent events. Time-associated differences in strength or vividness will be greater in the recent past, thus subjects should be better able to distinguish the times of occurrence of recent events. However, distance theories cannot explain many of the other findings that were summarized earlier. For example, they cannot explain why subjects sometimes recall times on fine scales more accurately than on grosser scales – a result that would be impossible if what was being recalled was a distance from the present – or why items from the beginning of a list are judged with greater accuracy than those in the middle. For this reason distance theories cannot provide a general account of humans' memory for time. But the question remains whether time-linked processes, such as a decline in the strengths of memories, play any role in humans' judgments of the times of past events.

Although adults rely on reconstruction in most experiments on memory for time, several recent studies suggest that humans can use distance information to gauge the times of past events when information about past locations is not available (Friedman, 1996). One line of evidence comes from the study of children. Children usually learn about long-scale time patterns, such as the order of months or seasons, at about six years of age or later. Younger children, who lack such knowledge, are unable to use locations in calendar time to judge the times of past events. For this reason they should be unable to answer accurately questions that are trivially easy for adults, such as whether one's birthday or Christmas was a shorter time ago. But children as young as four to five years of age can answer such questions with considerable accuracy when one of the events fell within the past several months and the other was substantially longer ago. Other studies show that adults can discriminate the temporal distances of events from many months before the test session even when required to respond so rapidly that reconstruction of past times is unlikely. Adults can also discriminate the times of stories presented on a television news magazine show over the past several months, even when few cues are available to permit them to reconstruct the temporal locations.

Presumably, some time-linked process, such as the decay of memory traces or the interference of subsequent processing, provides children and adults with information about the ages of memories. There is no clear evidence that shows what the critical attribute is, but it does appear to change in the way predicted by distance theories, declining most rapidly after an event occurs and more slowly with the passage of time. This is consistent with the findings of several studies that distance information is most useful in discriminating the times of events from the past few months and with the finding that when the distances of two events are compared, the ratio of their ages is a good predictor of subjects' accuracy. Probably, distance information is most often used by adults when we want to determine the approximate times of events from the recent past or when we compare the relative recency of two events that differ greatly in their ages.

A third theory of memory for time, order-code theory, explains humans' ability to judge which of two related events occurred more recently. For example, most people know that the Chernobyl nuclear accident occurred more recently than the Three Mile Island accident, even without locating either in time or comparing the strengths of memories of the two events. According to order-code theory, the occurrence of a new event activates the memory of earlier, related events, and lays down information about the temporal order of the pairs. This theory is supported by a final finding from the literature on memory for time. In several experiments participants were read lists of words and later judged which member of pairs of words had been seen more recently. The lists were made up of randomly ordered members of many different categories, such as animals or tools. Sometimes both members of a test pair were from the same category, and sometimes they were from different categories. If the order-code theory is correct, participants should be more accurate in the former condition, because the original presentation of a word when the list was read is more likely to have activated related than unrelated words. This is exactly what was found. Although order-code theory can explain these findings, it does not account for the fact that participants could also judge the order of unrelated words with some success; some other process must have been involved. Order-code theory also has difficulty explaining most of the other findings in the literature on memory for time. The process posited by order-code theory may be most relevant to one aspect of humans' chronological sense of the past, our ability to remember the sequence of meaningfully related events in our lives.

As is true for time perception, research shows that multiple processes contribute to humans' memory for the times of past events. Order-codes provide us with information about the temporal order of related events. Reconstruction is widely used to determine the locations of past events in time patterns. When neither of these sources of information is available, or when only very rough information about the time of a past event is needed, we rely on impressions of the temporal distances of remembered events.

15.4. Developmental approaches

Psychologists study the development of children's understanding of time to shed light on the components of adults' abilities and on how children come to adapt to the temporal features of the environment. One of the principal benefits of analyzing the gradual emergence of temporal abilities is that it demonstrates the multifaceted nature of humans' experience of time. Researchers have discovered that many other abilities besides time perception and memory for time underlie this experience. The general approach of developmental research is to study the performance of infants, children, or adolescents of one or more ages on tasks designed to measure their abilities as sensitively as possible. With infants this requires designing tasks that do not depend on verbal abilities, and with young children observations of behavior or relatively simple judgment tasks are most commonly used. As children grow older, they can be tested on procedures more nearly resembling those psychologists have used to study temporal abilities in adults. The research literature shows that infants begin to adapt to some of the temporal features of their environment in the early months of life, but important changes in temporal abilities continue through adolescence (see Levin and Zakay, 1989; Friedman, 1990; Macar et al., 1992; additional sources are cited below).

Within the first four months after birth, infants are sensitive to certain temporal properties of brief series of visual and auditory stimuli. For example, a rhythmic sequence consisting of two sound pulses, a longer pause, and four pulses can be distinguished from a 4-pause-2 sequence. In addition, distinctions between elementary speech sounds, or *phonemes*, can be made by infants of this age based on subtle differences in the timing of component sounds. By four months, infants notice when auditory and visual rhythms are synchronous, showing that they are able to integrate temporal information that is presented in two different sensory modalities. As a final example, infants of less than four months of age are able to learn to anticipate the appearance of visual stimuli that occur at regular temporal intervals (e.g. Haith et al., 1988). Clearly, infants are attuned to the short-term temporal structure of stimuli from a very early age.

Later in the first year of life, one can observe deliberate temporal sequencing in infants' own behavior. For example, eight-month-olds begin to order actions to attain a goal, such as a toy that an adult placed behind a barrier. By about 12 months, infants can learn to imitate a brief sequence of actions that is modeled for them, and they remember the sequence when tested many months later (Bauer, 1996). During the second and third years of life, children come to imitate progressively longer sequences of actions.

From about $1\frac{1}{2}$ years through five years of age, children master the most important aspects of time in language. Sixteen- to 18-month-olds use the order of words in a sentence to decipher its meaning (Hirsch-Pasek and Golinkoff, 1991). Between $1\frac{1}{2}$ and three years of age children learn the basics of the distinction between past, present, and future, as evidenced by their comprehension and use of terms for tense. Three- and four-year-old children understand temporal-relational terms, such as *before* and *after*, again revealing that the order in which events occur is a salient feature of young children's world. In fact, children of this age are aware of the temporal order of many familiar events in their lives. Researchers who have asked children to describe what happens when one bakes cookies or goes to a fast-food restaurant have found the three- and four-year-old children have a good understanding of the order in which many activities usually happen.

In the years that follow, children build up mental representations of much longer temporal patterns. Four-year-olds know the order of the main events of the waking day. Five- to six-year-olds can place in order a set of cards representing the seasons. Seven- and eight-year-old children are able to recite the days of the week and months in order. Adolescents show some appreciation of historical sequences.

The development of another kind of temporal knowledge has been the subject of a considerable amount of research. The influential developmental psychologist, Jean Piaget, was the first to study what we might call the homogeneity of time. Except for physicists, who understand time to be subject to the laws of relativity, most adults in modern societies conceive of time as a uniform dimension encompassing all events. Any two events can be uniquely ordered and any two intervals compared. Piaget's studies led him to believe that children begin with a more fragmentary view of time and must pass through a series of stages before they grasp the uniformity of time.

In one study he showed children two mechanical snails moving across a table side by side. The two snails started and stopped at the same times, but one moved more rapidly than the other and thus covered more distance. Piaget tested children of different ages, first ascertaining that they observed the basic information correctly, and then asked them whether the two snails moved for the same amount of time or one moved for more time. Children younger than about seven or eight years usually responded that the snail that moved faster and covered more distance had also moved for more time. Most of the older children recognized the logical necessity that two events that begin and end simultaneously must be of the same duration. Piaget concluded that for young children different actions can have their own localized times, times influenced by non-temporal factors such as distance covered or stopping point. According to his theory, children enter a new stage at about eight years of age, at which time they subordinate their perceptual impressions to logical principles. They also come to appreciate which kinds of information constrain duration and which do not. Children of this age also understand that different events can be integrated in the same temporal framework and that time is a measurable quantity.

Subsequent research has shown that the concepts that Piaget studied begin to develop earlier than he believed and are not fully mastered until later ages than he proposed. For example, in one study children were tested on problems similar to the snail problem, except that information was presented about the times that two dolls slept, and no motion was involved. Even five-year-olds were able to infer correctly the relative durations of the two intervals based on the starting and stopping times. Such findings reveal that the young child's notion of time is not always a matter of localized times. On the other hand, some of the non-temporal cues that misled Piaget's younger subjects have been found to continue to pose problems for children well beyond eight years of age. The correct integration of speed, distance, and time information can take until adolescence or later to develop.

Developmental research reveals the complexity of humans' understanding of time. Among the many facets of time that children must learn are the order of simple event sequences, the past-present-future distinction, the order of events in recurrent time patterns ranging in scale up to the yearly cycle, and the ways in which time does and does not relate to other dimensions. No single theory provides a good description of these diverse achievements. But a common theme of the findings is that as children grow older they develop more powerful conceptual tools for adapting to the temporal features of their environment.

15.5. The representation of time

The environment to which children learn to adapt is rich in recurrent temporal patterns. Some of these, such as the day and year, are determined by the earth's movement, whereas others, such as the week, are social constructions. But whether man-made or natural, humans must conceptualize the order of the parts of these patterns to be able to plan, remember the times of past events, and even understand one's present place in time. Psychologists attempt to describe the nature this knowledge through the study of mental representations. A relatively small body of evidence exists, but some progress has been made in characterizing the processes that underlie our grasp of time patterns (see Friedman, 1989, 1990, 1992 for summaries). In studies of adults' and children's representations of time patterns, participants are asked to make judgments about the order in which the elements of a pattern (e.g. the months of the year) occur. The relative difficulty of different problems can be compared to the predictions of different models of representation.

Two models seem to provide a good account of adults' and children's representations of days of the week and months of the year. The first, the image model, is an application of a general theory of representation to time patterns. A considerable body of research on mental representation supports the view that humans' knowledge of the appearance of objects and of locations in space depends, in part, on an image system. Image representations code spatial information and allow us to mentally 'see' spatial-like relations between different parts of the image. According to the image model of temporal representation, humans possess spatial-like images of time patterns, such as the days of the week and months of the year. These images capture information about the relative times that the elements occur, including their order and relative proximity.

Three kinds of evidence support the involvement of imagery in the representation of time patterns. First, when adults are asked whether they have visual images of times in a day, days of the week, months, and seasons, more than 70% report specific spatial arrangements of the parts of these time patterns, such as a line, circle, or oval. A second kind of evidence is the demonstration of similarities between judgments of the relative times of occurrence of days of the week or months of the year and the perception of pictures. In one study adults were read two intervals of months (e.g. December to May and July to October) and asked whether the first or the second interval is shorter. The lengths of the intervals to be compared varied across problems. In studies of visual perception, such as the comparison of the lengths of two lines, problems become more difficult as the magnitudes of the two stimuli become more similar. The image model predicts that the same will be true of the month task: Pairs of intervals that are close in length will be more difficult to compare than pairs that differ greatly. Response times and error rates supported this prediction. The third kind of evidence, selective interference with other concurrent tasks, will be described when evidence for the second model is presented.

According to another model, the order of the days of the week and the months of the year are represented as an ordered string of names. These verbal-list representations allow subjects to activate names of the elements sequentially in their usual order and count the number of elements that are activated. For example, most people have memorized the order of the letters of the alphabet and can covertly recite the letters to determine which letter is three after *h*. Verbal-list representations should be used when subjects are asked to judge the exact number of elements separating two days of the week or months of the year. In one experiment testing this model, subjects were asked to give the month that is *N* months after some stimulus month (e.g. "Which month is four months after August?"). On tasks like this one, subjects report using recitation, they show a substantial linear increase in response times as the number of months that must be counted increases, and they show an even steeper linear increase in response times when they must judge distances *backward* through the months. The verbal-list model predicts each of these properties.

A final kind of evidence bears on both of the models. If image representations process spatial-like information, one might expect subjects to have particular difficulty using imagery to solve temporal tasks while also keeping track of real positions in space. Similarly, verbal-list processing should be especially vulnerable to the concurrent recitation of numbers. In an experiment testing these predictions, participants solved a month task thought to rely on imagery. (A sample question is: "If it is October, which of the following pair of months would you come to first if you went *backward* through the months of the year, June or December?" [The correct answer is June.]). In another condition they were asked to perform a month task thought to rely on verbal-list processing (e.g. "What month is three months after June?" [September]). While they performed one or the other of these tasks, subjects were also required either to press a matrix of buttons in a particular pattern or to recite numbers aloud. The results showed that performance in the condition predicted to induce the use of imagery was disrupted more by the button-pressing task than the counting task, whereas in the condition thought to involve verbal-list processing, the opposite pattern was found. Taken together with the other findings, these results indicate that adults possess both image and verbal-list representations of certain time patterns and can use whichever kind of representation better provides the temporal information that is needed.

Developmental research shows that children acquire verbal-list representations of days of the week and months of the year by about seven or eight years of age, but image representations of these contents are usually not present until adolescence. Children who only possess verbal-list representations are unable to think about the time patterns in as flexible a way as adolescents and adults, who possess both kinds of representations. Because pre-adolescents have difficulty seeing the temporal relations between remote parts of these cycles, they probably have a more limited sense of orientation in time than adolescents and adults. However, image representations of another temporal cycle, the main activities of the waking day, appear at a much earlier age, by about four to five years. It seems likely that young children frequently think about and discuss with their parents and teachers which daily activities have already happened and which will happen, thus promoting the early acquisition of this flexible form of representation.

15.6. Conclusions

Psychological research over the past century has produced a substantial body of information about the experience of time. Humans and other animals can learn to anticipate the end of regular temporal intervals, probably using special biological timing mechanisms. Humans have impressions of the rate of time's passage and of quantities of time, and they are subject to a number of temporal illusions. Although biological timers may play a role in time perception, the registration of internal and external changes and memory for the contents of an interval probably explain most of humans' impressions of the magnitude of brief temporal durations. Our chronological sense of the past is the product of an ability to use what is remembered about an event to reconstruct its location in time; impressions of some quality of memories, such as their vividness; and a process that establishes temporal links between related events. Developmental research shows the numerous aspects of time that children gradually learn; including the order of parts of familiar activities; the past-present-future distinction; temporal patterns, such as the parts of a day, week, and year; and the homogeneity of time. Humans are able to have a sense of their place in time because they possess mental representations of recurrent time patterns. These representations can take several forms, including images and ordered verbal lists. By studying the nature of temporal experience, psychologists have been forced to recognize the many kinds of temporal information that inhere in the environment and the many ways that humans and animals adapt to time.

Acknowledgment

Preparation of this chapter was supported by a grant from the National Science Foundation, SBR 95-13881.

References

Allan, L. G., 1992, The internal clock revisited. In: F. Macar, V. Pouthas and W. J. Friedman (eds.), Time, action and cognition: Towards bridging the gap (Kluwer, Dordrecht, The Netherlands) pp. 191–202.

Allan, L.G. and Gibbon, J., 1991, Human bisection at the geometric mean. Learning and Motivation, **22**, 39–58.

Bauer, P.J., 1996, What do infants recall of their lives? Memory for specific events by one- and two-year-old children. American Psychologist, **51**, 29–41.

Block, R.A. (ed.), 1990, Cognitive models of psychological time (Erlbaum, Hillsdale, NJ).

Block, R.A. and Zakay, D., 1996, Models of psychological time revisited. In: H. Helfrich (ed.), Time and mind. Kirkland (Hogrefe and Huber, WA).

Brown, S.W., 1985, Time perception and attention: The effects of prospective and retrospective paradigms and task demands on perceived duration. Perception and Psychophysics, **38**, 115–124.

Church, R.M., 1984, Properties of the internal clock. In J. Gibbon and L. Allan (eds.), Timing and time perception (New York Academy of Sciences, New York), pp. 556–582.

Church, R.M., 1993, Human models of animal behavior. Psychological Science, **4**, 170–173.

Church, R.M., Meck, W.H. and Gibbon, J., 1994, Applications of scalar timing theory to individual trials. Journal of Experimental Psychology: Animal Behavior Processes, **20**, 135–155.

Eisler, H., 1976, Experiments on subjective duration 1868–1975: A collection of power function exponents. Psychological Bulletin, **83**, 1154–1171.

Fischer, R., 1966, Biological time. In J.T. Fraser (ed.), The voices of time (Braziller, New York), pp. 357–382.

Fraisse, P., 1963, The psychology of time. New York: Harper and Row.

Frankenhaeuser, M., 1959, Estimation of time: An experimental study (Almqvist and Wiksell, Stockholm).

Friedman, W.J., 1987, A follow-up to 'Scale effects in memory for the time of events': The earthquake study. Memory and Cognition, **15**, 518--520.

Friedman, W.J., 1989, The representation of temporal structure in children, adolescents and adults. In: I. Levin and D. Zakay (eds.), Time and human cognition: A life-span perspective (North-Holland, Amsterdam), pp. 259–304.

Friedman, W.J., 1990, About time: Inventing the fourth dimension (MIT Press, Cambridge, MA).

Friedman, W.J., 1992, The development of children's representation of temporal structures. In: F. Macar, V. Pouthas and W. J. Friedman (eds.), Time, action and cognition: Towards bridging the gap (Kluwer, Dordrecht, The Netherlands), pp. 67–75.

Friedman, W.J., 1993, Memory for the time of past events. Psychological Bulletin, **113**, 44–66.

Friedman, W.J., 1996, Distance and location processes in memory for the time of past events. To appear in D. L. Medin (ed.), The psychology of learning and motivation, **35** (Academic Press, Orlando, FL), pp. 1–41.

Gibbon, J. and Church, R.M., 1990, Representation of time. Cognition, **37**, 23–54.

Haith, M.M., Hazan, C. and Goodman, G.S., 1988, Expectation and anticipation of dynamic events by 3.5-month-old babies. Child Development, **59**, 467–479.

Hicks, R.E., 1992, Prospective and retrospective judgments of time: A neurobehavioral analysis. In: F. Macar, V. Pouthas and W.J. Friedman (eds.), Time, action and cognition: Towards bridging the gap (The Netherlands: Kluwer, Dordrecht), pp. 97–108.

Hirsch-Pasek, K. and Golinkoff, R.M., 1991, Language comprehension: A new look at some old themes. In: N.A. Krasnegor, D.M. Rumbaugh, R.L. Schiefelbusch and M.

Studdert-Kennedy (eds.), Biological and behavioral determinants of language development (Erlbaum: Hillsdale, NJ).

Leak, T.M. and Gibbon, J., 1995, Simultaneous timing of multiple interval: Implications of the scalar property. Journal of Experimental Psychology: Animal Behavior Proceses, **21**, 3–19.

Levin, I. and Zakay, D. (eds.) (1989, Time and human cognition: A life-span perspective (North-Holland, Amsterdam).

Macar, F., Grondin, S. and Casini, L., 1994, Controlled attention sharing influences time estimation. Memory and Cognition, **22**, 673–686.

Macar, F., Pouthas, V. and Friedman, W. J. (eds.), 1992, Time, action and cognition: Towards bridging the gap (Kluwer, Dordrecht, The Netherlands).

McGrath, J.E. (ed.), 1988, The social psychology of time: New Perspectives (Sage, Newbury Park, CA).

Meck, W.H., 1983, Selective adjustment of the speed of the internal clock and memory processes. Journal of Experimental Psychology: Animal Behavior Processes, **9**, 171–201.

Meck, W.H. and Angell, K.E., 1992, Repeated administration of pyrithiamine leads to a proportional increase in durations of remembered events. Psychobiology, **20**, 39–46.

Piaget, J., 1937, La construction du réel chez l'enfant (Delachaux and Niestlé, Paris).

Piaget, J., 1946, Le dévelopment de la notion de temps chez l'enfant (Presses Univ. de France, Paris).

Piaget, J., 1954, The construction of reality in the child. (Basic Books, New York).

Piaget, J., 1969, The child's conception of time (Routledge and Kegan Paul, London).

Poynter, D., 1989, Judging the duration of time intervals: A process of remembering segments of experience. In: I. Levin and D. Zakay (eds.), Time and human cognition: A life-span perspective (North-Holland, Amsterdam), pp. 305–331.

Roberts, S., 1983, Properties and function of an internal clock. In: R.L. Mellgren (ed.), Animal cognition and behavior (North-Holland, Amsterdam), pp. 345–397.

Rubin, D.C. (ed.) (1996, Remembering our past: Studies in autobiographical memory (Cambridge University Press, New York).

Tinklenberg, J.R., Roth, W.T. and Koppell, B.S., 1976, Marijuana and ethanol: Differential effects on time perception, heart rate, and subjective response. Psychopharmacology, **49**, 275–279.

Zakay, D., 1989, Subjective and attentional resource allocation: An integrated model of time estimation. In: I. Levin and D. Zakay (eds), Time and human cognition: A life-span perspective (Amsterdam, North-Holland), pp. 365–397.

Zakay, D., 1992, The role of attention in children's time perception. Journal of Experimental Child Psychology, **54**, 355–371.

Zakay, D., 1993, Relative and absolute duration judgments under prospective and retrospective paradigms. Perception and Psychophyics, **54**, 656–664.

Zakay, D. and Feldman, T., 1993, The role of segmentation and recallability in retrospective time estimation. The Psychological Record, **43**, 415–428.

Zakay, D., Tsal, Y., Moses, M. and Shahar, I., 1994, The role of segmentation in prospective and retrospective time estimation processes. Memory and Cognition, **22**, 344–351.

CONCLUSION: A TIME WHOSE IDEA HAS COME

ALAN SHIPMAN AND PATRICK BAERT

Time in Contemporary Intellectual Thought
Patrick Baert (Editor)

CONTENTS

To reflect on time is to arrest it. Captives of our own moment, we extend temporality as spatial captives extend territory, backwards by reinterpretation and forwards by anticipation. Views of the future are informed by present calculation, much of it rooted in past experience. Images of the past are altered by present reflection, in the light of what was later learnt or earlier hoped for. This crossing of boundaries when evaluating the present situation leads to crossing of purposes when subsequent action is taken. Expectation can become self-fulfilling or self-frustrating, as action repeats or repudiates the conditions that gave rise to it. Recollection can be distorted by imagination or preference, as witnessed by courtroom battles over retrieved memory and diplomatic storms over rewritten national history.

By importing past and future into the present for inspection, we invite the same misunderstanding as naturalists who study caged animals, or pollsters who quiz focus groups. Our sample may not be representative, or may change its ways when taken out of context. But whereas isolation helps the study of spatial phenomena by controlling for irrelevant interactions, the study of temporal phenomena becomes impoverished without such interactions. Time is an elusive concept not because it keeps on getting away, but because it keeps on returning, however cunningly clock and calendar try to keep it in its place. This book has forsaken the unified pounce on a hollow carcase in favour of snapshots in the field, from a variety of angles. In this concluding commentary, some common themes are identified and some tentative lessons drawn.

The extended present

Stretching the present, to encompass the future and assess the past, is an instinctive defence against time's relentless passage. Life in an instantaneous present would be, even 350 years after Hobbes' Leviathan, 'solitary, poore, nasty, brutish and short', deprived even of a decent memorial if its best moments faded and only the worst returned to haunt. In contrast, the extended present can savour promised pleasures before they arrive, and bask in reflected glories after they have gone. Inventing a present may, indeed, be one of the traits that makes us human. Physical and chemical processes slide seamlessly from future into past, and it takes a mental process to stand against and analyse the flow.

The extended present, as a zone of 'real' time in which action informed by knowledge of loose causal process becomes an active agent for change, takes centre stage in 'Austrian' economics (as remarked by Rizzo in Chapter 9) and plays an important role in several versions of continental philosophy (examined by Sandbothe in Chapter 2). The incentive to retrieve the past is, as Lane notes in her discussion of time in politics (Chapter 12), especially strong when we believe that golden ages reside there and authorities' legitimacy (or lack of it) derives from there. The incentive to anticipate the future is enhanced if we believe that to foresee it is to control it – although business for prophets and soothsayers can be equally brisk in communities convinced that their destiny is preordained. Whether it is gauged through the pulse of atomic clocks or the passage of seasons, whether viewed as linear progression or eternal return, there seems to be consensus that time goes by. Recollection and rational calculation, drawing on what traces and trailers can be summoned to today's screen, are our protection from having to take each moment as it comes.

Before contemporary physics brought them together in one continuum, the asymmetry of time and space seemed a matter of common sense. Whereas ground can be stood and steps retraced in space, experience of time is ephemeral, and movement through it irreversible. Our quest for a platform in time that is big enough to walk around on is motivated by the knowledge that it has only limited steering, and no reverse gear. Remembering and reasoning, with their respective gifts of precedent and prediction, embed current action in a template of stored experience and studied expectation. With the past second-thought and the future second-guessed, passing events need no longer overwhelm us, and can instead yield up their secrets to the storehouses of knowledge and culture, thereby equipping us to deal more effectively with what comes next.

Newtonian physics appeared to tilt towards determinism, a prospect notoriously explored by Laplace. It stepped back from the brink with the assumption of absolute time, passing independently of physical events. Einsteinian physics risked a similar brush with predestination by unifying time and space geometrically, dethroning simultaneity so that clocks might strike and observations arrive at different moments for those moving with different velocities. But Einstein's troublesome offspring, quantum mechanics, has turned the Newtonian threat on its head. If the universe's initial conditions were known absolutely, then its end-point and all intermediate points appear (at least in principle) predictable if known causal laws are not to be violated. But if, as in quantum theory, the initial conditions point only a probabilistic way forward, then the future must be 'open': it is the specification of an end-point that will now defy those causal laws. The inaugural 'Big Bang', and its apparent asymmetry with the terminal 'Big Crunch', enshrine such open-endedness – in a way its own theorists may not fully have acknowledged, as Newton-Smith argues powerfully in Chapter 4.

This new characterisation of relative openness challenges determinism at cultural as well as cosmic level. Concepts of time may differ between peoples as profoundly as between people, depending on which path-dependent technologies a community has encountered, and which choices were taken (and by whom) at each historical 'crossroads'. Societies differ profoundly in their characterisation and treatment of time, as Gell's anthropological overview (Chapter 13) makes clear. The tendency of traditions to take root in it (explored by Lane in the political context in Chapter 12), and of circumstances to alter the perception of it (instanced by Friedman in Chapter 15), make time a significantly more variable social construct than space.

Assessing other times

Despite the heightened sense of its achievement, the extended present must remain a conceptual fiction. A past rewritten from the present could go on to subvert that present. A future foretold in too much detail would already have been ushered in, or headed off. While space travel now borders on the commonplace, time travel remains a speculative flash in physicists' equation systems – the speculation lapsing easily into backdoor metaphysics, as Newton-Smith contends (Chapter 4). Equation of time and space in relativity stops a long way short of their equation in everyday logic. In practice we remain confined to one point in time just as much as to one point in space. This being so, there are dangers in the asymmetry with which our senses tend to treat the two.

Stuck at a particular point in space, we make intuitive correction for our displaced perspective. Practicality and past experience tell us it's the train that's pulled away, not the station platform sliding backwards. We credit parallax for our departure seeming to be on time by the overhead clock, and refraction for the spoon appearing bent when dipped into aqueous inter-city tea. Temporal illusion is, however, harder to circumvent than optical illusion – if only because, compared with the deceptively visible spatial present, the dimly-glimpsed past holds so many more surprises in store. Stuck at a particular point in time, we may be less alive to the distortions of our present perspective, or aware of the action needed to correct them. Friedman (Chapter 15) explores some of the cognitive challenges to, and responses by, beings condemned to live in time without any sense specifically assigned to its apprehension and interpretation. (Or perhaps, unlike their focused binocular outlook on space, confused by a collection of biological, cultural, historical and wall-mounted clocks which rarely if ever run in synch).

The dangers of judging the past by present standards, and gauging the future by present trends, have sparked arguments which range from cultural relativism in historical sociology to data-mining in econometrics, and which rage unabated by the discovery of 'legal' temporal imports ('stone-age' tribes unearthed in the modern jungle, or leading indicators in this year's statistics). Misunderstanding of what some past authors have said about time, as identified in particular by Sandbothe (Chapter 2) and Baert (Chapter 11), adds a recursive dimension to an already complicated debate. Time may, as Gell (Chapter 13) argues, be a universal concept (akin to space and number), with only its interpretation varying between cultures, but this offers little clarification unless we can step outside such cultures – a privilege, Quie reminds us, that is denied by the postmodern critique (Chapter 14). Moreover, as Baert also notes (following Mead), 'deep' structures can come back to the surface if present surprises force us to review the past more critically. After which (re-evaluating Foucault) surprises from the reconstrued past can disrupt our established views of the present, inverting the cognitive convention by which past familiarity forms a template on which to assimilate unfamiliar events.

Even if annals can be scanned and crystal balls scried without temporal prejudice, the process of reporting such observations may well reintroduce it. Not all languages make historic, continuous and anticipative uses of the present tense in quite the way as does English. Some get along without a future tense, as Gell (Chapter 13) observes. But all languages, whether saving time with shorthand or killing it with chat, betray their users' attachment to the present through a temporal acquisitiveness of their own. A lexicon which can make Nixon Flies To China a headline before, during or after the president's plane ride may be too flexible to convey conclusions that depend on precise demarcation of the here and now.

The language of formal modelling offers only limited escape, since it still conflates past and future into the present, even if occurrences are formally sequenced and their interconnections one-way. Assigning events to separate periods labelled $t=1,2\ldots n$ does not overcome the fact that all are being considered at a single point in time. Nor does it avoid the problems of imposing discrete units on what is logically a continuous flow. Formalisation of the (Bergsonian) intuition that a full understanding of time's dynamics cannot be constructed from the snapshots provided by static descriptions is a central theme in Lofgren's assessment of the understanding of time (Chapter 3).

The impossibility of separating description (syntax) from interpretation (semantics) when speaking or writing about time is, as Lofgren observes, a restatement of the dichotomy observed by McTaggart almost a century ago (glossing over how a century might be defined). McTaggart's A series captures the constant shading of future into past via a transitory present, while his B series presents the three series as discrete. The first concept has an intuitive appeal and operational usefulness, but depends on a conflation of incompatible states which deprives it of realism. The second restores that realism, but deprives it of any practical use – the past being irretrievable and the future inconceivable from where we are now. Philosophers' subsequent battle to bridge that gap, re-establishing past, present and future as objective features of reality, is chronicled by Teichmann (Chapter 1) and Sandbothe (Chapter 2). Human groups' efforts to survive on that bridge, resurrecting the past for those cyclical processes needed to sustain (biological and cultural) life while learning enough from the past to keep life's developmental processes moving forward, are discussed in Lane in the political context (Chapter 12) and by Gell in that of anthropology (Chapter 13).

The unreality of time when reflection brings past and future into the present, and the futility of reflecting on a past and future which the present can do nothing about, seem at first to be problems confined to human sciences and philosophy. Time's suspension between truth and tractability is less obviously troubling when the subject matter has no power to reflect on its self or surroundings, or to learn from what observers have concluded about it. Arguments for a basic distinction between 'natural' and 'historical' remain strong, despite attempts to unify the concept, as Sandbothe notes with particular reference to the philosophy of Ricoeur (Chapter 2). Confidence that objects and the cause-effect relations between them were as separable across time as across space underlay the approach to the physical world adopted by 'Enlightenment' physics and chemistry – and even exported to parts of the social world by early attempts to reduce biology to chemistry, psychology to physiology, and history to divine or dialectical grand design.

Timeless reasoning

Enlightenment reductionism, breaking the whole into parts which tended to 'equilibrium' regardless of their spatial and temporal starting-point, can in retrospect be cast as a captive of a Newtonian mechanics: a world-view subjecting movement in space to passage of mechanical time, which runs at a constant speed beyond the control of those inhabiting the space. In its Laplacean extrapolation, the natural time of Newton's clockwork renders historical time irrelevant, precise measurement of initial conditions and prediction of all consequent motion and energy transfer delivering the ultimate in determinism. Once universal laws instil the present with rigid links to the past and a map of its own future, any control exerted by the extended present becomes an illusion. In its place is an atemporality that stretches to eternity. Societies seem relegated from making time to marking time, from actively registering its passage through new events and new ideas to passively watching the clock as someone else's design unfolds around and through them.

Einsteinian physics has subsequently challenged the notion of time as existing independently of space, or of being measurable independently of those who occupy the space. Like an imperial power restoring usurped local monarchs to their thrones, relativity rescues time from omnipresence's oblivion by ending its independent existence. As a complementary dimension of space, which is itself re-drawn on lines far removed from those of Euclid, time's passage is no longer invariant with respect to the position and motion of those who observe it.

Quantum theory has since revealed subatomic processes in which predictive certainty gives way, at best, to probability. More recently, chaos theory has scaled up the resultant indeterminacy to complex systems as big as our own.

Micro-world events that could go either way, and macro-world developments that could have been vastly different with imperceptible changes in initial conditions, present significant conceptual and practical difficulties to natural science. Newton-Smith (Chapter 4) reveals the magnitude of the challenge in his study of the precursor to all initial conditions, the spectacular (but not necessarily singular) Big Bang. Relativity's white knight restores the fun to betting, but it rides a Trojan horse. The new physics leaps from the atemporal frying-pan into omnidirectional fire, across a quantum world that may consist of many more than four dimensions. After philosophers' battle to re-establish the reality of time, physicists have been similarly challenged to re-assert its ireversibility.

Coveney (Chapter 5) draws on insights from physics and chemistry to show how complex systems and dissipative structures might resolve the inconsistency between the irreversibility of macro-time and the apparent reversibility of micro-time. De Hemptinne (Chapter 6) focuses on one particular system – gas expanding from a burst balloon – to show how irreversibility depends on a system being open to its environment. Isolation begins to break down when solid walls reflect the sound wave at its edges, but only when permeable boundaries prevent them bouncing back are the initial conditions fully lost, and changes at the edge allowed to creep back into the core.

At first glance, natural sciences' tribulations appear to be social sciences' liberation, restoring the freedom Newton's apple had threatened to knock on the head. Growing up in the shadow of Newton and Descartes, social science's enduring problem lay less in keeping time moving forwards than in making it move at all. Newton's laws still hold on spatio-temporal scales between those of quark and quasar – if not perfectly, then well enough for planes to land safely and towerblocks to take our weight. Initially in a hurry to turn Newton's calculus and mechanical analogies into a new 'social physics', its pioneers were to discover they had all the time in the world.

In one of the boldest feats of atemporal theorising, neoclassical economists demonstrated the possibility of free decentralised trade producing an optimal transaction pattern with welfare-maximising properties. But this result is at risk as soon as anyone trades at a non-equilibrium price, or sinks capital into a project whose (normal) profit expectations are not precisely fulfilled. As Hahn, one of its pioneers, readily concedes (in Chapter 10), the basic general equilibrium model requires that exchanges either take place all at once (allowing state-contingent trades on a full set of forward markets), or unfold according to a sequence predetermined by preferences, technology and initial income distribution. (Perhaps endorsing the dictum that time is money, the basic model has no obvious place for money either).

Models that dispense with time can still be defended instrumentally – as yielding predictions whose accuracy justifies their lack of descriptive value, or as promoting understanding even if a causal explanation is not on offer. Atemporal models may, indeed, gain added strength by abstracting from messy reality, and some theorists take pride in the 'unrealism' of their models. Gone, in one clean swing of the axiom (or Occam's razor?), are the observation errors due to mismeasurement and time lags, and background noise that obscures the main causal relation or gets it the wrong way round. At the macro level, our human (if not Humean) tendency to assume that causes precede events can invert the true sequence, if a change that produced the observed result only becomes observable after it. Viewed from a distance, stock markets always seem to crash before the bad profit figures get published, and sprinters always seem to jump the gun. Recent neurological research suggests that intention sometimes stays ahead of action only because the brain delays our sensory signals until it has sent its own. At the micro level, atemporal (axiomatic or statistical) theory may be the only way to view a 'reality' that changes once under the experimenter's lens.

The problem remains, however, that when the lessons of such models come to be applied, it is to a world where time still passes; or whose inhabitants like to believe it does, and act accordingly. Newtonian mechanics addresses the shapes, sights and sounds of direct human experience, but robs them of logical (chronological) order by turning epoch-hopping thought-experiment into real intertemporal intrusion. If we actually rewrote the past from the present, we would no longer arrive at that present. And if we could fully anticipate the future, we would surely have got there already. Or, if it didn't look so rosy, redirected things to be sure of ending up somewhere else. In the very act of liberating individuals, by identifying the conditions and regularities that informed and followed from their actions, Newtonian social science seemed to find new shackles for those actions. In the teleological leg-irons of remorseless thesis, antithesis and synthesis, if not the ontological straitjacket of unarguable rational choice.

Introduced as a device to stop everything happening at once, time was in danger of stopping anything happening at all. It was almost with relief that economists looked to the persistence of unemployment, and of the illogically configured Qwerty keyboard, to remind themselves that real economic processes exhibit 'path-dependency' which may give rise to more than one equilibrium. Blessed with experimental opportunities which economists (perhaps fortunately) lack, natural science is now helping to relieve the dullness of equilibrium by revealing the exciting things that can happen if a system moves away from it. Coveney (Chapter 5) describes how thermodynamic systems displaced sufficiently far from 'equilibrium' can shift to new arrangements with fundamentally different patterns of behaviour. Echoing the 'catastrophe theory' with which sociologists explained compositional 'tipping' of neighbourhoods (and the decline of tipping in restaurants) in the 1960s, such nonlinear observations have recently given economists new hope of finally explaining why nations differ so persistently in their rates of unemployment, inflation and growth.

Irreversibility of much physico-chemical change equates to asymmetry of time in the chemical systems described by Coveney and de Hemptinne (Chapters 5 and 6). A similar relation between time and change has been evoked in relation to dissipative structures and autopoiesis in biology, now finding their metaphorical way into the study of organisations and other social institutions. But re-asserting the possibility of change does not necessitate the reintroduction of time. While history's iron laws may have rusted, and psychology's Skinner boxes succumbed to woodworm that didn't wait to press a button, timelessness still commands a conceptual premium. Abstraction from time is especially valued in disciplines where the mists of time engulf past understanding more quickly than the pedestals of time cah enshrine it, and where cultural borders limit comparison across space.

From Plato onwards, as Lane recounts in her review of time in political theory (Chapter 12), observers have been tempted to equate time with decay, bringing events which drag good life and good government down. Rawls' *Theory of Justice* has, in our own time, resurrected the quest for a timeless approach to rulemaking and power-sharing by 'social contract'. Its critics' attempts to re-introduce history as 'authorising power' (notably by Nozick over the unequal distribution of property) have not been to everyone's taste. Baert notes in his assessment of time in social theory (Chapter 11) a similar resistance to the ravages of time among structuralists (for whom synchronic relations across space are the key to general explanation of social order and its development), and those ontologists who would completely subordinate the relative time-invariance of institutions to spontaneous individual will.

Economists' attachment to the statics of general equilibrium despite its aversion to change, and to dynamic equilibrium models which fail to identify the main sources of growth, reflects a deeply-held faith in the modelbuilding usefulness of Newton's 'instantaneous present'. As Rizzo observes (in Chapter 9), even models that incorporate 'expectations' ground them in a generalised model calibrated on unambiguous past data – which happen to be the same as those deployed by the model-builder – and so represent nothing more than the playing-out of a sequence determined at $t=0$. Econometricians' long-held faith in time-series estimation, already under fire from the mixed record of its forecasts, has recently been shaken from several methodological angles – notably the 'Lucas critique' that the structural parameters might vary across time. One cause of this variation, people's ability to learn about the model that policymakers are using and predict their reaction to current events, has a much wider resonance for social sciences, which were quick to seize on classical experimenters' assumed separation of observer from reality, and have been much slower to incorporate the more recent acknowledgement of subject-object interaction – even in the physical world – with its related indeterminacies.

Evolving explanations

Reintroducing time, via the Bergsonian concepts of novelty and emergence, has become a central task for the contemporary social sciences. Its necessity is established by the observation that social systems change, and the aspiration of social actors to change them (even if the two processes are rarely consonant, despite the best-laid revolutionary plans). Its pursuit has drawn increasing strength from another revolution in natural science, almost contemporaneous with those of Einstein and Planck but temporally and spatially on starkly contrasting planes. Dealing with time-spans as big as cosmology's distances, by digging for artefacts as fragmentary as sociology's deep structures, evolutionary theory has risen from the 'softest' of the natural sciences to challenge some of its firmest beliefs about predictability, time and change.

Delving into a more distant past throws a very different light on the foreseeable future, as emerges from Friday's account of the evolutionary story so far (Chapter 7). By emphasising the indeterminateness of variation and selection, the multi-level interaction between organism and environment, and their resultant co-development, evolution's time-bound reconstruction restores the creative licence which mechanisms' 'liberating' construction inadvertently closed down. Panglossian superlatives give way to Darwinian comparatives as change ceases to be equated with inevitable progress and rising complexity, path dependency knots the lines of least resistance, and multiple equilibria dethrone the one best way.

After its initial injection of historical sense into Linnaeus' 'natural history' classification, evolution encountered a practical problem related to elapsed time. The smallness and low pass rate of natural variation could not easily account for the rate at which new species had formed and 'family' trees branched off; or for the rate at which one hairless ape had sprinted away from the tree of its prehensile progenitors. Once Darwin's 'random' variation displaced the environmentally-determined version of Lamarck, a more philosophical problem also intruded. Evolution fills the animate past with path-dependency, while emptying the future of predictability. Can a species so shaped by past selection, so suffused by genetic inheritance, really be free to defy its 'code' or modify its environment? Unsurprisingly, molecular determinism from within has been quick to fill the void left by material determinism from without. While Lamarckian directed variation remains out of favour, teleological tweaks in the selection process threaten to reintroduce the grand designer Darwin expressly overthrew. As the genome map unfolds, evolutionary theory slips easily from essential open-endedness into the belief that what survives is of special significance in terms of environmental 'fitness', and that growing complexity gives at least a hint of where the process may head next.

'Punctuated equilibrium', as outlined in Chapter 7, rebuffs both constraining tendencies. In place of the short strides and slow pace of incremental change, it traces speciation to sudden ruptures interspersed with long phases of stability. It also defies the simple equation of change with progress. While the environment may not induce variation, it can be changed by the action of the more successful variants, resulting in a constantly changing selection test which denies any simple notion of direction, progress, or improvement according to any absolute standard of fitness. Even if organisms do tend towards increasing complexity, this can stiffen the selection test (much of which arises from biotic competition with other organisms) so that relative performance stays unchanged. After their initial breakthrough, bipeds are left running to stand still. Rejected by 'selfish' genetics but supported by the fossil record, punctuated equilibrium offers a sedimentary echo of the complex chemical systems examined in Chapters 5 and 6.

Tattersall's appraisal of the evolution of evolution (Chapter 8) shows that the story is far from complete, even if genetic 'history' now promises to fill the fossil gaps and finesse ontogeny's refusal to recapitulate phylogeny. But evolution's ultimate message – that there may not be any ultimate message – fires a broadside across mechanistic modelling's over-choreographed bows. Away from the world of the weak anthropic principle (dismissed by Newton-Smith in Chapter 4) and static general equilibrium (defended, but with the circumspection of one a designer who knows its limits, by Hahn in Chapter 10), there need be no grand designs except those we construct or choose to infer. What we observe is only one version of what might have been, and what comes next is at least partly our decision. In a world where chance processes shape development, irreversibly, at the level of cells and societies as well as splitting atoms, counterfactuals count and genuine choices re-emerge.

Evolution restores the future as a place where the extant can take new turnings, and even the extinct acquire new meanings. Yet it also ensures that the scope for novelty is never absolute. We can only build on what we have – lost species cannot be retrieved, however they might have thrived in today's conditions – and we cannot predict which design will be next to take its place. Selfish genes may not dictate our actions, but nor does culture entirely reprieve us from selection pressure, or avoid an often crude 'survival of the fittest' at the level of artefact and artwork. Not as helpless as we act, we are also not as powerful as we think.

Baert (Chapter 11) shows how several strands of social theory have reappropriated the concept of a 'relatively open' future, treading between the twin determinisms of overarching social framework and optimising individual choice. People follow rules but can also break them, extrapolate from trends but can also buck them, pursue change but only by reinforcing much of the context in which their action occurs. Aware of what they are doing (perhaps by overhearing the social scientists discussing it), people can refine their knowledge and redesign their strategies in the light of experience. If evolution is absent from, or even repudiated by, certain versions of this new orthodoxy, it is largely through an oversensitivity to its earlier, cruder sociobiological uses, and an overliteral use of the biological metaphor that fails to acknowledge the interplay between variation and competitive selection in the social realm. With 'agency' restored to interaction with 'structure', past and future enter a dual relation with the extended present. We can reassess our history, but only using knowledge and beliefs whose acquisition has been shaped by that history. We can anticipate the future, but in a way which generates knowledge and beliefs just as likely to frustrate it as to fulfil it.

Structuralist enquiry was never fatalistic. One of the principal aims of acquainting us with the past was to unshackle us from it. Historical sociologists, as much as psychoanalysts, sought to dig out the deep structures so that we could give them a decent burial, and move ahead without their unacknowledged pull. Similarly, social forecasting has never been entirely deterministic. Futurologists have characteristically warned of impending disaster so that we can act to avert it, rather than hailing approaching paradise and goading us to give history a push. But re-opening the future has its existential downsides, as quickly emerged from the Heideggerian considerations of being and time reviewed by Sandbothe (Chapter 2). It was, after all, Sartre who conceded that a consciousness projecting itself into past and future confronts itself as not-being, revealing 'a nihilating structure of temporality'; and who recognised the anguish that results from continual free choice with no objective validation. When useful structures and conventions mutate or break down, it is little comfort to be told we played a part in undermining them. When well intentioned plans go wrong, taking responsibility for them soon ceases to be liberating. With no external divine or dialectical agent to dictate our next action, weak will propels us all too easily into the internalised determinisms of class position, social status, oafish parents or selfish genes.

In our own time

The new empowerment of ordinary people and their 'daily-life time' also has a theoretical pitfall. If everyone can monitor their own and others' practices, using their findings to explain past occurrence and inform future action then what is so special about the social scientist? How can those who profess this title step sufficiently far outside society to reach 'objective' conclusions about it? How can they publicise those judgements without undermining them, by influencing and changing the way people think and behave? Generalised self-reflection slips easily into reflexivity, under which the answer to how we can comprehend and improve the social world is swamped by the question of how we can presume any privileged insights into it.

The postmodern critique, as Quie wryly observes (Chapter 14), represents the extended present taken to its insufferable extreme. Time expands to conquer space, pronouncing an end to history by annihilating the potential for change. Action and reaction, sequential under Hume's science, equal and opposite under Newton's, become instantaneous as airwaves and cables spin out their legacy to span the globe (a realisation owed to Marshall McLuhan when technicolor and electronic networking were still in their infancy). Suddenly time is tangible, tradeable, intelligible, and it is space that is rendered unreal. The manipulation of signs, which can run backwards as easily as hitting a rewind button, replaces any attempt to change things on the tarmac-buried, tree-denuded, screen-reconstituted ground.

Given semiotics' apparent power to stop (or lose touch with) the clock, this book's progression from natural science (Chapters 3–8) to social science (Chapters 9–15) has a superficial directional logic. If not a slippery slope from order into chaos, then at least the melting of solid models and measurements into more confusing feedback and flux. But as even the harder sciences' certainties soften on closer inspection, this apparently linear course from investigative rigour to interpretive rigor mortis becomes more of a circular tour. The cutting edge of contemporary cosmology shows itself just as open to paradox, conjecture, distorting subject-object interaction and destabilising self-reflection as the blunter instruments of contemporary philosophy, as Newton-Smith (Chapter 4) makes clear early on (as early, indeed, as rival Big Bang theories are willing to allow).

Circular tours can still be worth it for the scenery, and we hope this absence of certainty about the nature and meaning of time does not detract from the usefulness of the search. Consideration of respective disciplines' outstanding problems with time, as well as their possibilities for comprehending or transcending it, has not been a destructive exercise. While some scientists still look ahead to the day when atoms, apes and asteroids are subsumed beneath a single Theory of Everything, the immensity of the task has led others to lower their sights towards a 'bootstrap' view in which every subdiscipline has something to say, and none can say it all. Although more than a passive backdrop against which events unfold, time stops short of being a padded cell into which all human life can be crammed.

With so multifaceted and malleable a target, many close approaches may be preferable to one direct but disorientating hit. In this spirit, *Time in Modern Intellectual Thought* set out to throw some interdisciplinary light on the two-way flow between old orthodoxies and new paradoxes. Moving on from long-running debates on where time goes, and whether it even exists, it has shown that no discipline has a monopoly on temporal wisdom, but that some time-related concepts – notably of quantum decay and evolutionary change – may usefully transcend their original subject boundaries. The arrow may not have run entirely straight, but its sidetracks pass through some challenging new territory.

Millennium celebrations have shown how an arbitrary date can still excite new assessments of the past and new visions of the future. Their worldwide telecasting also highlighted the use of technology to truncate the lapse of time, as a complement to its application for expanding opportunities in space. Watching the clock has long permitted a lack of concern with what lies behind it. But once it is sufficiently saved, stored and shifted, rationing of time no longer excuses us from rationalising time. Instant, continuous communication, superimposing action and reaction, does more than allay post-industrial frustration of wealth outgrowing the time available for its consumption. It encourages a holistic, synchronous perspective on events once clearly distinguished by time and place. Such a perspective is breathing new life into the natural and social sciences, as these chapters have tried to show. Whether scientists' apparent success in the struggle against compartmentalisation will spur historians to a new push against periodisation, only time will tell. But cross-disciplinary space suggests a rethink is already bringing rewards.

INDEX

www.ingramcontent.com/pod-product-compliance
Ingram Content Group UK Ltd.
Pitfield, Milton Keynes, MK11 3LW, UK
UKHW050947290926
15470UKWH00005B/1695

9 780444 829030